NEUROPROTECTION IN AUTISM, SCHIZOPHRENIA AND ALZHEIMER'S DISEASE

NEUROPROTECTION IN AUTISM, SCHIZOPHRENIA AND ALZHEIMER'S DISEASE

Edited by

Illana Gozes

Joseph Levine

ELSEVIER

ACADEMIC PRESS

An imprint of Elsevier

Academic Press is an imprint of Elsevier
125 London Wall, London EC2Y 5AS, United Kingdom
525 B Street, Suite 1650, San Diego, CA 92101, United States
50 Hampshire Street, 5th Floor, Cambridge, MA 02139, United States
The Boulevard, Langford Lane, Kidlington, Oxford OX5 1GB, United Kingdom

Library of Congress Cataloging-in-Publication Data
A catalog record for this book is available from the Library of Congress

British Library Cataloguing-in-Publication Data
A catalogue record for this book is available from the British Library

ISBN 978-0-12-814037-6

For information on all Academic Press publications
visit our website at https://www.elsevier.com/books-and-journals

Publisher: Nikki Levy
Acquisition Editor: Melanie Tucker
Editorial Project Manager: Kristi Anderson
Production Project Manager: Paul Prasad Chandramohan
Cover Designer: Miles Hitchen

Typeset by SPi Global, India

Contents

Contributors ix
About the editors xi
Introduction xiii

I

Overview

1 Activity-dependent neuroprotective protein (ADNP)/NAP (CP201): Autism, schizophrenia, and Alzheimer's disease
Illana Gozes

Introductory remarks: Activity-dependent neuroprotective protein (ADNP) discovery through vasoactive intestinal peptide (VIP) neuroprotective mechanism 3
ADNP macromolecular interactions 4
The ADNP syndrome in patients and models: A CP201 (NAP) connection 7
The schizophrenia connection 11
The Alzheimer's disease (AD) connection 13
Epilogue: Autism, schizophrenia, and AD shared mechanism through the ADNP prism 13
Selected Internet resources 14
Acknowledgments 14
Conflict of interest 15
References 15

2 Clinical convergence of autism, schizophrenia, and Alzheimer's disease: The case of social cognition
Joseph Levine

Introduction 21
Clinical characteristics and symptomatology of schizophrenia, ASD, and Alzheimer's disease 21
Clinical convergence of schizophrenia, Alzheimer's disease and ASD 23
Social cognition 24
Social cognition and ASD 26
Social cognition and schizophrenia 30

Social cognition and Alzheimer's disease 35
Pharmacological treatment of social cognition impairments 37
References 43

II

Autism

3 A contemporary view on the molecular basis of neurodevelopmental disorders

Ilse M. van der Werf, Geert Vandeweyer, R. Frank Kooy

Neurodevelopmental disorders 57
Genetic causes of NDDs 61
Gene networks in neurodevelopmental disorders 70
References 73

4 Autism spectrum disorder: A clinical path to early diagnosis, evaluation, and intervention

Lidia V. Gabis

Diagnosis 79
Evaluation of ASD in infancy 79
Recognition of signs and symptoms neurodevelopmental workup 80
About the "red flags" 85
Diagnostic methods 86
Comprehensive evaluation and planning of treatment 89
The significance of early diagnosis 93
Summary of Challenges 94
Acknowledgment 95
References 95

5 Neuroinflammation and neuroprotection in schizophrenia and autism spectrum disorder

Ehud Mekori-Domachevsky, Hadar Segal-Gavish, Raz Gross

Part I: Schizophrenia **102**
Microglia 102
Cytokines 103
Oxidative stress 104
BDNF 107
References 109
Part II: Autism spectrum disorder (ASD) **111**
Microglial dysfunction in ASD 111
Cytokine alterations in ASD 112
Immune-modulating drugs for ASD 114

Brain-derived neurotropic factor (BDNF) dysregulation in ASD 116
Oxidative stress in ASD 117
References 118

III

Neuropsychiatric disorders

6 Neuroprotective roles of neurotrophic factors in depression

Christina Gross, Kim B. Seroogy

Introduction 125
The need for novel treatment strategies in depression 126
Evidence suggestive of impaired neuroprotection and neurotrophic factors signaling in depression 127
Evidence for enhanced neuroprotection as an underlying mechanism of current successful therapeutic strategies in depression 134
The potential of neurotrophic factors in disease-modifying therapies for mood disorders and beyond 136
Acknowledgments 137
References 137

7 Neuroprotective roles of neurotrophic growth factors in mood disorders

Cheng Jiang, Stephen R. Salton

Neurotrophic growth factors and their receptors 145
Contributions of BDNF to the pathophysiology of mood disorders 147
Mechanisms by which BDNF and other neurotrophins regulate mood disorders and antidepressant efficacy 154
Targets of BDNF/TrkB signaling with potential roles in mood disorders 159
Overview 160
References 160

8 Heme metabolism, mitochondria, and complex I in neuropsychiatric disorders

Lee S. Ifhar, Dorit Ben-Shachar

Introduction 173
Heme 174
Heme in neurons 179
Alterations in heme metabolism in neuropsychiatric disorders 180
Alzheimer's disease and the mitochondria 182
Heme metabolism in Alzheimer's disease 184
Parkinson's disease and the mitochondria 185
Heme metabolism in Parkinson's disease 186
Schizophrenia and The mitochondria 187

Heme metabolism in schizophrenia 188
Conclusions 189
Acknowledgment 191
References 191

9 Neuroprotective effects of lithium in neuropsychiatric disorders
Galila Agam, Joseph Levine

Introduction 209
Lithium's effect on autophagy, BDNF, inflammation, mitochondrial function, and apoptosis 210
Lithium and neuropsychiatric disorders 215
Summary 227
References 228

IV

Alzheimer and neurodegenerative diseases

10 Tau-based therapies for Alzheimer's disease: Promising novel neuroprotective approaches
Carmen Laura Sayas

Introduction 245
Modulators of tau posttranslational modifications 247
MT stabilizers 254
Inhibitors of tau aggregation 255
Immunotherapy 257
Reduction of tau levels 263
Concluding remarks and future directions 264
References 264

11 Acetylation of tubulin: A feasible protective target from neurodevelopment to neurodegeneration
Cappelletti Graziella, Cartelli Daniele

Introduction 273
Acetylation of tubulin in neurodevelopmental disorders 278
Acetylation of tubulin in neurodegenerative diseases 281
Open questions and future perspective 285
Acknowledgments 287
References 288

Index 295

Contributors

Galila Agam Department of Clinical Biochemistry and Pharmacology and Psychiatry Research Unit, Faculty of Health Sciences, Ben-Gurion University of the Negev and Beer-Sheva Mental Health Center, Beer-Sheva, Israel

Dorit Ben-Shachar Laboratory of Psychobiology, Department of Psychiatry, Rambam Health Care Campus, B. Rappaport Faculty of Medicine, Technion IIT, Haifa, Israel

Cartelli Daniele Neuroalgology Unit, IRCCS Foundation, "Carlo Besta" Neurological Institute, Milan, Italy

Lidia V. Gabis Director of the Child Development Center at Edmond and Lilly Safra Children's Hospital, Sheba, Tel Hashomer; Sackler School of Medicine, Tel Aviv University, Tel Aviv, Israel

Illana Gozes The Lily and Avraham Gildor Chair for the Investigation of Growth Factors, Dr. Diana and Zelman Elton (Elbaum) Laboratory for Molecular Neuroendocrinology, Department of Human Molecular Genetics and Biochemistry, Sackler Faculty of Medicine, Sagol School of Neuroscience and Adams Super Center for Brain Studies, Tel Aviv University, Tel Aviv, Israel

Cappelletti Graziella Department of Biosciences; Center of Excellence on Neurodegenerative Diseases, Università degli Studi di Milano, Milan, Italy

Christina Gross Division of Neurology, Cincinnati Children's Hospital Medical Center; Department of Pediatrics, University of Cincinnati College of Medicine, Cincinnati, OH, United States

Raz Gross Sackler School of Medicine, Tel Aviv University, Ramat-Aviv; Division of Psychiatry, The Chaim Sheba Medical Center, Tel Hashomer, Israel

Lee S. Ifhar Laboratory of Psychobiology, Department of Psychiatry, Rambam Health Care Campus, B. Rappaport Faculty of Medicine, Technion IIT, Haifa, Israel

Cheng Jiang Fishberg Department of Neuroscience, Icahn School of Medicine at Mount Sinai, New York, NY, United States

R. Frank Kooy Department of Medical Genetics, University of Antwerp, Antwerp, Belgium

Joseph Levine Beersheva Mental Health Center, Beersheva; Ben Gurion University of the Negev, Beer-Sheva, Israel

Ehud Mekori-Domachevsky Division of Child and Adolescent Psychiatry, The Chaim Sheba Medical Center, Tel Hashomer; Sackler School of Medicine, Tel Aviv University, Ramat-Aviv, Israel

Stephen R. Salton Fishberg Department of Neuroscience; Brookdale Department of Geriatrics and Palliative Medicine; Friedman Brain Institute, Icahn School of Medicine at Mount Sinai, New York, NY, United States

Carmen Laura Sayas Institute of Biomedical Technologies (ITB), University of La Laguna (ULL), Tenerife, Spain

Hadar Segal-Gavish Division of Child and Adolescent Psychiatry, The Chaim Sheba Medical Center, Tel Hashomer; Sackler School of Medicine, Tel Aviv University, Ramat-Aviv, Israel

Kim B. Seroogy Department of Neurology, University of Cincinnati College of Medicine, Cincinnati, OH, United States

Ilse M. van der Werf Department of Medical Genetics, University of Antwerp, Antwerp, Belgium

Geert Vandeweyer Department of Medical Genetics, University of Antwerp, Antwerp, Belgium

About the editors

Illana Gozes

Dr. Illana Gozes is a professor emerita at the Sackler Faculty of Medicine and is a faculty of the Sagol School of Neuroscience and Adams Super Center for Brain Studies at Tel Aviv University in Tel Aviv, Israel. She is the first incumbent of the Lily and Avraham Gildor Chair for the Investigation of Growth Factors and heads the Elton Laboratory for Molecular Neuroendocrinology. Prof. Gozes won many awards of excellence (including Bergmann Award; Fogarty Scholar in Residence; Tel Aviv University's Vice President Award, best applied scientist; Olson Prize; Julodan Prize; Teva Prize; Neufeld Award; Hanse-Wissenschaftskolleg (HWK) fellowship; Humboldt Award; Landau Prize for Life Achievements; and Champion of Hope—Science International—2016, Global Genes, the United States. She is an ex-president of the Israel Society for Neuroscience; served on the governing committee and the Board of Governors of Tel Aviv University; is currently serving on the Council of the European Society for Neurochemistry (Secretary Elect), the Israeli Ministry of Education, Council of Higher Education; is the Israeli coordinator of the International Brain Bee Competition for high school students. Professor Gozes serves also as the Editor-in-Chief of the Journal of Molecular Neuroscience; and serves on many journal editorial boards. Professor Gozes is the inventor of numerous patents including CP201 (NAP, davunetide), a clinical drug candidate targeted at the rare disease indication, the ADNP syndrome (founded Allon Therapeutics and currently the chief scientific officer at Coronis Neurosciences). Professor Gozes discovered ADNP, an essential protein for brain formation implicated in autism, schizophrenia, Alzheimer's disease, and cancer. CP201 is a snippet of ADNP, enhancing ADNP's protective activity. Professor Gozes published >320 papers and has an h-index of 73 (Google Scholar).

Joseph Levine

Prof. Emeritus Joseph Levine is a dedicated researcher and a specialist in psychiatry with more than 35 years of experience in psychiatry. Prof. Levine is a former director of adult and forensic psychiatric and psychogeriatric wards and outpatient clinic in Israel. Prof. Levine has a doctor of medicine degree from the Sackler School of Medicine, University of Tel Aviv (Israel, 1982), and a MA Sci. (cum laude) degree in physiology and pharmacology from the University of Tel Aviv (1985). Between the years 1997 and 2000, Prof. Levine was undergoing 3 years of research in the bipolar unit and the Neurophysics Laboratory of the psychiatric department of the University of Pittsburgh, United States. Since 2002, Prof. Levine is an associate professor in the psychiatric division of the Faculty of Health Sciences at Ben-Gurion University of the Negev. Recently, in 2018, Prof. Levine spent a 1-year sabbatical at the neuroscientific laboratory of Prof. Illana Gozes laboratory at Tel Aviv University. Since 1987, Prof. Levine has published more than 136 scientific manuscripts in the international scientific literature with an h-index of 48.

Introduction

Illana Gozes[a], *Joseph Levine*[a,b]

[a]The Lily and Avraham Gildor Chair for the Investigation of Growth Factors, Dr. Diana and Zelman Elton (Elbaum) Laboratory for Molecular Neuroendocrinology, Department of Human Molecular Genetics and Biochemistry, Sackler Faculty of Medicine, Sagol School of Neuroscience and Adams Super Center for Brain Studies, Tel Aviv University, Tel Aviv, Israel, [b]Division of Psychiatry, Faculty of Health Sciences, Ben-Gurion University of the Negev, and Mental Health Center, Be'er Sheva, Israel

Neuroprotection is the consequence of chemical, genetic, or biological/physiological/behavioral intervention leading to a protective effect in the nervous system against neurodegeneration or brain malfunctioning. Such intervention may be applied either prior to the breakout of a neurodegenerative disease or at different phases along the course of such disease or brain malfunctioning. Such interventions may affect either pathophysiological or compensatory adaptive brain mechanisms. Neuroprotection may be also linked to neurotrophic effects, impacting slow or aberrant developmental processes. Interestingly, as evident in the following chapters, interventions may be frequently relevant for several neurodegenerative or neurodevelopmental disorders (NDDs) rather than to a single one.

Our book brings together latest findings, encompassing basic and clinical aspects pertaining to neuroprotection in autism spectrum disorders (ASD), schizophrenia and related neuropsychiatric disorders, Alzheimer's disease (AD), and other neurodegenerative disorders. The compiled chapters are authored by leading researchers and clinical key opinion leaders representing basic neuroscience disciplines such as neurochemistry, neuroimmunology, molecular neuroendocrinology, and molecular genetics or from medical and clinical disciplines such as clinical biochemistry and pharmacology, medical genetics, biomedicine, neurology, child development and pediatrics, geriatrics, and psychiatry.

The first section is an overview. This overview contains two chapters. The first chapter by one of us (coeditors), Illana Gozes, discusses activity-dependent neuroprotective protein (ADNP)/NAP (CP201) in autism, schizophrenia, AD, and beyond. ADNP is a protein essential for brain formation and function, key to synapse development. Importantly, this chapter also discusses the discovery of an active eight amino acid neuroprotective ADNP snippet named NAP (NAPVSIPQ), a drug candidate also named davunetide or CP201. The NAP target-interacting molecules are microtubule end-binding proteins essential for microtubule dynamics, axonal transport, and synapse formation and function. Interestingly, microtubule end-binding proteins positively impact the microtubule-associated protein tau binding to the microtubule shaft, and this tau-microtubule binding is dramatically enhanced by NAP, protecting against tauopathy, a major hallmark of AD. The chapter discusses the macromolecular interactions attributed to ADNP followed by a description of the ADNP syndrome (with de novo ADNP mutations being a leading cause for syndromic autism). The chapter continues with the discovery of a key role for autophagy in schizophrenia, showing a direct ADNP involvement. The chapter further provides updates on ADNP in AD with ADNP levels in the serum correlating with intelligence. The ADNP description in this chapter is entwined by updates on CP201 (NAP) protection against ADNP deficiencies, enhancing ADNP activities at the microtubule/autophagy levels; encompassing ASD, schizophrenia, and AD; and representing diseases of aberrant synapses.

The second chapter in the overview section is by the second coeditor of this book, Joseph Levine. This chapter discusses the clinical convergence of autism, schizophrenia, and AD: the case of social cognition. This chapter first briefly reviews some key clinical characteristics and symptomatology of each of these diagnoses. Second, it suggests that the clinical convergence of these three diagnoses lies mainly in the cognitive sphere, especially in the realm of social cognition. Third, social cognition, a complex series of highly phylogenetically preserved processes that are used to interpret social interaction, is discussed. Fourth, research studies and metaanalyses regarding impairments in social cognition, along with brain areas associated with social cognition, are reviewed for each of the three diagnoses. Finally, the chapter reviews the existing pharmacological research treatment studies, along with metaanalysis of these studies, targeting social cognitive impairments in ASD and schizophrenia, suggesting that such pharmacological treatments based on the repurposing of existing drugs may transcend diagnostic boundaries.

The focus of the second book section is on the ASD. The first chapter in this section by Ilse M. van der Werf, Geert Vandeweyer, and R. Frank Kooy discusses a contemporary view on the molecular basis of NDDs. The authors stress that identifying the genetic cause of a disease within the spectrum of NDDs is important on multiple levels as it teaches about normal and disturbed cellular processes in the development of the brain and can yield potential targets that can be used in drug development. On a more individual scale, they continue; it can be helpful in determining the most appropriate support for patients and can inform parents of patients on recurrence risk for next pregnancies. The authors also point out that in the era where rapid technological advances promised to drive the identification of the genetic cause of disease in the vast majority of individuals with NDDs, we are still left with "unsolved cases." Therefore, it is concluded that new research strategies have to be applied in the field of NDD genetics. One such strategy may be to focus on gene networks and the global deregulation of these networks in NDDs, rather than on individual genes within the networks.

The second chapter in this section by Lidia V. Gabis discusses ASD as a clinical path to early diagnosis, evaluation, and intervention. In this chapter, the early signs and symptoms of ASD are discussed, as well as etiologies and specific links to neurodegenerative disorders. Despite major advances in autism research, this author stresses that there is still a significant gap between the current "state of the art" research, targeting early signs of ASD and the common practice associated with the age of diagnosis and management. Research so far has established a direct connection between early diagnosis, early intervention, and subsequent outcomes. Yet, this path is neither linear nor available for all diagnosed children.

The third chapter in this section by Ehud Mekori-Domachevsky, Hadar Segal, and Raz Gross discusses neuroinflammation and neuroprotection in schizophrenia and ASD. Schizophrenia and ASD—these authors state—are both severe, complex psychiatric disorders with a heterogeneous clinical presentation. There are some common features to both disorder, such as aberrant social interaction, peculiar thought process, and cognitive deficits, and while ASD is considered a "true" NDD, current knowledge also suggests a significant etiological role for early-life antecedents in the evolution of schizophrenia. Despite decades of research, the multifactorial etiology of both disorders is still mostly unknown. The inflammatory (or neuroinflammatory) hypothesis that suggests that pathological functioning of the immune system plays a central role in schizophrenia and in ASD is not new. Nevertheless, mounting evidence from current advanced laboratory studies has moved the immune system to the center stage of psychiatric or more specifically immunopsychiatric research. The aim of this chapter is to outline the different aspects of the immune system that are implicated in schizophrenia and ASD from a neuroprotective perspective.

The third section deals with neuropsychiatric disorders in general and contains four chapters. The first chapter in this section by Christina Gross and Kim B. Seroogy discusses the neuroprotective roles of neurotrophic factors in depression. These authors state that increasing evidence suggests that impaired neurotrophic factor expression or signaling may be a shared pathological mechanism of depression and neurodegenerative disorders associated with depression and a potential treatment target. Altered neurotrophic signaling may also contribute to comorbid depression in neurodegenerative disorders including AD and Parkinson's disease (PD). This chapter provides a concise overview of impaired neurotrophic factor signaling in depression and comments on how impaired neuroprotection may contribute to depressive disorders in neurodegenerative diseases. The authors also discuss some of the ongoing efforts to target these defects in neurotrophic factor signaling for therapeutic treatment and explain what is needed to fully exploit altered neurotrophic factor signaling as a treatment target in depression and neurodegenerative disorders associated with depression.

The second chapter in this section by Cheng Jiang and Stephen R. Salton discusses neuroprotective roles of neurotrophic growth factors in mood disorders. These authors state that neurotrophic growth factors (neurotrophins), other growth factors, their receptors, and downstream signaling pathways have long been appreciated to play critical roles in the modulation of depressive behavior and the regulation of antidepressants efficacy. Mechanisms by which neurotrophin actions impact depression-like behavior and antidepressant responses in animal models include regulation in the CNS of neuroprotection, neurogenesis, synaptic plasticity and stability, and neuronal cytoarchitecture, including dendritic length and spine density, with correlative genetic association and big data studies in humans. The chapter reviews the role that neurotrophins in general and brain-derived neurotrophic factor (BDNF) in particular play in the modulation of mood disorders and antidepressant efficacy. The chapter further discusses the neurotrophin protein family and relevant receptors; reviews evidence that associates neurotrophin gene polymorphisms, transcript levels, and circulating protein levels with susceptibility or resilience to mood disorders; and finally focus on the specific mechanisms by which neurotrophins regulate depressive behavior and antidepressant actions.

The third chapter in this section by Lee S. Ifhar and Dorit Ben-Shachar discusses heme metabolism, mitochondria, and complex I in neuropsychiatric disorders. Heme plays a role in major cellular processes such as signal transduction, protein synthesis, and complex assembly and regulation of transcription and translation. As such, it is essential for brain metabolism, oxygen sensing, and neuronal survival. Heme synthesis is composed of eight distinct steps; the first and last three steps take place in the mitochondria. Accumulating evidence points at mitochondrial dysfunction and altered heme metabolism in neuropsychiatric disorders. This chapter reviews the growing experimental data demonstrating deficits in heme and mitochondria and specifically in the first complex (Co-I) of the electron transport chain in AD, PD, and schizophrenia. Furthermore, these authors provide evidence for an association between mitochondrial Co-I dysfunction and heme metabolism and suggest a potential for mitochondrial Co-I dysfunction and heme metabolism to become therapeutic targets that will improve the efficiency of the current treatments.

The fourth chapter in this section by Galila Agam and Joseph Levine (coeditor) discusses neuroprotective effects of lithium in neuropsychiatric disorders. The chapter points to the fact that the recognition of lithium is not only as a mood stabilizer but also as an inducer of a broad neuroprotective effects (burst at the dawn of the previous millennium). The myriad of effects culminating in neuroprotection are explained by the discoveries of several key cell signaling-related enzymes being lithium inhibitable. These include glycogen synthase kinase-3, inositol monophosphatase, phosphoadenosylphosphate phosphatase, and Akt/beta-arrestin 2 (Akt). Processes associated with these enzymes include autophagy, BDNF cellular signaling, neuroinflammation, mitochondrial function, and apoptosis. The involvement of these processes in a variety of brain disorders beyond bipolar disorders, such as stroke, Huntington's disease, AD, PD, fragile X syndrome, amyotrophic lateral sclerosis, and multiple sclerosis, raised the interest in exploring lithium's potential neuroprotective property in these neurodegenerative and NDDs. The first part of the chapter discusses lithium's effects on the cellular processes detailed earlier. The second part deals with data regarding lithium's effects on cellular and animal models of each of the aforementioned disorders which, by and large, encouraged clinical trials of lithium in these brain disorders. Finally, the chapter discusses lithium human studies in each of these CNS disorders.

The fourth section focusses on AD and neurodegenerative diseases containing two chapters. The first chapter by Carmen Laura Sayas discusses tau-based therapies for AD as a promising novel neuroprotective approaches. Tau neurofibrillary tangles and amyloid-beta plaques are the two main hallmarks of AD. In AD, tau undergoes conformational changes and subsequent seeding, aggregation, and transneuronal spreading, contributing to the propagation of tauopathy in the brain in a prion-like fashion. Notably, the progression of cognitive decline in AD correlates much better with tauopathy rather than with deposition of amyloid-beta plaques. For decades, the development of potential therapies for AD has been centered on counteracting the formation of amyloid-beta plaques, with complete failure in clinical trials. As a consequence, the focus of AD drug discovery has been switched toward tau. In this chapter, current tau-targeting future therapies for AD are summarized, from tau-kinase inhibitors to acetylation inhibitors, microtubule stabilizers, aggregation inhibitors, monoclonal anti-tau antibodies, or active tau vaccines. Special emphasis is placed on the most promising therapeutic agents that have reached clinical trials. This chapter touches back to the first chapter of the book, dealing with ADNP/NAP (CP201) protecting against tauopathy.

The second chapter in this section and last in the book is by Cappelletti Graziella and Cartelli Daniele discusses the acetylation of tubulin as a possible feasible protective target from neurodevelopmental aberrations to neurodegeneration. These authors state that previously increasing evidence linked the defective regulation of microtubules to a spectrum of disorders from neurodevelopmental to neurodegenerative diseases. Acetylation of tubulin determines the biochemical and biophysical diversity of microtubules, regulates microtubule function, and has been recently related to the pathogenic events in different disorders including autism, schizophrenia, and AD. The chapter critically examines the experimental data emerging from in vitro to in vivo disease models and from patients with the aim of understanding whether targeting tubulin acetylation could be a promising strategy for neuroprotection. However, a more comprehensive analysis of molecular details at the base of tubulin acetylation and, most importantly, the consequent determination of novel compounds targeting this process are *the condition* sine qua non for moving toward therapeutic interventions targeting tubulin acetylation.

We hope that our book will advance the understanding of neuroprotection in ASD, schizophrenia/neuropsychiatric disorders, AD, and other neurodegenerative disorders hopefully paving the path to better disease management and novel therapeutics.

Acknowledgments

Professor Illana Gozes is supported by the following grants, ISF 1424/14 and 2340/1 ERA-NET neuron AUTISYN and ADNPinMED and BSF-NSF 2016746, AMN Foundation, as well as by Drs. Ronith and Armand Stemmer, Mr. Arthur Gerbi (French friends of Tel Aviv University), Spanish friends of Tel Aviv University, and the Alicia Koplowitz Foundation.

Conflict of interest

Professor Illana Gozes is the Chief Scientific Officer of Coronis Neurosciences developing CP201 for the ADNP syndrome.

SECTION I

Overview

CHAPTER

1

Activity-dependent neuroprotective protein (ADNP)/NAP (CP201): Autism, schizophrenia, and Alzheimer's disease

Illana Gozes

The Lily and Avraham Gildor Chair for the Investigation of Growth Factors, Dr. Diana and Zelman Elton (Elbaum) Laboratory for Molecular Neuroendocrinology, Department of Human Molecular Genetics and Biochemistry, Sackler Faculty of Medicine, Sagol School of Neuroscience and Adams Super Center for Brain Studies, Tel Aviv University, Tel Aviv, Israel

Introductory remarks: Activity-dependent neuroprotective protein (ADNP) discovery through vasoactive intestinal peptide (VIP) neuroprotective mechanism

Deciphering how neuropeptides regulate brain development and functioning, we focused on vasoactive intestinal peptide (VIP)[1] as a gene/protein highly expressed at the time of synapse formation,[2] which was translated to VIP-associated neuroprotection[3] and VIP-related synaptogenesis, through astrocyte activation.[4] Importantly, VIP is tightly associated with brain development, with VIP blockade during embryonic development resulting in severe microcephaly,[5,6] and VIP blockade during postnatal development resulting in impaired dendritic spines[7] and impaired acquisition of developmental milestones.[8] VIP receptors have been further linked to schizophrenia.[9] Additionally, the VIP-related peptide, pituitary adenylate cyclase-activating polypeptide (PACAP), was linked

https://doi.org/10.1016/B978-0-12-814037-6.00001-X

to posttraumatic stress syndrome.[10] VIP has also been implicated in neuroprotection in Alzheimer's disease.[11] The question then arose as to common disease mechanisms underlying VIP activity.

Original studies indicated that VIP neuroprotection is mediated through astrocyte activation, which entailed secretion of protein growth factors, leading to the cloning/discovery of activity-dependent neurotrophic factor (ADNF)[12] and activity-dependent neuroprotective protein (ADNP) and its active neuroprotective site, NAP (NAPVSIPQ).[13, 14] The VIP-related peptide PACAP was also discovered as an ADNP regulator.[15–18] Knocking out (KO) the mouse *Adnp* gene in vivo resulted in neural tube closure defects, halted brain formation, and embryonic death.[19] At the single-cell level, Adnp expression in astrocytes is required for neuroprotection, and while in nonneuronal cells, ADNP is found in the nucleus, in neurons, upon neuronal maturation, the protein is found also in the cytoplasm with specific RNA silencing resulting in the loss of microtubules/loss of neurites.[20] Contrasting KO embryos, haploinsufficient (heterozygous) *Adnp* mice survive while exhibiting cognitive impairment.[21] The *Adnp*-deficient mouse is characterized by microtubule insufficiency, reduced axonal transport,[22] and reduced dendritic spines.[23] These findings are in line with patient results showing intellectual disabilities in case of ADNP gene heterozygous microdeletion or truncating mutation.[24–26] ADNP is a large protein, presenting key signature motifs for macromolecular interactions linked with cognition, with emphasis on social cognition, impaired, for example, in autism, schizophrenia, and Alzheimer's disease (AD). The connection to AD was recently reviewed.[27] Here, I will mention macromolecular interactions attributed to ADNP followed by a description of the ADNP syndrome (within the autism spectrum disorders—ASD), the schizophrenia connection of ADNP, and finally the updates on ADNP and AD. Importantly, when we cloned ADNP, we discovered an active eight-amino acid neuroprotective snippet that we have named NAP (NAPVSIPQ), drug candidate name, davunetide, or CP201.[13] Thus ADNP description will be entwined by updates on CP201 (NAP) protection against ADNP deficiencies and beyond.

ADNP macromolecular interactions

ADNP macromolecular interactions and tight links with the regulation of cognition have been recently reviewed.[28] Fig. 1.1 reveals the multiple interactions of ADNP with key regulatory proteins in the cell nucleus and in the cytoplasm. Fig. 1.1 further depicts the distribution of currently reviewed ADNP de novo mutations[29] resulting in the ADNP syndrome as discussed later.

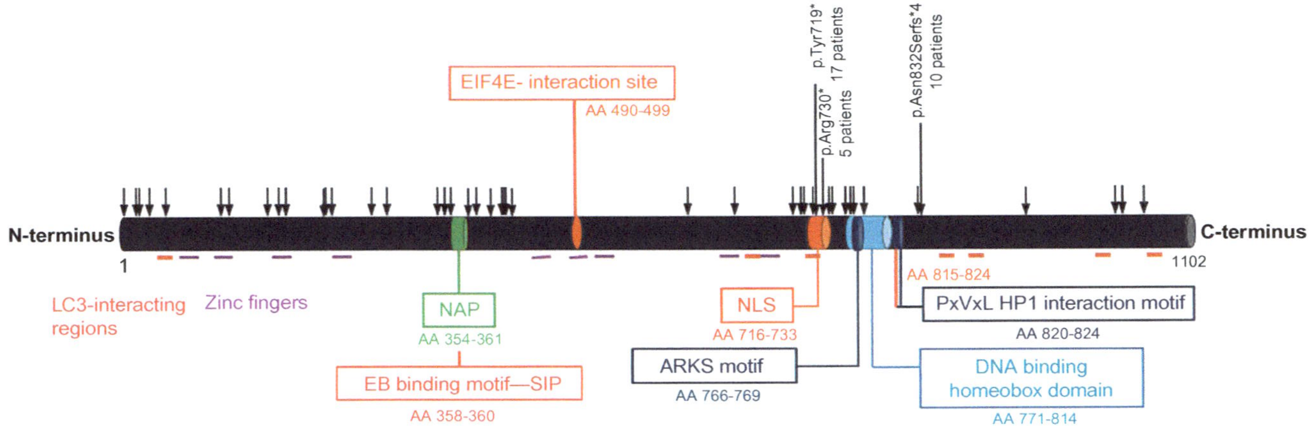

FIG. 1.1 ADNP—protein motifs and ADNP syndrome mutations. Selected motifs in the ADNP human protein are shown as described in our previous publications (e.g., Refs.13, 14, 36, 73, 97), we also described additional interactions[77, 98, 104] and were corroborated and extended by others,[105, 106] as recently reviewed.[107] The picture is a modification of the figure appearing in a recent extensive description of a large ADNP syndrome cohort.[29] The different ADNP mutations have been recently suggested to affect ADNP's cellular distribution.[108] It should be noted that LC3 binding motifs were not specifically shown before and those are added in the figure (contribution of Ms. Yanina Ivashko-Pachima) from the ELM database.[109]

(Continued)

FIG. 1.1 continued

ELM name	Matched sequences	ADNP positions	Elm description
LIG_LIR_Gen_1	TTWEDV	51–56 [A]	Canonical LIR motif that binds to Atg8 protein family members to mediate processes involved in autophagy
	TYKCI	661–665 [A]	
	EFPLL	723–727 [A]	
	SDSFENL	876–882 [A]	
	DSFENL	877–882 [A]	
	SPFDPV	891–896 [A]	
	SKYETI	934–939 [A]	
	SSYGKV	1031–1036 [A]	

Functional site class: Atg8 protein family ligands. **Functional site description:** The autophagy-related protein Atg8 and its homologs LC3 and GABARAP play an important role in selective autophagy. During autophagy, Atg8 proteins get directly conjugated to phosphatidylethanolamine (PE) lipids to mediate membrane fusion events involved in autophagosome biogenesis such as phagophore formation and elongation. In addition, different Atg8 protein family members can recruit specific adaptors bound to ubiquitylated proteins, organelles, or pathogens for degradation. Many of these adaptor proteins contain an **LC3-interacting region (LIR)** that mediates binding to Atg8 and Atg8-related proteins. These LIR:Atg8/LC3/GABARAP interactions are essential for cellular cell homeostasis and the control of intra- and extracellular stress conditions.

The ADNP syndrome in patients and models: A CP201 (NAP) connection

In a recent survey of 78 ADNP syndrome individuals, which followed up the original ADNP identification of ADNP in autism[30] and syndrome description (Ref. 31 reviewed in Refs. 32, 33), individuals were recognized as sharing similar facial features, including a prominent forehead with a high anterior hairline; a wide and depressed nasal bridge; and a short nose with full, upturned nasal tip. Most individuals had a thin upper lip, often combined with an everted lower lip, and a pointed chin, more pronounced at younger age.[29] About 25% of the children present short stature, with the animal model of *Adnp*$^{+/-}$ also presenting shorter length.[23]

In general the disease is characterized as a rare syndromic intellectual disability encompassing global developmental delay, gastrointestinal problems, hypotonia, delayed speech, behavioral and sleep problems, pain insensitivity, seizures, structural brain anomalies, dysmorphic features, visual problems, early tooth eruption, and autistic features.

In detail (Table 1.1), about 80% of the children present early deciduous dentition, with almost a full mouth of teeth by the age of 1 year.[34] This is coupled to the identification of ADNP bone regulation in mice and men.[34] Further characterizing 78 ADNP children identified developmental delay in all tested individuals with late age of independent walking, after

TABLE 1.1 ADNP syndrome symptoms (selected)[29, 34]

Symptom	Prevalence (%)
Intellectual disabilities	100% (>50% severe)
Motor deficits	(100%); ~87% delayed walking
Speech delays	~98%
Autistic traits	~93%
Feeding or gastrointestinal problems	~83%
Early tooth eruption	~80%
Delay in bladder training	>80%
Hypotonia	~78%
Brain abnormalities (MRI)	~75%
Visual problems	~74%
Sensory process disorders	~67%
Sleep problems	~65%
Cardiac problems	~38%

18 months of age (~87% of the children, average age of walking about 2.5 years).[29] ADNP syndrome patients with a p.Tyr719* mutation[35] start walking significantly later (average 3.5 years).[29] The delayed walking may be linked to hypotonia, as 78% of the children had hypotonia, while only 4% of the children present hypertonia. Standing unassisted for long periods of time or walking long distances is difficult for many of the children, with abnormal walking. A minority was not able to walk at the time of last evaluation.[29] These findings were also seen in our unique *Adnp*$^{+/-}$ mouse model exhibiting sex-dependent reduced muscle strength and gait abnormalities, with partial correction by CP201 treatment.[23]

In the ADNP syndrome-affected individuals, muscle problems were also associated with feeding. Thus, 83% of the affected individuals have feeding or gastrointestinal problems, mainly gastroesophageal reflux, frequent vomiting, and constipation. A few have excessive appetite. At the age of assessment, 20.9% of the individuals were overweight, and 7.5% were obese, according to standard WHO classification. Oral movement problems, with implications for feeding and speech, are common (45.6%) and significantly more common in individuals with mutations in the nuclear localization signal (NLS) and C-terminal of this domain.[29]

Thirty-eight percent of the ADNP patients had one or more congenital cardiac defect,[29, 32] which may be linked with ADNP regulation of heart formation.[36]

All ADNP syndrome individuals present intellectual disabilities (ID), 52% severe, 36% moderate, and 12% mild ID. This is coupled with speech delay (98.6% of individuals). Nineteen percent have no language development at all.[29] The connection of ADNP with cognition was first discovered in our mouse model with ADNP being essential for brain formation[19] and with *Adnp*$^{+/-}$ mice displaying marked cognitive impairments, reversed by intranasal NAP (CP201) treatment.[21, 23] Furthermore, the *Adnp*$^{+/-}$ mouse pups displayed reduced ultrasonic vocalizations (a mean of social contact), and NAP treatment enhanced vocalization.[23] Back to the children cohort, 81% of them have a considerable delay in bladder training, and many are still not toilet trained when approaching puberty.[29]

Apparent loss of acquired abilities has been reported in 15% of the children language and for skills like counting, riding a bicycle, or being toilet trained.[29] In this respect the *Adnp*$^{+/-}$ mice exhibit age-associated neurodegeneration measured by immunohistochemical methods.[21]

The *Adnp*$^{+/-}$ mouse model also displays delayed maturation of dendritic spines and reduced spine density, which could be translated to brain abnormalities.[23] Furthermore, ADNP has been linked to white matter development.[37, 38] In the ADNP syndrome patients, MRI (performed in 75.6% of the individuals) showed cerebral abnormalities, including atypical white matter lesions,[39] delayed myelination, cortical dysplasia or atrophy, perinatal hypoxic ischemic encephalopathy, hydrocephalus, and hippocampal hypoplasticity in 56% of the observed cases,[29] including delays in hippocampal maturation.[35]

Comparing cognition with autistic features, ADNP patients present higher rates of ID but less severe social affect symptoms and high levels of stereotyped motor behaviors compared with high levels of restricted interests in idiopathic ASD.[40] Thus, 93% of the ADNP affected individuals present with autistic features,[29] and 65%–67% of them have been reported to have a clinical diagnosis of ASD.[29, 40] Parental reports suggest that 88% of the children are overall happy and friendly, while behavioral problems are reported in 77.6% of them. Several present with obsessive compulsive behavior, mood disorder, a high anxiety level, temper tantrums, self-injurious, and (verbally) aggressive behavior. In this respect 44% are hyperactive or easily distracted. About one-third of them have a diagnosis of attention deficit hyperactivity disorder (ADHD). Several individuals are on behavior-regulating medication like methylphenidate or atypical antipsychotics like risperidone or olanzapine to help control behavioral disturbances, particularly aggression. In this respect, NAP (CP201) mechanism has been linked to risperidone,[41] and NAP treatment protects against anxiety[42] and against autistic-like features in the *Adnp*$^{+/-}$ mouse.[23] Furthermore, NT4-NAP/AAV protected against depression in rodents.[43, 44]

Back to the human, 67% ADNP patients have also been diagnosed with sensory processing disorder. A high pain threshold is reported in 63.6% of individuals. Interestingly, all individuals with a p.Tyr719* mutation are included in this group.[29, 35] In terms of sensory problems, the *Adnp*$^{+/-}$ mouse showed accelerated development of the ear twitch response, attesting to a sensory process disorder and potentially increased irritability.[23]

Sleep problems are present in 65.2%. Some of them are extremely anxious, with struggles falling asleep and frequent nighttime awakenings. Some were treated with melatonin. Many individuals have a low daytime activity level or excessive daytime sleepiness; a minority has sleep apnea.[29]

Other sensory problems include visual and auditory problems. Thus visual problems were present in 73.6% of the individuals, not only hypermetropia (40.3%) and strabismus (49.2%) but also myopia and astigmatism. Many of them are prescribed glasses. Forty-one percent of the individuals have a diagnosis of cerebral visual impairment (CVI). Some have an everted or notched eyelid or mild ptosis, the latter particularly in individuals with mutations in the NLS and C-terminal of this domain.[29] Interestingly, NAP treatment has been linked to retinal protection,[45–49] and ADNP controls PAX6, an important regulator of cortical and eye development.[19]

Regarding the auditory system, many of the children experience chronic otitis media requiring ventilation tubes. Some of them (11.7%) were diagnosed with mild hearing loss in childhood. Two children have hearing aids for sensorineural hearing loss. Ear-nose-throat problems, including narrow ear canals, laryngomalacia, and sleep apnea, were present in 32.1% of the individuals.[29] Furthermore, we have recently shown atypical auditory brainstem response and protein expression aberrations related to ASD and hearing loss in the *Adnp*$^{+/-}$ mouse brain.[50] Ear infections in the children are also associated with general recurrent infections (51% of the individuals) and with the ADNP/NAP association with protection against inflammation.[51–55]

Additional problems (selected)

Thirty-four percent of the males had unilateral or bilateral cryptorchidism, 12.5% were born with renal anomalies (narrow ureters, bilateral vesicoureteral reflux, which was surgically repaired). Twenty-five percent had nail abnormalities, and 8% children had plagiocephaly of whom three wore a cranial molding helmet. Sixteen percent have seizures.

Routine symptomatic care by primary care professionals is recommended including speech, occupational, and physical therapy, specialized individualized learning programs, treatment of neuropsychiatric features including sleep disorders, behavioral problems, and/or seizures. Nutritional/hormonal support, routine treatment of cardiac, ophthalmologic, and auditory findings is also recommended.

Additional selected articles regarding the ADNP gene in autism[56–58] including case reports are available,[59–61] placing ADNP as one of the leading genes in syndromic autism.[62] The connection of NAP (CP201) with development has been also established, as illustrated earlier, and in a number of publications including multiple studies in fetal alcohol syndrome models (e.g., Refs. 63–67), Down's syndrome neurons,[68] hypoxia,[69] excitotoxicity (implicated in cerebral palsy),[38] and hematopoiesis.[70]

Regarding case reports, we have most recently published a case reporting a novel activity of ADNP in the skin that may serve to define the clinical phenotype of patients with ADNP syndrome and provide an attractive therapeutic option for skin alterations in these patients, which is mimicked by our *Adnp* heterozygous (haploinsufficient) mouse and treatable by NAP.[71]

Regarding fetal alcohol syndrome, we have recently shown that alcohol consumption regulates ADNP expression and *Adnp* haploinsufficient female mice showed higher alcohol consumption and preference, compared with *Adnp* intact females, whereas no genotype difference was observed in males. Importantly, daily intranasal administration of the NAP normalized alcohol consumption in *Adnp* haploinsufficient females. Finally, female *Adnp* haploinsufficient mice showed a sharp increase in alcohol intake after abstinence suggesting that ADNP is a potential novel biomarker and regulator of alcohol-drinking behaviors (also regulating gene associated with alcohol consumption).[72]

The schizophrenia connection

The drug candidate NAP (CP201) provides the ADNP microtubule interacting site, enhancing microtubule dynamics and protection.[73] As schizophrenia was tightly associated with microtubule abnormalities, we tested NAP in a microtubule-deficient mouse model, the stable tubule-only polypeptide (STOP) model, or microtubule associated protein 6 (Map6) haploinsufficient model.[74] NAP (generic name, davunetide or CP201) is an active fragment of activity-dependent neuroprotective protein (ADNP). The haploinsufficient mice showed schizophrenia-like symptoms (hyperactivity) that were ameliorated by chronic treatment with the antipsychotic drug, clozapine. Daily intranasal NAP treatment significantly decreased hyperactivity in the mice and protected visual memory.

An independent study comparing the stability of microtubules in olfactory neuroepithelial cells between schizophrenia cases and matched nonpsychiatric comparison subjects subjected to nocodazole enhancing microtubule depolymerization showed that the mean percentages of cells with intact microtubules were significantly greater for schizophrenia cases than for the matched comparison subjects, suggesting an aberrant microtubule control mechanism.[75]

As ADNP regulates microtubule dynamics, through its NAP motif,[73] we then asked whether ADNP is dysregulated in schizophrenia postmortem brains, comparing gene expression of *ADNP* with its family member gene, *ADNP2*. Quantitative real-time polymerase chain reaction in postmortem hippocampal specimens from normal control subjects exhibited a significant ADNP to ADNP2 transcript level correlation ($r = 0.931$, $P < .001$), while in the hippocampus of matched schizophrenia patients, this correlation ($r = 0.637$, $P = .014$) was drastically decreased in a statistically significant manner ($P = .03$), mirroring disease-associated increased ADNP2 transcripts.[76] Further dysregulation was discovered in patient-derived lymphocytes[77] showing a significantly increased expression in ADNP and ADNP2, similar to increases in the antiapoptotic, beclin1-interacting, Bcl2. The increase in ADNP was associated with the initial stages of the disease, possibly reflecting a compensatory effect. The increase in ADNP2 might be a consequence of neuroleptic treatment, as seen in rats subjected to clozapine treatment.[77] These results were further corroborated in an independent study, with blood ADNP mRNA levels found to be significantly increased in the patients with schizophrenia when compared with controls.[78]

Mechanistically, as indicated in Fig. 1.1, ADNP interacts with microtubule associated protein 1 light chain 3 (LC3), which is the building block of the autophagosome.[77] In this respect, it should be mentioned that autophagy is a process preserving the balance between synthesis, degradation, and recycling of cellular components and is therefore essential for neuronal survival and function. Several key proteins govern the autophagy pathway including beclin1 and LC3. At the protein level, we showed that ADNP coimmunoprecipitated with LC3B, suggesting a direct association with the autophagy process. We further showed a brain-specific reduction in beclin1 expression in postmortem hippocampus of schizophrenia patients, not detected in peripheral lymphocytes. This is in contrast with *ADNP*. Furthermore, ADNP haploinsufficiency in mice, which results in age-related neuronal death or cognitive and social dysfunction,[21] exhibited reduced hippocampal beclin1 and increased Bcl2 expression (mimicking schizophrenia and normal human aging).[77] Further connection of ADNP with the autophagy process was revealed in the STOP (Map6)-deficient mouse model with NAP protecting against reduction in hippocampal beclin1 (Becn1) mRNA but with no protection by the antipsychotic clozapine. Normalization of Becn1 expression by NAP was coupled with behavioral protection against hyperlocomotion and cognitive deficits measured in the object recognition test. Clozapine reduced hyperlocomotion below control levels and did not significantly affect object recognition. The combination of clozapine and NAP resulted in normalized outcome behaviors.[79]

It should be further emphasized that microtubule dynamics and autophagy are closely linked and ADNP/NAP are important for both processes.[80] In the case of microtubules, we also chose to study a mouse model expressing the mutated disrupted in schizophrenia 1 (DISC1) protein,[41] a microtubule-regulating protein that when mutated leads to schizophrenia and other psychiatric illnesses. To elaborate on the mechanism, the NAP (NAPVSIPQ) sequence contains the SxIP motif, microtubule end-binding (EB) protein target, which is critical for microtubule dynamics leading to synaptic plasticity and neuroprotection.[73] Our further bioinformatics prediction for FDA-approved drugs mimicking SxIP-like motif that displace NAP-EB binding identified Risperidone.[41] Risperidone or NAP effectively ameliorated object recognition deficits in the mutated DISC1 mouse model. NAP but not Risperidone reduced anxiety in the mutated mice. Transcripts of Forkhead-BOX P2 (Foxp2), a gene regulating DISC1 and associated with human ability to acquire a spoken language, were increased in the hippocampus of the DISC1 mutated mice and were significantly lowered after treatment with NAP, Risperidone, or the combination of both.[41] These findings are in agreement with our recent observations of ADNP controlling social vocalization in mice and NAP protecting against ADNP deficiencies—linked vocalization impediments.[23]

Importantly, phase II clinical studies have shown NAP-dependent augmentation of functional activities of daily living coupled with brain protection,[81, 82] translating laboratory data to a clinical development impact.

The Alzheimer's disease (AD) connection

I have recently described the ADNP-cognition connection,[28] with interesting findings correlating cytoskeleton and chromatin interactions. Our laboratory has shown a tight correlation with ADNP/NAP and tau phosphorylation, tangle formation[21] as most lately reviewed and extended in our paper "NAP protects against tau hyperphosphorylation through GSK3."[83] Our original studies showing NAP protection against tau hyperphosphorylation in AD models and increases in amyloid beta[84, 85] are now extended to protection against amyloid associated changes in AKT,[86] with AKT being central to ADNP/NAP activity.[23] Regarding tau aggregation, our original studies showing NAP protection[87, 88] have now been extended to attenuation of human tau granular-shaped oligomers.[89] These findings explain the NAP (davunetide/AL-108/CP201) protection of short-term memory in amnestic mild cognitive impairment patients.[90, 91] Most interestingly, complete proteomics of serum from AD patients has identified ADNP as the only protein decreasing in content compared with controls,[92] while our studies identified a correlation between ADNP serum levels and intelligence in elderly individuals at risk for AD.[93] Furthermore, while NAP (davunetide/AL-108/CP201) provided protection in amnestic mild cognitive impairment patients, expressing all forms of tau,[91, 94] it did not protect in progressive supranuclear palsy (PSP) patients, known as a four-repeat tau pathology (tau containing four microtubule-binding repeat domains, tau 3R, in contrast to Alzheimer's disease with tau pathology containing both tau 3R and 4R).[94, 95] In this respect, our most recent experiments showed that NAP through binding to EB1/EB3 proteins, which in turn bind tau, dramatically enhanced tau microtubule interaction,[96] specifically for the 3R tau.[94] Importantly, the ADNP/NAP AD connection has been thoroughly reviewed in our recent book chapter,[27] and new findings are discussed later within the current book focus on autism, schizophrenia, and AD.

Epilogue: Autism, schizophrenia, and AD shared mechanism through the ADNP prism

Interestingly and somewhat paralleling, reduced cognition is found in autism with ID, in cognitive impairment associated with schizophrenia, and in AD, the most common dementia. While ADNP is mutated and loses part of its function in the ADNP syndrome, it is deregulated in schizophrenia[76, 77] and ADNP serum level are correlated with cognition in elderly patients at risk for AD.[93] Furthermore, ADNP/NAP regulate genes/proteins common and differentiating among these diseases.[22, 23, 36, 97, 98] The shared pathway that is emerging is that all three diseases are abnormalities of synapse, with ADNP/NAP interacting with microtubules essential for axonal transport[22, 99] and dendritic spine formation.[23, 73] Furthermore, ADNP/NAP regulate the glutamatergic synapse,[100] a key regulatory synapse in brain function. The mechanism by which ADNP/NAP regulate cellular microtubules is through the interaction with the microtubule end-binding proteins EB1 and EB2, accelerating microtubule dynamics at the growing tip of the microtubule and recruiting tau to the microtubule shaft, thus protecting against tau hyperphosphorylation and protecting axonal transport.[28, 83, 96] Microtubule health is tightly linked with the autophagy pathway, which is regulated by ADNP/NAP,[77, 79] and has been implicated in all three diseases. Importantly, NAP (CP201) has shown direct influence on the autophagic flux, microtubule integrity, and mitochondrial dynamics in models of Parkinson's disease,[101, 102] associated also with tau pathology,[103] suggesting new horizons beyond autism/schizophrenia and AD. Lastly, ADNP is crucial for brain formation[19]/synapse formation,[23] and ADNP deficiency is key to neuronal survival[21] and cognitive/motor/sensory functions,[23, 50] underlying the activity of the healthy brain and dysregulated in disease situations, partly addressed by NAP (CP201, davunetide) treatment.[27]

Selected Internet resources

http://omim.org/entry/611386
http://omim.org/entry/615873
https://www.adnpkids.com/
https://www.adnpfoundation.org/
http://humandiseasegenes.nl/adnp/graph-and-chart/
http://www.coronisns.com/
https://www.orpha.net/consor/cgi-bin/SupportGroup_Search.php?lng=EN&data_id=125978&title=Association-Amis-de-ADNP-France
https://www.orpha.net/consor/cgi-bin/OC_Exp.php?Lng=EN&Expert=404448

Acknowledgments

Professor Illana Gozes is supported by the following grants, ISF 1424/14 and 2340/1 ERA-NET neuron AUTISYN, BSF-NSF 2016746, AMN Foundation as well as Drs. Ronith and Armand Stemmer, Mr. Arthur Gerbi (French Friends of Tel Aviv University), and Spanish Friends of Tel Aviv University, Alicia Koplowitz Foundation.

Conflict of interest

Professor Illana Gozes is the chief scientific officer of Coronis Neurosciences developing CP201 for the ADNP syndrome.

References

1. Bodner M, Fridkin M, Gozes I. Coding sequences for vasoactive intestinal peptide and PHM-27 peptide are located on two adjacent exons in the human genome. *Proc Natl Acad Sci U S A*. 1985;82:3548–3551.
2. Gozes I, Shani Y, Rostene WH. Developmental expression of the VIP-gene in brain and intestine. *Brain Res*. 1987;388:137–148.
3. Brenneman DE, Eiden LE. Vasoactive intestinal peptide and electrical activity influence neuronal survival. *Proc Natl Acad Sci U S A*. 1986;83:1159–1162.
4. Blondel O, Collin C, McCarran WJ, et al. A glia-derived signal regulating neuronal differentiation. *J Neurosci*. 2000;20:8012–8020.
5. Gressens P, Hill JM, Gozes I, Fridkin M, Brenneman DE. Growth factor function of vasoactive intestinal peptide in whole cultured mouse embryos. *Nature*. 1993;362:155–158.
6. Gressens P, Hill JM, Paindaveine B, Gozes I, Fridkin M, Brenneman DE. Severe microcephaly induced by blockade of vasoactive intestinal peptide function in the primitive neuroepithelium of the mouse. *J Clin Invest*. 1994;94:2020–2027.
7. Hill JM, Mervis RF, Politi J, et al. Blockade of VIP during neonatal development induces neuronal damage and increases VIP and VIP receptors in brain. *Ann N Y Acad Sci*. 1994;739:211–225.
8. Hill JM, Gozes I, Hill JL, Fridkin M, Brenneman DE. Vasoactive intestinal peptide antagonist retards the development of neonatal behaviors in the rat. *Peptides*. 1991;12:187–192.
9. Vacic V, McCarthy S, Malhotra D, et al. Duplications of the neuropeptide receptor gene VIPR2 confer significant risk for schizophrenia. *Nature*. 2011;471:499–503.
10. Ressler KJ, Mercer KB, Bradley B, et al. Post-traumatic stress disorder is associated with PACAP and the PAC1 receptor. *Nature*. 2011;470:492–497.
11. Gozes I, Bardea A, Reshef A, et al. Neuroprotective strategy for Alzheimer disease: intranasal administration of a fatty neuropeptide. *Proc Natl Acad Sci U S A*. 1996;93:427–432.
12. Brenneman DE, Gozes I. A femtomolar-acting neuroprotective peptide. *J Clin Invest*. 1996;97:2299–2307.
13. Bassan M, Zamostiano R, Davidson A, et al. Complete sequence of a novel protein containing a femtomolar-activity-dependent neuroprotective peptide. *J Neurochem*. 1999;72:1283–1293.
14. Zamostiano R, Pinhasov A, Gelber E, et al. Cloning and characterization of the human activity-dependent neuroprotective protein. *J Biol Chem*. 2001;276:708–714.
15. Nakamachi T, Ohtaki H, Yofu S, et al. Pituitary adenylate cyclase-activating polypeptide (PACAP) type 1 receptor (PAC1R) co-localizes with activity-dependent neuroprotective protein (ADNP) in the mouse brains. *Regul Pept*. 2008;145:88–95.
16. Nakamachi T, Li M, Shioda S, Arimura A. Signaling involved in pituitary adenylate cyclase-activating polypeptide-stimulated ADNP expression. *Peptides*. 2006;27:1859–1864.
17. Li M, David C, Kikuta T, Somogyvari-Vigh A, Arimura A. Signaling cascades involved in neuroprotection by subpicomolar pituitary adenylate cyclase-activating polypeptide 38. *J Mol Neurosci*. 2005;27:91–105.
18. Zusev M, Gozes I. Differential regulation of activity-dependent neuroprotective protein in rat astrocytes by VIP and PACAP. *Regul Pept*. 2004;123:33–41.

19. Pinhasov A, Mandel S, Torchinsky A, et al. Activity-dependent neuroprotective protein: a novel gene essential for brain formation. *Brain Res Dev Brain Res*. 2003;144:83–90.
20. Mandel S, Spivak-Pohis I, Gozes I. ADNP differential nucleus/cytoplasm localization in neurons suggests multiple roles in neuronal differentiation and maintenance. *J Mol Neurosci*. 2008;35:127–141.
21. Vulih-Shultzman I, Pinhasov A, Mandel S, et al. Activity-dependent neuroprotective protein snippet NAP reduces tau hyperphosphorylation and enhances learning in a novel transgenic mouse model. *J Pharmacol Exp Ther*. 2007;323:438–449.
22. Amram N, Hacohen-Kleiman G, Sragovich S, et al. Sexual divergence in microtubule function: the novel intranasal microtubule targeting SKIP normalizes axonal transport and enhances memory. *Mol Psychiatry*. 2016;21:1467–1476.
23. Hacohen-Kleiman G, Sragovich S, Karmon G, et al. Activity-dependent neuroprotective protein deficiency models synaptic and developmental phenotypes of autism-like syndrome. *J Clin Invest*. 2018;128:4956–4969.
24. Borozdin W, Graham Jr. JM, Bohm D, et al. Multigene deletions on chromosome 20q13.13-q13.2 including SALL4 result in an expanded phenotype of Okihiro syndrome plus developmental delay. *Hum Mutat*. 2007;28:830.
25. Vandeweyer G, Helsmoortel C, Van Dijck A, et al. The transcriptional regulator ADNP links the BAF (SWI/SNF) complexes with autism. *Am J Med Genet C: Semin Med Genet*. 2014;166C:315–326.
26. Huynh MT, Boudry-Labis E, Massard A, et al. A heterozygous microdeletion of 20q13.13 encompassing ADNP gene in a child with Helsmoortel-van der Aa syndrome. *Eur J Hum Genet*. 2018;26:1497–1501.
27. Gozes I. Neuroprotective drug development: the story of ADNP, NAP (davunetide) and SKIP. In: Gozes I, ed. *Neuroprotection in Alzheimer's Disease*. Academic Press an Imprint of Elsevier; 2017:253–270.
28. Gozes I, Ivashko-Pachima Y, Sayas CL. ADNP, a microtubule interacting protein, provides neuroprotection through end binding proteins and tau: an amplifier effect. *Front Mol Neurosci*. 2018;11:151.
29. Van Dijck A, Vulto-van Silfhout AT, Cappuyns E, et al. Clinical presentation of a complex neurodevelopmental disorder caused by mutations in ADNP. *Biol Psychiatry*. 2019;85:287–297.
30. O'Roak BJ, Vives L, Fu W, et al. Multiplex targeted sequencing identifies recurrently mutated genes in autism spectrum disorders. *Science*. 2012;338:1619–1622.
31. Helsmoortel C, Vulto-van Silfhout AT, Coe BP, et al. A SWI/SNF-related autism syndrome caused by de novo mutations in ADNP. *Nat Genet*. 2014;46:380–384.
32. Gozes I, Helsmoortel C, Vandeweyer G, Van der Aa N, Kooy F, Sermone SB. The compassionate side of neuroscience: Tony Sermone's undiagnosed genetic journey--ADNP mutation. *J Mol Neurosci*. 2015;56:751–757.
33. Gozes I, Yeheskel A, Pasmanik-Chor M. Activity-dependent neuroprotective protein (ADNP): a case study for highly conserved chordata-specific genes shaping the brain and mutated in cancer. *J Alzheimers Dis*. 2015;45:57–73.
34. Gozes I, Van Dijck A, Hacohen-Kleiman G, et al. Premature primary tooth eruption in cognitive/motor-delayed ADNP-mutated children. *Transl Psychiatry*. 2017;7:e1043.
35. Gozes I, Patterson MC, Van Dijck A, et al. The eight and a half year journey of undiagnosed AD: gene sequencing and funding of advanced genetic testing has led to hope and new beginnings. *Front Endocrinol (Lausanne)*. 2017;8:107.
36. Mandel S, Rechavi G, Gozes I. Activity-dependent neuroprotective protein (ADNP) differentially interacts with chromatin to regulate genes essential for embryogenesis. *Dev Biol*. 2007;303:814–824.
37. Malishkevich A, Leyk J, Goldbaum O, Richter-Landsberg C, Gozes I. ADNP/ADNP2 expression in oligodendrocytes: implication for myelin-related neurodevelopment. *J Mol Neurosci*. 2015;57:304–313.

38. Sokolowska P, Passemard S, Mok A, Schwendimann L, Gozes I, Gressens P. Neuroprotective effects of NAP against excitotoxic brain damage in the newborn mice: implications for cerebral palsy. *Neuroscience*. 2011;173:156–168.
39. Pescosolido MF, Schwede M, Johnson Harrison A, et al. Expansion of the clinical phenotype associated with mutations in activity-dependent neuroprotective protein. *J Med Genet*. 2014;51:587–589.
40. Arnett AB, Rhoads CL, Hoekzema K, et al. The autism spectrum phenotype in ADNP syndrome. *Autism Res*. 2018;11:1300–1310.
41. Vaisburd S, Shemer Z, Yeheskel A, Giladi E, Gozes I. Risperidone and NAP protect cognition and normalize gene expression in a schizophrenia mouse model. *Sci Rep*. 2015;5.16300.
42. Alcalay RN, Giladi E, Pick CG, Gozes I. Intranasal administration of NAP, a neuroprotective peptide, decreases anxiety-like behavior in aging mice in the elevated plus maze. *Neurosci Lett*. 2004;361:128–131.
43. Liu F, Liu YP, Lei G, et al. Antidepressant effect of recombinant NT4-NAP/AAV on social isolated mice through intranasal route. *Oncotarget*. 2017;8:10103–10113.
44. Ma XC, Chu Z, Zhang XL, et al. Intranasal delivery of recombinant NT4-NAP/AAV exerts potential antidepressant effect. *Neurochem Res*. 2016;41:1375–1380.
45. Belokopytov M, Shulman S, Dubinsky G, Gozes I, Belkin M, Rosner M. Ameliorative effect of NAP on laser-induced retinal damage. *Acta Ophthalmol*. 2011;89:e126–e131.
46. Jehle T, Dimitriu C, Auer S, et al. The neuropeptide NAP provides neuroprotection against retinal ganglion cell damage after retinal ischemia and optic nerve crush. *Graefes Arch Clin Exp Ophthalmol*. 2008;246:1255–1263.
47. Lagreze WA, Pielen A, Steingart R, et al. The peptides ADNF-9 and NAP increase survival and neurite outgrowth of rat retinal ganglion cells in vitro. *Invest Ophthalmol Vis Sci*. 2005;46:933–938.
48. D'Amico AG, Maugeri G, Rasa D, et al. NAP modulates hyperglycemic-inflammatory event of diabetic retina by counteracting outer blood retinal barrier damage. *J Cell Physiol*. 2019;234:5230–5240.
49. D'Amico AG, Maugeri G, Rasa DM, et al. NAP counteracts hyperglycemia/hypoxia induced retinal pigment epithelial barrier breakdown through modulation of HIFs and VEGF expression. *J Cell Physiol*. 2018;233:1120–1128.
50. Hacohen-Kleiman G, Yizhar-Barnea O, Touloumi O, et al. Atypical auditory brainstem response and protein expression aberrations related to ASD and hearing loss in the Adnp haploinsufficient mouse brain. *Neurochem Res*. 2019;44:1494–1507.
51. Heimesaat MM, Fischer A, Kuhl AA, Gobel UB, Gozes I, Bereswill S. Anti-inflammatory properties of NAP in acute toxoplasma gondii-induced ileitis in mice. *Eur J Microbiol Immunol (Bp)*. 2015;5:210–220.
52. Heimesaat MM, Giladi E, Kuhl AA, Bereswill S, Gozes I. The octapetide NAP alleviates intestinal and extra-intestinal anti-inflammatory sequelae of acute experimental colitis. *Peptides*. 2018;101:1–9.
53. Escher U, Giladi E, Dunay IR, Bereswill S, Gozes I, Heimesaat MM. Anti-inflammatory effects of the octapeptide NAP in human microbiota-associated mice suffering from subacute ileitis. *Eur J Microbiol Immunol (Bp)*. 2018;8:34–40.
54. Quintana FJ, Zaltzman R, Fernandez-Montesinos R, et al. NAP, a peptide derived from the activity-dependent neuroprotective protein, modulates macrophage function. *Ann N Y Acad Sci*. 2006;1070:500–506.
55. Braitch M, Kawabe K, Nyirenda M, et al. Expression of activity-dependent neuroprotective protein in the immune system: possible functions and relevance to multiple sclerosis. *Neuroimmunomodulation*. 2010;17:120–125.
56. Coe BP, Witherspoon K, Rosenfeld JA, et al. Refining analyses of copy number variation identifies specific genes associated with developmental delay. *Nat Genet*. 2014;46:1063–1071.

57. De Rubeis S, He X, Goldberg AP, et al. Synaptic, transcriptional and chromatin genes disrupted in autism. *Nature*. 2014;515:209–215.
58. Deciphering Developmental Disorders S. Large-scale discovery of novel genetic causes of developmental disorders. *Nature*. 2015;519:223–228.
59. Gale MJ, Titus HE, Harman GA, et al. Longitudinal ophthalmic findings in a child with Helsmoortel-Van der Aa syndrome. *Am J Ophthalmol Case Rep*. 2018;10:244–248.
60. Pascolini G, Agolini E, Majore S, Novelli A, Grammatico P, Digilio MC. Helsmoortel-Van der Aa syndrome as emerging clinical diagnosis in intellectually disabled children with autistic traits and ocular involvement. *Eur J Paediatr Neurol*. 2018;22:552–557.
61. Alkhunaizi E, Walkiewicz MA, Chitayat D. Mutation in the ADNP gene associated with Noonan syndrome features. *Clin Dysmorphol*. 2018;27:53–57.
62. Larsen E, Menashe I, Ziats MN, Pereanu W, Packer A, Banerjee-Basu S. A systematic variant annotation approach for ranking genes associated with autism spectrum disorders. *Mol Autism*. 2016;7:44.
63. Spong CY, Abebe DT, Gozes I, Brenneman DE, Hill JM. Prevention of fetal demise and growth restriction in a mouse model of fetal alcohol syndrome. *J Pharmacol Exp Ther*. 2001;297:774–779.
64. Chen S, Charness ME. Ethanol disrupts axon outgrowth stimulated by netrin-1, GDNF, and L1 by blocking their convergent activation of Src family kinase signaling. *J Neurochem*. 2012;123:602–612.
65. Chen S, Charness ME. Ethanol inhibits neuronal differentiation by disrupting activity-dependent neuroprotective protein signaling. *Proc Natl Acad Sci U S A*. 2008;105:19962–19967.
66. Wilkemeyer MF, Chen SY, Menkari CE, Brenneman DE, Sulik KK, Charness ME. Differential effects of ethanol antagonism and neuroprotection in peptide fragment NAPVSIPQ prevention of ethanol-induced developmental toxicity. *Proc Natl Acad Sci U S A*. 2003;100:8543–8548.
67. Sari Y, Gozes I. Brain deficits associated with fetal alcohol exposure may be protected, in part, by peptides derived from activity-dependent neurotrophic factor and activity-dependent neuroprotective protein. *Brain Res Rev*. 2006;52:107–118.
68. Busciglio J, Pelsman A, Helguera P, et al. NAP and ADNF-9 protect normal and Down's syndrome cortical neurons from oxidative damage and apoptosis. *Curr Pharm Des*. 2007;13:1091–1098.
69. Rotstein M, Bassan H, Kariv N, Speiser Z, Harel S, Gozes I. NAP enhances neurodevelopment of newborn apolipoprotein E-deficient mice subjected to hypoxia. *J Pharmacol Exp Ther*. 2006;319:332–339.
70. Dresner E, Malishkevich A, Arviv C, et al. Novel evolutionary-conserved role for the activity-dependent neuroprotective protein (ADNP) family that is important for erythropoiesis. *J Biol Chem*. 2012;287:40173–40185.
71. Mollinedo P, Kapitansky O, Gonzalez-Lamuno D, et al. Cellular and animal models of skin alterations in the autism-related ADNP syndrome. *Sci Rep*. 2019;9:736.
72. Ziv Y, Rahamim N, Lezmy N, et al. Activity-dependent neuroprotective protein (ADNP) is an alcohol-responsive gene and negative regulator of alcohol consumption in female mice. *Neuropsychopharmacology*. 2019;44:415–424.
73. Oz S, Kapitansky O, Ivashco-Pachima Y, et al. The NAP motif of activity-dependent neuroprotective protein (ADNP) regulates dendritic spines through microtubule end binding proteins. *Mol Psychiatry*. 2014;19:1115–1124.
74. Merenlender-Wagner A, Pikman R, Giladi E, Andrieux A, Gozes I. NAP (davunetide) enhances cognitive behavior in the STOP heterozygous mouse—a microtubule-deficient model of schizophrenia. *Peptides*. 2010;31:1368–1373.
75. Brown AS, Borgmann-Winter K, Hahn CG, et al. Increased stability of microtubules in cultured olfactory neuroepithelial cells from individuals with schizophrenia. *Prog Neuro-Psychopharmacol Biol Psychiatry*. 2013;48:252–258.

76. Dresner E, Agam G, Gozes I. Activity-dependent neuroprotective protein (ADNP) expression level is correlated with the expression of the sister protein ADNP2: deregulation in schizophrenia. *Eur Neuropsychopharmacol*. 2011;21:355–361.
77. Merenlender-Wagner A, Malishkevich A, Shemer Z, et al. Autophagy has a key role in the pathophysiology of schizophrenia. *Mol Psychiatry*. 2015;20:126–132.
78. Li XS, Wu HT, Yu Y, et al. Increased serum FGF2 levels in first-episode, drug-free patients with schizophrenia. *Neurosci Lett*. 2018;686:28–32.
79. Merenlender-Wagner A, Shemer Z, Touloumi O, et al. New horizons in schizophrenia treatment: autophagy protection is coupled with behavioral improvements in a mouse model of schizophrenia. *Autophagy*. 2014;10:2324–2332.
80. Sragovich S, Merenlender-Wagner A, Gozes I. ADNP plays a key role in autophagy: from autism to schizophrenia and Alzheimer's disease. *Bioessays*. 2017;39(11).
81. Javitt DC, Buchanan RW, Keefe RS, et al. Effect of the neuroprotective peptide davunetide (AL-108) on cognition and functional capacity in schizophrenia. *Schizophr Res*. 2012;136:25–31.
82. Jarskog LF, Dong Z, Kangarlu A, et al. Effects of davunetide on N-acetylaspartate and choline in dorsolateral prefrontal cortex in patients with schizophrenia. *Neuropsychopharmacology*. 2013;38:1245–1252.
83. Ivashko-Pachima Y, Gozes I. NAP protects against tau hyperphosphorylation through GSK3. *Curr Pharm Des*. 2018;24(33):3868–3877.
84. Matsuoka Y, Jouroukhin Y, Gray AJ, et al. A neuronal microtubule-interacting agent, NAPVSIPQ, reduces tau pathology and enhances cognitive function in a mouse model of Alzheimer's disease. *J Pharmacol Exp Ther*. 2008;325:146–153.
85. Matsuoka Y, Gray AJ, Hirata-Fukae C, et al. Intranasal NAP administration reduces accumulation of amyloid peptide and tau hyperphosphorylation in a transgenic mouse model of Alzheimer's disease at early pathological stage. *J Mol Neurosci*. 2007;31:165–170.
86. Zhang J, Wei SY, Yuan L, et al. Davunetide improves spatial learning and memory in Alzheimer's disease-associated rats. *Physiol Behav*. 2017;174:67–73.
87. Gozes I, Iram T, Maryanovsky E, et al. Novel tubulin and tau neuroprotective fragments sharing structural similarities with the drug candidate NAP (Davuentide). *J Alzheimers Dis*. 2014;40(Suppl 1):S23–S36.
88. Gozes I, Schirer Y, Idan-Feldman A, David M, Furman-Assaf S. NAP alpha-aminoisobutyric acid (IsoNAP). *J Mol Neurosci*. 2014;52:1–9.
89. Mokhtari F, Riazi G, Balalaie S, et al. Peptides NAP and SAL attenuate human tau granular-shaped oligomers in vitro and in SH-SY5Y cells. *Neuropeptides*. 2016;59:21–31.
90. Gozes I, Stewart A, Morimoto B, Fox A, Sutherland K, Schmeche D. Addressing Alzheimer's disease tangles: from NAP to AL-108. *Curr Alzheimer Res*. 2009;6:455–460.
91. Morimoto BH, Schmechel D, Hirman J, et al. A double-blind, placebo-controlled, ascending-dose, randomized study to evaluate the safety, tolerability and effects on cognition of AL-108 after 12 weeks of intranasal administration in subjects with mild cognitive impairment. *Dement Geriatr Cogn Disord*. 2013;35:325–336.
92. Yang MH, Yang YH, Lu CY, et al. Activity-dependent neuroprotector homeobox protein: a candidate protein identified in serum as diagnostic biomarker for Alzheimer's disease. *J Proteome*. 2012;75:3617–3629.
93. Malishkevich A, Marshall GA, Schultz AP, Sperling RA, Aharon-Peretz J, Gozes I. Blood-borne activity-dependent neuroprotective protein (ADNP) is correlated with premorbid intelligence, clinical stage, and Alzheimer's disease biomarkers. *J Alzheimers Dis*. 2016;50:249–260.
94. Ivashko-Pachima Y, Maor-Nof M, Gozes I. NAP (davunetide) preferential interaction with dynamic 3-repeat tau explains differential protection in selected tauopathies. *PLoS ONE*. 2019;14. e0213666.

95. Boxer AL, Lang AE, Grossman M, et al. Davunetide in patients with progressive supranuclear palsy: a randomised, double-blind, placebo-controlled phase 2/3 trial. *Lancet Neurol*. 2014;13:676–685.
96. Ivashko-Pachima Y, Sayas CL, Malishkevich A, Gozes I. ADNP/NAP dramatically increase microtubule end-binding protein-tau interaction: a novel avenue for protection against tauopathy. *Mol Psychiatry*. 2017;22:1335–1344.
97. Malishkevich A, Amram N, Hacohen-Kleiman G, Magen I, Giladi E, Gozes I. Activity-dependent neuroprotective protein (ADNP) exhibits striking sexual dichotomy impacting on autistic and Alzheimer's pathologies. *Transl Psychiatry*. 2015;5.e501.
98. Mandel S, Gozes I. Activity-dependent neuroprotective protein constitutes a novel element in the SWI/SNF chromatin remodeling complex. *J Biol Chem*. 2007;282:34448–34456.
99. Jouroukhin Y, Ostritsky R, Assaf Y, Pelled G, Giladi E, Gozes I. NAP (davunetide) modifies disease progression in a mouse model of severe neurodegeneration: protection against impairments in axonal transport. *Neurobiol Dis*. 2013;56:79–94.
100. Sragovich S, Malishkevich A, Piontkewitz Y, et al. The autism/neuroprotection-linked ADNP/NAP regulate the excitatory glutamatergic synapse. *Transl Psychiatry*. 2019;9:2.
101. Esteves AR, Gozes I, Cardoso SM. The rescue of microtubule-dependent traffic recovers mitochondrial function in Parkinson's disease. *Biochim Biophys Acta*. 2014;1842:7–21.
102. Melo TQ, van Zomeren KC, Ferrari MF, Boddeke HW, Copray JC. Impairment of mitochondria dynamics by human A53T alpha-synuclein and rescue by NAP (davunetide) in a cell model for Parkinson's disease. *Exp Brain Res*. 2017;235:731–742.
103. Magen I, Ostritsky R, Richter F, et al. Intranasal NAP (davunetide) decreases tau hyperphosphorylation and moderately improves behavioral deficits in mice overexpressing alpha-synuclein. *Pharmacol Res Perspect*. 2014;2.e00065.
104. Schirer Y, Malishkevich A, Ophir Y, Lewis J, Giladi E, Gozes I. Novel marker for the onset of frontotemporal dementia: early increase in activity-dependent neuroprotective protein (ADNP) in the face of tau mutation. *PLoS ONE*. 2014;9. e87383.
105. Mosch K, Franz H, Soeroes S, Singh PB, Fischle W. HP1 recruits activity-dependent neuroprotective protein to H3K9me3 marked pericentromeric heterochromatin for silencing of major satellite repeats. *PLoS ONE*. 2011;6.e15894.
106. Ostapcuk V, Mohn F, Carl SH, et al. Activity-dependent neuroprotective protein recruits HP1 and CHD4 to control lineage-specifying genes. *Nature*. 2018;557:739–743.
107. Gozes I. ADNP regulates cognition: a multitasking protein. *Front Neurosci*. 2018;12:873.
108. Cappuyns E, Huyghebaert J, Vandeweyer G, Kooy RF. Mutations in ADNP affect expression and subcellular localization of the protein. *Cell Cycle*. 2018;17:1068–1075.
109. Dinkel H, Van Roey K, Michael S, et al. ELM 2016—data update and new functionality of the eukaryotic linear motif resource. *Nucleic Acids Res*. 2016;44:D294–D300.

CHAPTER

2

Clinical convergence of autism, schizophrenia, and Alzheimer's disease: The case of social cognition

Joseph Levine

Beersheva Mental Health Center, Beersheva, Israel

Introduction

Schizophrenia, ASD, and Alzheimer's disease are the main topic of this book. This chapter will first briefly review some key clinical characteristics and symptomatology of each of these diagnoses. Second, it will be suggested that the clinical convergence of these three diagnoses lies mainly in the cognitive sphere, especially in the realm of social cognition. Third, social cognition—a complex series of highly phylogenetically preserved processes that are used to interpret social interaction—will be discussed. Forth, research studies and their metaanalyses regarding impairments in social cognition—along with brain areas associated with them—will be reviewed for each of the three diagnoses. Finally the chapter will review the existing pharmacological research treatment studies—along with metaanalysis of these studies—targeting social cognitive impairments in ASD and schizophrenia, suggesting that such pharmacological treatments, based on the repurposing of existing drugs may transcend diagnostic boundaries.

Clinical characteristics and symptomatology of schizophrenia, ASD, and Alzheimer's disease

Schizophrenia is a severe psychotic remitting relapsing mental disorder with a life time prevalence of 0.3%–0.7%.[1] The disorder is characterized

https://doi.org/10.1016/B978-0-12-814037-6.00002-1

by disturbances in thought, perception, and behavior demonstrating a chronic course. The diagnosis is based on a constellation of positive, negative, and cognitive signs and symptoms. The positive symptoms include delusions, hallucinations, disorganized speech, and grossly disorganized or catatonic behavior. Its negative symptoms include paucity of speech (alogia), the lack of motivation (avolition), anhedonia, and diminished emotional expression (affective flattening), while its cognitive symptoms include impairments of memory, language, and executive functions as well as in social cognition.[2] According to the Diagnostic and Statistical Manual of Mental Disorders V,[3] some signs and/or symptoms of the disorder must last for a continuous period of at least 6 months. This 6-month period must include at least 1 month of active positive symptoms (or less if treated) and may include periods of residual symptoms. During residual periods, mainly negative and cognitive symptoms are present. The onset of schizophrenia commonly occurs in the second or third decades of life. It is of note that cognitive impairments and negative symptoms, even in the lack of positive symptoms, usually lead to social and occupational dysfunction as well as to deteriorated functioning in interpersonal relationships.[1]

Alzheimer's disease is named after Alois Alzheimer, the physician who first described a case of the disease in 1906. The DSM-5 presents diagnostic criteria for either major or mild neurocognitive disorder due to Alzheimer's disease (AD). This neurological disorder is characterized by a progressive form of dementia that leads to cognitive deficits including declining of memory functions accompanied by time-dependent deteriorated thinking, as well as behavior, social, and daily-life functioning. It is also characterized by the accumulation of beta amyloid plaques and neurofibrillary tangles in the brain. Its prevalence according to the DSM-5 is 5%–10% of the general population in the eighth decade of life and 25% in the ninth decade and above.[3, 4] It is expected that the number of people with AD dementia will increase considerably in the next few decades. Finally, risk factors for Alzheimer's disease include old age, family history, and traumatic brain injury.[3, 4]

Autism spectrum disorder (ASD) is defined by the DSM-5 as having persistent deficits in social communication and social interaction as shown by deficits in social-emotional reciprocity: deficits in nonverbal communicative behaviors needed for social interaction and deficits in developing, understanding, and maintaining relationships. There are also restricted, repetitive patterns of behavior, interests, or activities, as manifested by stereotyped or repetitive motor movements, use of objects, speech, and/or insistence on sameness. Such patterns are also manifested by inflexible adherence to routines, or rituals. These behavioral patterns may be associated with highly restricted and fixated interests that are abnormal in intensity. They may also demonstrate hyper- or hyporeactivity to sensory input that may be manifested by unusual interests in sensory aspects of the environment. It is of note that symptoms are present in the early phase of development but may become fully manifest to caregivers only after the second year of life. These symptoms cause impairment in social, occupational, or other important areas of daily-life functioning. In this regard, Gross in 2004 and Hill and Frith in 2003 showed that patients with autism have considerable difficulty evaluating social signals and difficulties in emotion recognition, leading to greatly impaired interpersonal interactions.[5, 6] Finally, these disturbances are not better explained by an intellectual disability or global developmental delay. However, it is of note that intellectual disability and autism spectrum disorder may frequently cooccur. In such a case, social communication is expected to be lower than that of a subject suffering from the same degree of intellectual disability.[3]

ASD is mainly idiopathic with unknown cause; however, in about 10%–20% of ASD cases, a specific and recognized genetic cause can be identified such as tuberous sclerosis or fragile X syndrome.[7] Richards et al.[8] who screened a group of 21 genetic syndromes that may present ASD symptomatology—by the use of a systematic review and metaanalysis—showed that ASD phenomenology varied between genetic syndromes but was considerably higher than that expected for the general population.

Clinical convergence of schizophrenia, Alzheimer's disease and ASD

Interestingly, schizophrenia, ASD, and Alzheimer's disease although apparently different do share some clinical features. Obviously, Alzheimer's disease presents a deteriorated course and so may schizophrenia and certain cases of ASD. Besides positive symptoms such as delusions and hallucinations, which characterize schizophrenia and may appear in some stages of Alzheimer's disease as well as in certain cases of ASD, these disorders may also present negative symptoms as emotional blunting, impaired motivation, paucity of speech, and even the lack of speech in some cases of ASD, late stage of Alzheimer's disease, and during the acute phase of schizophrenia as characterized by catatonic stupor.

However, I suggest that a pivotal clinical commonality between these three disorders lies in the cognitive sphere as all these three disorders present a variety of impairments in cognitive abilities. Especially important in this regard is the complex area of social cognition and its impairments shown to be shared by a variety of different mental and brain disorders.[9–11] Therefore, this chapter will focus on social cognition and its impairments in schizophrenia, Alzheimer's disease, and ASD.

Social cognition

Social skills are a set of behaviors critical for successful communication between individuals and consequently for mental health and general well-being. Social skills are based on a set of social cognitive processes titled social cognition, being one of the six core components of neurocognitive functions according to the American Psychiatric Association's Diagnostic and Statistical Manual for Mental Disorders.[3]

Social cognition is a complex set of processes enabling the individual to perceive and interpret the behavior and intentions of others to enable him to act with accordance to accepted social norms.[12] Such processes include (a) social perception: the perceiving and processing of social cues including emotional expressions of facial and body language; (b) social understanding: the understanding of affective states of other people along with their cognitive aspects; and (c) social decision-making: acting appropriately according to social norms and one's goals, taking into consideration the understanding of the behaviors and intentions of others.[12] Another definition of social cognition stressing its phylogenetic roots is that of Frith and Frith[13] who stated that "Social cognition refers to a series of highly phylogenetically preserved processes that are used to interpret social interaction."

The domains that comprise social skills are theorized to include four categories: (1) emotional processing, (2) social perception, (3) attribution style, and (4) theory of mind (ToM).[14]

Emotional processing enables the perception, understanding, and acting appropriately to the emotional stimuli of other people. It was suggested that this entity includes three subdomains: emotion perception, emotion understanding, and emotion managing.

Social perception is the initial step of evaluating the intentions and state of mind of others by analyzing their gaze direction, body movements, and other social cues.

Attribution style reveals how individuals interpret and understand the causes of a specific event and whether the cause is personal or situational, internal (related to the individual), or external (related to the environment).

Finally, theory of mind (ToM) is the ability to understand or conceive the beliefs, motivations, and intentions of others. Theory of mind is a process of taking the other/s perspective and understanding their motivations and intentions. In general, ToM is a complex set of abilities regulating the comparison of multiple perspectives. In this regard, it was suggested that to successfully navigate the human social world, one needs to realize that behavior is guided by mental states that relate to goals and beliefs.

It is of note that one important facet of social cognition is reciprocity. This refer to joint attention, a recently introduced component of social cognition enabling the coordinated attention between the individual and others.[15]

One may next ask what are the brain structures and networks that may be involved in social cognition?

Human functional magnetic resonance imaging (fMRI) studies suggested that social cognition seem to involve brain structures that include the prefrontal cortex (PFC), amygdala, cingulate gyrus, fusiform gyrus, insula and regions in the somatosensory cortex, superior temporal sulcus, and supramarginal gyrus.[16]

It was further suggested[16] that of the previously mentioned structures, the amygdala, alongside the ventromedial prefrontal cortex, is of great interest to social cognition researchers as they share rich functional connections and have been found to play a role in autism, schizophrenia, and other mental disorders.[16]

It was also pointed out that the neural brain structures and networks involved in the different social cognition domains not only are related but also have each distinctive features. Face processing, for example, seems to activate the fusiform face area,[17, 18] while the ventromedial PFC plays a role in facial emotion recognition and the ability to process ToM.

As will be discussed later, there are now research data demonstrating social cognitive impairments in neurodevelopmental, psychiatric, and neurodegenerative disorders.[19–21]

In this regard, social withdrawal was suggested to be one of the first and common signs of early social dysfunction in a number of important neuropsychiatric disorders including schizophrenia, ASD, and Alzheimer's disease, due to the complexity of the brain processes required to initiate and maintain social relationships.[22]

It is also of note that social cognition was suggested to have a predictive value as ASD or psychotic patients with relatively low social cognitive abilities will demonstrate low social functioning.[23, 24]

Interestingly, Cotter et al.[25] recently identified 31 metaanalyses comparing facial emotion recognition and/or ToM task performance of subjects with psychiatric, neurological, or neurodevelopmental disorders with healthy controls. Based on data obtained for schizophrenia, ASD, and Alzheimer's disease along with data on other neurological and mental brain disorders, Cotter et al.[25] suggested that social cognitive deficits seem to be a core cognitive phenotype of many clinical disorders. They stressed the need to understand the "real world" effect of these social cognitive impairments, to explore whether they can serve as a marker of neurological abnormality, and to find appropriate effective trans-diagnostic interventions.

To sum up this section, social cognitive impairments seem to transcend clinically determined diagnostic boundaries[26] for ASD, schizophrenia, and Alzheimer's disease raising a question of some shared mechanisms in the base of these disorders.

Social cognition and ASD

The first to demonstrate impairments of social cognition in ASD was Baron-Cohen who showed in the 1980s that autistic children lack a normal theory of mind.[27]

In general, ASD subjects demonstrate a range of deficits in social cognition and in its various domains, including the domains of emotional processing, theory of mind (ToM), social perception, and attributional style.

Many studies researched these domains in ASD using a variety of operational variables relating to face processing (i.e., face identity, face recognition, facial emotion recognition,[28–34]), body motion perception,[35–38] and joint attention,[39–41] and a variety of brain areas as well as biological processes were reported to be associated with the different aspects of social cognition in ASD.[12, 42–52]

Given the broad range of research studies conducted on the different aspects of social cognition, several metaanalyses were held to evaluate general trends among groups of studies, each of them focusing on similar aspects of social cognition. Such metaanalyses are grouped later to one of five different groups: (a) metaanalyses of ASD subjects in general, (b) metaanalysis of subjects with specific genetic diagnosis of ASD, (c) metaanalyses comparing ASD and schizophrenia, (d) metaanalyses comparing ASD and ADHD; (e) metaanalyses of studies detecting brain areas that are associated with given domains of social cognition in ASD. Since more recent metaanalyses tend to include more research studies and thus more data and since in certain cases they also contain improved method\s of analysis, these metaanalyses are arranged chronologically according to their year of publication (although it does not always reflect exactly the upper time range of studies included in the meta analyses).

Metaanalyses of ASD subjects in general

In this section, we will briefly mention two metaanalyses.

The first is by Mendelson et al. published in 2016 that conducted a metaanalysis of literature regarding school age friendships (suggested to be dependent on social cognition among other factors) among boys with ASD.[53] The authors report that youth with ASD do make friends according to their peers and parents. However, self-reported friendship quality and parent- and peer-reported quantity were poorer than typically developing peers.

The second metaanalysis is by Velikonja et al. published in 2019 that conducted a metaanalysis of studies assessing at least one domain of neurocognitive functioning or social cognition using standard measures.[54] Seventy-five studies were included. Adults with ASD showed large impairments in theory of mind and emotion perception and processing.

The authors conclude that adults with ASD show impairments in social cognitive domains and in specific nonsocial cognitive domains. Both of these metaanalyses suggest either directly or indirectly impairments in social cognition domains in autism.

Metaanalysis of subjects with specific genetic diagnosis of ASD

Chisholm et al. in 2018 conducted a metaanalysis of social functioning in ASD children and adults with neurofibromatosis Type 1.[55] Twenty-two publications were included. Findings from this review and metaanalysis provide evidence that children and adults with neurofibromatosis Type 1 exhibit significantly higher prevalence and severity of social dysfunction and ASD symptomatology.

Although only one metaanalysis of a specific genetic disease related to ASD was published, it may show that social dysfunction and possibly social cognition are impaired not only in idiopathic ASD but also in the specific genetic diseases demonstrating ASD symptomatology.

Metaanalyses comparing ASD with schizophrenia

Three metaanalyses comparing ASD and schizophrenia were published thus far.

The first is by Sugranyes et al. published in 2011 that reviewed 33 studies using fMRI to compare ASD or schizophrenia subjects with healthy control group, comparing standardized facial emotion recognition or theory of mind (ToM) paradigms.[56] Both schizophrenia and ASD showed medial prefrontal hypoactivation, which was more pronounced in ASD, while ventrolateral prefrontal dysfunction was associated mainly with schizophrenia. Amygdala hypoactivation was observed in schizophrenia patients during facial emotion recognition and in ASD during more complex ToM tasks. Both disorders were associated with hypoactivation within the superior temporal sulcus during ToM tasks, but activation in these regions was increased in ASD during affect processing. Disease-specific differences were noted in somatosensory engagement, which was increased in schizophrenia and decreased in ASD. Reduced thalamic activation was uniquely seen in schizophrenia.

These authors conclude that reduced frontolimbic and superior temporal sulcus engagement emerges as a shared feature of social cognition deficits in both schizophrenia and ASD. However, there were disease- and stimulus-specific differences. These findings the authors suggest may aid future studies on schizophrenia and ASD and facilitate the formulation of new hypotheses regarding their pathophysiology.

The second metaanalysis is by Chung et al. published in 2014 that conducted a metaanalysis of 37 studies assessing verbally or visually components of mentalizing (ToM abilities) in adults with schizophrenia or ASD reporting similar large estimated effect sizes of impairments in verbal and visual mentalizing tasks for both schizophrenia and ASD subjects.[57] Adults with schizophrenia showed a trend toward larger impairments on a verbal than on a visual mentalizing task, while adults with ASD did not show different levels of impairment on the verbal compared with the visual tasks. The authors suggest that their results demonstrate that both schizophrenia and ASD share, at least in part, some common cognitive processing deficits associated with mentalizing impairments.

Finally, Fernandes et al. in 2018 aiming at reviewing studies comparing social cognitive performance between subjects with schizophrenia and ASD performed a metaanalysis of 19 research studies.[58] Eight studies providing data on facial emotion perception demonstrated better performance in schizophrenia versus ASD. No significant differences were found between the groups as to ToM tasks, emotional intelligence, and social skills. The authors suggest that instruments that are currently available for the evaluation of social cognition poorly differentiate between individuals with schizophrenia and ASD. On the other hand, it may be argued that such results are due to common dysfunction of the social cognitive domains between the both diagnoses.

Taken together, these metaanalyses seem to suggest some common elements between both disorders regarding social cognition as manifested by (a) reduced frontolimbic and superior temporal sulcus engagement as a shared feature of social cognition deficits in schizophrenia and ASD; (b) both schizophrenia and ASD share some common cognitive processing deficits associated with ToM and perhaps other features of social cognition as well.

Metaanalysis comparing ASD with ADHD

Bora and Pantelis published in 2016 conducted a metaanalysis of social cognition, including emotion recognition and ToM studies in attention-deficit/hyperactivity disorder (ADHD) subjects compared with healthy controls and ASD patients.[59] Forty-four studies comparing ADHD with healthy controls and 17 studies comparing ADHD with ASD were included. Deficits in social cognition, especially ToM, were significantly more pronounced in ASD compared with ADHD. The authors suggest that performance of individuals with ADHD on social cognition lies in an intermediate stage between ASD and healthy controls.

Metaanalyses of studies detecting brain areas associated with given domains of social cognition in ASD

At least five metaanalyses aimed at detecting brain regions and networks associated with social cognition in ASD.

First, Di Martino et al. published in 2009 conducted 2 separate voxel-wise activation likelihood estimation metaanalyses of 39 functional neuroimaging studies consisting of 24 studies examining social processes (e.g., theory of mind and face perception) and 15 studies examining nonsocial processes.[60] Compared with control subjects, ASD patients showed greater likelihood of hypoactivation in two medial wall regions: the perigenual anterior cingulate cortexes (ACC) in social tasks only and the dorsal ACC in nonsocial studies. Furthermore, the right anterior insula was more likely to be hypoactivated in ASD according to the analyses of social studies. The authors conclude that their results showed ASD-related patterns of hypofunction and aberrant activation that depends on the specific cognitive domain, that is, social versus nonsocial.

Second, Sugranyes et al. (mentioned earlier) in 2011 reviewed 33 studies using fMRI to compare ASD or schizophrenia subjects with healthy control group, as to standardized facial emotion recognition or theory of mind (ToM) paradigms. As mentioned earlier, these authors reported reduced frontolimbic and superior temporal sulcus involvement as a shared feature of social cognition deficits in schizophrenia and ASD. However, the authors also note that there were disease- and stimulus-specific differences.

Third, van Veluw and Chance in 2014 conducted a metaanalysis of fMRI studies that assessed self-awareness and awareness of others (ToM).[61] The authors confirmed that self-recognition in contrast to recognition of others' faces and awareness of others not only involves a network that consists of separate, distinct neural pathways but also includes overlapping regions of higher order in the prefrontal cortex where these processes may be combined.

Forth, Yang and Hofmann published in 2016 conducted a metaanalysis of functional magnetic resonance imaging studies that investigated the mirror neuron system (a possible link between action observation and social skills; see Ref. 62) that is reflected by action observation and imitation in ASD and typically developing subjects.[63] Thirteen studies were selected in the analysis. The results suggested that ASD individuals compared with typically developing subjects showed stronger effects in the anterior inferior parietal lobule, a part of the putative human mirror neuron system. In addition, the ASD group demonstrated altered effects in the occipital cortex, dorsolateral prefrontal cortex, cingulate cortex, and insula. The authors suggest that ASD individuals demonstrate dysfunction of the mirror neuron system during action observation and imitation. The authors suggested that brain regions involved in visual processing, executive function, and social cognitive function might also show dysfunction during action task performance.

Finally, Patriquin et al. published in 2016 conducted a metaanalysis of 50 neuroimaging studies of social cognition in children and adults with

ASD demonstrating consistent differences in activation in the posterior superior temporal sulcus at the temporoparietal junction, middle frontal gyrus, fusiform face area, inferior frontal gyrus, amygdala, insula, and cingulate cortex between ASD and typically developing control individuals.[64] Surface-based cortical morphometry analysis showed alterations in the thickness, volume, and surface area in individuals with ASD in the superior temporal sulcus, insula, and fusiform face area. Increased cortical thickness was found in individuals with ASD in the inferior frontal gyrus. The authors conclude that their results provide functional and anatomical bases for social cognition abnormalities in ASD by identifying common signatures from a large pool of neuroimaging studies.

To sum up, there are clearly data providing functional and anatomical basis of social cognition abnormalities in ASD.

Social cognition and schizophrenia

As stated previously, schizophrenia patients demonstrate negative symptoms among which is reduced interest in social interaction.[3] Thus, interest in social interactions and social functioning is impaired in subjects with schizophrenia, many of them demonstrating increased social withdrawal.[65] Importantly in this regard, poor social functioning and social withdrawal in schizophrenia were suggested to be caused, at least in part, by behavioral deficits due to impaired social cognition.

Four domains of social cognition skills are frequently assessed in schizophrenia: emotional processing, theory of mind (ToM), social perception, and attributional style and schizophrenia seem to be associated with impairments in all four of these domains of social cognition.[66–76] Such impairments in social cognition seem to appear prior to the schizophrenia onset, at the first episode, and during its course.[77–82]

Most importantly, impairments in social cognition were reported to be associated with the disease functional outcome and overall recovery, an outcome associated with gross work disabilities and dysfunctional daily living activities.[83–87]

Moreover, social cognition was suggested by some to be an important mediating link between neurocognition and social functioning, and it was suggested that impairment in social cognition is more important than impairment in neurocognition of a general nature as to the functional outcome of schizophrenia.[86]

Interestingly, in this regard, Mancuso et al.[88] examined a sample of outpatients with psychosis showing that lower-level social cue detection along with other variables (higher-level inferential and regulatory processes) correlated with functional outcome but not with clinical symptoms. Moreover, lower-level social cue detection accounted for unique incremental variance in functional capacity, above and beyond nonsocial neurocognition and negative symptoms.

In the same line, Mike et al.[89] performed an exploratory factor analysis on patients within the early course of schizophrenia assessing various domains of social cognitive skills to investigate the relationships between these distinct domains of social cognition skills and (a) neurocognitive performance, (b) clinical symptoms, and (c) social functioning satisfaction. The results revealed the presence of a three-factor solution, representing the domains of emotion management, emotion recognition, and ToM, together accounting for above 50% of the variance. Moreover, higher scores on the ToM factor were significantly related to higher social functioning satisfaction measures as well as with lower clinical symptoms severity.

Finally, impairments in social cognition in schizophrenia seem to be associated with neural dysfunction in a variety of brain regions and networks involved in social processing—that seem to include the amygdala, somatosensory-related cortex, ventral striatum, and medial orbitofrontal cortex; cognitive control (used in self-regulatory processes)—the lateral prefrontal cortex, anterior cingulate cortex, and superior parietal lobe; and theory of mind—superior temporal sulcus, temporoparietal junction, posterior cingulate, and medial prefrontal cortex.[71]

The degree of dysfunction of one or more of these brain regions was suggested to be related to the specific social cognitive performance and real-world social behavior.[14, 71, 90–93]

Given the broad range of research studies exploring social cognition domains in schizophrenia, several metaanalyses tried to evaluate general trends among group of studies focusing on similar aspects of social cognition.

Such metaanalyses are grouped later to one of seven groups:

- **(a)** metaanalysis of social cognition domains in high risk for psychosis,
- **(b)** metaanalysis of social cognition domains in first-degree relatives of people with schizophrenia,
- **(c)** metaanalyses of social cognition domains and processes in schizophrenia subjects,
- **(d)** metaanalysis on the relationship between neurocognition and social cognition with functional outcomes in nonaffective psychosis,
- **(e)** metaanalyses comparing social cognition domains and processes between schizophrenia and ASD,
- **(f)** metaanalysis comparing social cognition domains and processes between schizophrenia and bipolar disorder,
- **(g)** metaanalyses of studies detecting brain areas associated with given domains of social cognition in schizophrenia.

As stated earlier, since more recent metaanalyses tend to include more research studies and thus more data and in certain cases also improved method/s of analysis, the metaanalyses in each group are arranged chronologically according to the year of publication.

Metaanalysis of social cognition domains in high risk groups for psychosis

Lee et al. published in 2015 conducted a metaanalysis of 20 studies examining social cognitive domains (theory of mind, social perception, attributional bias, and emotion processing) in individuals with clinical high risk for psychosis reporting that individuals at clinical high risk for psychosis exhibited significant impairments in all domains of social cognition compared with healthy controls, with the largest effect size identified for attributional bias.[81]

Metaanalysis of social cognition domains in first-degree relatives of people with schizophrenia

Lavoie et al. in 2013 conducted a metaanalysis of studies investigating at least one domain of social cognition (mentalizing or ToM, emotional processing, social perception, social knowledge, and/or attributional style) in adult first-degree relatives of patients with schizophrenia.[94] Twenty-nine studies were included of which 11 evaluated mentalizing, 20 emotional processing, and 2 social perception. Moderate mean effect sizes were obtained for these three components. Overall, the results from this metaanalysis highlight that social cognition is globally affected in first-degree relatives of people with schizophrenia, suggesting that social cognition deficits in schizophrenia may be related to a genetic vulnerability for the disorder.

Metaanalysis of social cognition domains and processes in schizophrenia subjects

Five metaanalyses were conducted in this regard:

First, Kohler et al. published in 2010 conducted a metaanalysis of 53 studies demonstrating a large deficit in emotion perception in schizophrenia, irrespective of task type.[95] Several factors moderated the observed impairment including illness-related clinical factors and patients' demographics.

Second, Savla et al. published in 2013 conducted a metaanalysis on 112 studies of social cognition (ToM, social perception, social knowledge, attributional bias, emotion perception, and emotion processing) in schizophrenia.[96] Schizophrenia subjects performed worse than normal controls across all domains, with large effects for social perception, emotion perception, and emotion processing. Greater deficits in social

and emotion perception were associated with inpatient status, and greater deficits in emotion processing were associated with longer illness duration. The authors conclude that the evidence suggest deficits across multiple social cognitive domains in schizophrenia. Third, Song et al. published in 2015 stated, "Previous meta-analytic studies conducted in Western countries have consistently revealed impairments in ToM in schizophrenia.[97] However, there is no systematic meta-analytic review of ToM deficits in individuals with schizophrenia in non-Western countries." This led them to conduct a metaanalytic study of 13 studies comparing ToM performance of adults with schizophrenia to healthy controls in South Korea. Large impairments in affective and cognitive ToM and verbal and visual ToM were found in Korean subjects with schizophrenia. There were no differences in magnitude between affective and cognitive ToM or between verbal and visual ToM. The authors suggest that their results demonstrate that Korean subjects with schizophrenia experience substantial impairments in various ToM domains.
Forth, Bonfils et al. published in 2016 conducted a metaanalysis of studies examining empathy (an important social skill and an element of social cognition) in schizophrenia subjects. Individuals with schizophrenia exhibited statistically significant, medium deficits in affective empathy.[98] Measurement type moderated the affective empathy deficit such that performance-based measures showed larger schizophrenia group deficits than self-report measures. The authors concluded that consistent significant deficits in affective empathy were found when comparing people with schizophrenia with healthy controls, especially when using performance-based assessments.
Finally, Bonfils et al. in 2017 conducted a metaanalysis of 32 studies examining self-reported empathic tendencies in schizophrenia subjects compared with healthy controls.[99] Compared with healthy controls, schizophrenia subjects reported significantly reduced tendencies for empathic concern, perspective-taking, and fantasy but significantly greater tendencies for personal distress.

Thus, these metaanalyses clearly demonstrate deficits across multiple social cognitive domains in schizophrenia in both western and nonwestern countries.

Metaanalysis on the relationship between neurocognition and social cognition with functional outcomes in nonaffective psychosis

Fett et al. published in 2011 conducted a metaanalysis of 52 studies examining cross-sectional associations between neurocognition, social cognition, and functional outcome in individuals with nonaffective psychosis.[86] The authors concluded that social cognition was more strongly associated with functioning within one's community than neurocognition, with the strongest associations being between theory of mind and functional outcomes. However, the authors stated that three-quarters of variance in outcome were still left unexplained.

Metaanalyses comparing social cognition domains and processes between schizophrenia and ASD

Here, three metaanalyses by Sugranyes et al.,[56] Chung et al.,[57] and Fernandes et al.[58] were conducted; please see the section on metaanalyses comparing social domains between ASD and schizophrenia appearing earlier.

Metaanalysis comparing social cognition domains and processes between schizophrenia and bipolar disorder

Bora and Pantelis in 2016 conducted a metaanalysis of 26 studies comparing facial emotion recognition and theory of mind (ToM) abilities in schizophrenia and bipolar disorder.[100] Schizophrenia patients significantly underperformed compared with bipolar patients in both facial emotion recognition and ToM. Neurocognitive deficits significantly contributed to schizophrenia-bipolar disorder group differences for ToM. However, between-group differences for social cognition were not statistically more severe than neurocognition. The authors conclude that (1) the social cognitive impairment is more severe in schizophrenia, (2) between-group differences were modest and are comparable with other neurocognitive differences between schizophrenia and bipolar disorder, and (3) there is a significant overlap in social cognitive performance deficits observed in both disorders.

Metaanalyses of studies detecting brain areas associated with given domains of social cognition in schizophrenia

Kronbichler et al. published in 2017 conducted a metaanalysis of 21 studies researching functional brain abnormalities in schizophrenia trying to identify regions exhibiting convergent under- or overactivation during ToM tasks.[101] Common regions associated with ToM like medial prefrontal cortex and temporoparietal junction revealed abnormal activation in schizophrenic patients: Underactivation was identified in the medial prefrontal cortex, left orbitofrontal cortex, and a small section of the left posterior temporoparietal junction, while marked overactivation was identified in a more dorsal, bilateral section of the temporoparietal junction. Also, abnormal activation was identified in medial occipitoparietal cortex, right premotor areas, left cingulate gyrus, and lingual gyrus. The authors suggest that their findings demonstrate that schizophrenia patients simultaneously show over- and underactivation in ToM-related regions. Interestingly, temporoparietal junction reveals diverging activation patterns with an underactivating left posterior and an overactivating bilateral dorsal section. The authors conclude that schizophrenia patients show less specialized brain activation in regions linked to ToM and increased activation in attention-related networks suggesting compensatory effects.

Finally, Jáni and Kašpárek in 2017 used a metaanalytic method to explore brain activation during facial emotion recognition and theory of mind tasks in schizophrenia patients.[102] The authors report that schizophrenia patients showed lesser recruitment of the facial emotion processing network; behavioral performance was associated with the activation of the precentral gyrus. They also report abnormal activation of the mentalizing network in schizophrenia patients during a task concerning reasoning about other people's mental states; patients with worse performances showed lesser activation in the right insula and superior temporal gyrus. Multimodal metaanalysis showed overlaps of brain-related abnormalities for both modalities in schizophrenia, with reduced recruitment of the right insula, anterior cingulate, and medial prefrontal cortex and increased activation in the bilateral parietal cortex. The authors conclude that their findings suggest a complex impairment of social cognition demonstrated by neural-related circuit disruptions during facial emotion processing and theory of mind tasks in schizophrenia.

To sum up the previously mentioned section about metaanalyses of social cognition in schizophrenia, (a) there are data demonstrating deficits across multiple social cognitive domains in schizophrenia in both western and nonwestern countries; (b) there are data providing functional and anatomical basis of social cognition abnormalities in schizophrenia, some of this anatomical basis appearing also in other diagnoses; and (c) such abnormalities are found even prior to the disease breakout as demonstrated in high-risk groups for psychosis and may have genetic background as demonstrated by findings showing dysfunctional social cognition in first-degree relatives of patients with schizophrenia.

Social cognition and Alzheimer's disease

Alzheimer's disease (AD) is the most frequent neurocognitive disorder characterized by progressive decline of cognitive abilities[103] among them decline of social cognition, while mild cognitive impairment (MCI) is a disorder that causes mild measurable decline in cognitive abilities posing the individual with MCI at an increased risk of developing dementia mainly Alzheimer's.[103, 104]

We will discuss later four aspects of social behavior in AD and MCI subjects. Two of them are dependent at least partially on social cognition (social awkwardness and social withdrawal) and the other two (ToM and emotion recognition) constituting important components of social cognition.

Thus, AD patients may present social awkwardness[105, 106]—a multifactorial entity that seem to be in part secondary to impairments in social cognition. Lindau et al.[105] reported that social awkwardness seems to appear as the first symptoms in AD in about 5% of patients. However, they stressed that its appearance as a first symptom is more characteristic to frontotemporal dementia.

Another symptomatology of Alzheimer's disease includes negative symptoms among them social withdrawal[103, 107] that may be manifested even prior to AD diagnosis and seem to increase gradually during the disease's course.[108, 109]

Yet, another important component of social cognition is ToM. Several studies assessing ToM abilities have demonstrated poor ToM skills in AD patients[110] as well as in MCI patients[111] compared with healthy controls. Interestingly, it seems that advanced ToM abilities are impaired first in AD, while lower level of ToM abilities are impaired later in the course of the disease.[112] Recently, Bora and Yener in 2017[113] performed a metaanalysis of theory of mind (ToM) and emotion recognition deficits in MCI incorporating 17 studies. Mild cognitive impairment was associated with significant impairments in ToM and in the facial emotion recognition of fear and sadness. Social cognitive deficits were more severe in multidomain MCI.

The same group, Bora et al. in 2015 performed a metaanalysis comparing facial emotion recognition performances between the behavioral variant of frontotemporal dementia, AD, and healthy controls.[114] Their metaanalysis included 19 studies; behavioral variant of frontotemporal dementia was associated with more severe deficits than AD in total emotion recognition and with the emotion recognition score of each of the negative emotions studied.

Several authors tried to explain the dysfunctional brain areas associated with different aspects of ToM in MCI and AD. Thus, Lindau et al. suggested that the ToM deficits in MCI and AD patients may be secondary to deficits in working memory and executive functions that are required in more complex ToM tasks,[105] while Dermody et al. suggested that empathic (dependent on ToM abilities) impairments may be caused by early atrophy in AD of the left temporoparietal junction, temporal fusiform cortex, inferior temporal gyrus, angular gyrus, and bilateral middle and superior posterior temporal gyrus.[115]

In this regard, some authors suggested that brain atrophy associated with ToM reflect the level of impairment in ToM abilities.[114, 116, 117] Also, it was reported that brain structures suggested to be associated with social functioning show abnormal changes that tend to progress with the years of illness.[109, 112]

Finally, there is the question of whether ToM and other domains of social cognition are solely dependent on general neurocognition or perhaps constitute a separate entity. In this regard, Cosentino et al. monitoring 517 subjects with probable AD for 5.5 years suggested that social abilities and general neurocognition in AD seems to consist of separate entities and that social cognition seem to be related to clinical outcomes associated with the patient's disability.[118]

To sum up this section, I will cite Porcelli et al.[14] who summarized the findings as related to the impairments of social cognition in AD stating that "AD physiopathology seems to progressively impair the social brain at multiple sites, starting from the structures that sustain motivation to the mentalizing network…resulting *(in)* poor social functioning."

Pharmacological treatment of social cognition impairments

As stated earlier, social cognition is reported to be impaired in ASD, schizophrenia, and Alzheimer's disease, and one may ask—in the lack effectiveness of the current drugs used to treat these diagnoses—the question whether there are existing pharmacological agents that may be repurposed to alleviate the impairment of social cognition domains.

Green[84] recently wondered, "With so many drugs available to improve the symptoms of schizophrenia, why is the disorder still associated with such severe disability?" (see also Ref. 119 in this regard). Green[84] suggested that "researchers and clinicians have begun to realize that a hindrance to widespread recovery (*in schizophrenia*) is that (*the*) available antipsychotic medications have been effective in treating the positive symptoms (*hallucinations, delusions*) of schizophrenia but no other features of this illness such as (*social*) cognitive impairment," concluding that impairments in social cognition play a significant role on the functional status of these patients.

So at least for schizophrenia, the standard treatment with antipsychotic drugs does not seem to alleviate social cognition impairment as well as daily living functioning shown to be mediated by such impairment.

In this regard, Gabay et al. conducted a metaanalysis of antipsychotic treatment effects on facial affect processing deficits in schizophrenia, the latter being an important component of social cognition.[120] Nine studies met their criteria for entry into the metaanalysis that included data from

1162 patients and 6 antipsychotic medications. The authors reported that antipsychotic medications are poor at improving facial affect processing compared with reducing symptoms.

Such example of the lack of effectiveness of the current drugs used to treat schizophrenia on impairments associated with social cognition open the door for research as to effectiveness of other pharmacological agents.

Oxytocin: A candidate treatment for dysfunctional social cognition

One possible candidate drug to alleviate impairment of social cognition domains is oxytocin. Oxytocin is a peptide hormone and neuropeptide produced by the paraventricular nucleus of the hypothalamus and released by the posterior pituitary. It plays an important role in the regulation of affiliative behavior and social bonding in both nonhuman mammals and humans, among other functions. Thus oxytocin plays a modulatory role in a wide range of socially relevant behaviors, including pair bonding, anxiety, stress reactivity, aggression, maternal behavior, sexual behavior, social interaction behavior, and social recognition. These properties of oxytocin make it a possible candidate drug for the treatment of a variety of impairments in social cognition.[121–124]

In this regard, a recent metaanalysis of studies using healthy participants found that intranasal oxytocin (the most common route of administering this drug) enhanced emotion recognition of human faces, in particular faces that demonstrated happiness or fear.[125]

Oxytocin treatment in ASD

Since deficits in social cognition affiliative behavior and social bonding are pervasive and characteristic of neurodevelopmental disorders including ASD, oxytocin was researched as a potential drug to alleviate various aspects of social cognition in ASD.

Many studies examining the effect of oxytocin in ASD were conducted (see, for example, oxytocin studies by Lim et al.,[126] Hollander et al.,[127] Andari et al.,[128] Guastella et al.,[129] Domes et al.,[130] Domes et al.,[131] Veening and Olivier,[132] Lukas and Neumann,[133] and Gordon et al.[134]) reporting that the effects of oxytocin on brain activity could vary depending on the social significance of the stimuli being encountered. The results of the various trials of oxytocin in ASD (most of them with intranasal route of administration) were inconsistent, with mixed results and thus inconclusive. Trying to sort out if such treatment is relevant to ASD, four metaanalyses were held in the recent years. From the reasons mentioned in the sections earlier, I will mention them chronologically according to the year of publication.

First, LoParo and Waldmanin stating that "Inconsistency in the effect sizes of the oxytocin receptor (OXTR) variants included in association studies render it unclear whether OXTR is truly associated with ASD, and, if so, which OXTR single-nucleotide polymorphisms (SNPs) are associated" conducted a metaanalysis published in 2015 of 16 oxytocin receptor SNPs that included 3941 individuals with ASD from 11 independent samples.[135] They found significant associations between ASD and the several SNPs (rs7632287, rs237887, rs2268491, and rs2254298). Oxytocin receptor was also significantly associated with ASD in a gene-based test.

Second, Ooi et al. published in 2017 reviewed randomized controlled trials examining the effects of oxytocin administration on social cognition and the restricted, repetitive behaviors in individuals with ASD.[136] Twelve trials were included in this review. Seven out of the 11 studies that examined social cognition reported improvements following oxytocin administration. However, the results as a whole suggested that oxytocin has no significant effect on social cognition. The authors concluded that previous evidence revealed mixed findings about the effects of oxytocin on ASD. However, they added that given the limited number of trials, the effectiveness of oxytocin on ASD should still be considered tentative.

Third, Keech et al. published in 2018 held a metaanalysis of randomized controlled trials considering the effect of intranasal oxytocin on social cognitive domains across a range of neurodevelopmental disorders. Seventeen studies met inclusion criteria, comprising 466 participants with neurodevelopmental disorders.[137] This metaanalysis reported no significant effect of oxytocin on emotion recognition; a moderate but nonsignificant effect on empathy; and a small, significant effect on ToM. Metaregression indicated that the effect of intranasal oxytocin on social cognition was not moderated by the diagnosis, the age of participant, the dose, or the frequency of intranasal oxytocin administration. The authors called for more well-designed randomized controlled trials and added that until then the promise of intranasal oxytocin should be considered tentative.

Finally, Cai et al. published in 2018 conducted a systematic review and metaanalysis of the reported adverse events of long-term intranasal oxytocin treatment for autism spectrum disorder.[138] There were 223 participants in 5 trials. The common side effects included nasal discomfort (14.3%), tiredness (7.2%), irritability (9.0%), diarrhea (4.5%), and skin irritation (4.5%). The authors conclude that their results support that intranasal oxytocin is a tolerated and safe drug for use in ASD subjects.

Thus these metaanalyses suggest that (a) the promise of intranasal oxytocin (suggested to have relatively tolerated and safe profile) should be still considered tentative until more data are provided and (b) that the oxytocin genetic variants (i.e., single nucleotide polymorphisms) should be considered in future well-designed randomized controlled trials.

Oxytocin treatment in schizophrenia

Oxytocin has been also examined as a possible pharmacological tool to enhance social cognition in schizophrenia (see Shilling and Feifel[139]; Bukovskaya and Shmukler[140] for review). However, as in the case of ASD, despite considerable efforts to study this drug in schizophrenia, its effects on social cognition are not consistent across studies,[141–147] and it was suggested that a variety of variables may mediate the response of subjects with schizophrenia to this drug. Thus Davis et al.[148] reported that high dose of oxytocin administration improved relatively high-grade social cognition assessments, including empathy and detection of sarcasm, but not lower-level assessments such as facial affect perception and social perception. In general, it was also suggested that some improvements in schizophrenia appears after relatively high dosage of oxytocin, rather than low ones.[149] In the same line of thought, Davis et al.[150] aiming at advancing the understanding of the genetic variations associated with social cognitive deficits in schizophrenia genotyped schizophrenia outpatients assessing seven single nucleotide polymorphisms (SNPs) of the oxytocin receptor (OXTR). For one of the seven SNPs, rs2268493, the "T" allele was significantly associated with poorer performance on a composite social cognition index, as well as with specific tests of ToM and social perception. The authors suggest that although the sample size was small, their findings seem to point to an involvement of genetic variants of the oxytocin receptor in social cognitive impairments in schizophrenia. Perhaps, such variants should be taken into account in future studies of oxytocin for social cognition impairments in schizophrenia.

Interestingly, Watanabe et al.[151] studied oxytocin receptor gene and risk of schizophrenia in case-control and family-based analyses reporting no significant associations between any of the 14 SNPs examined and schizophrenia in both the case-control and trio samples. A metaanalysis using both samples indicated only a nominally significant association between rs9840864 and schizophrenia.

To try and sort out the efficacy of oxytocin in schizophrenia symptomatology, four metaanalyses were held in recent years. Three studied oxytocin effect on positive and negative symptomatology, while one studied its effects on social cognition. We will describe them here chronologically:

Oya et al. published in 2016 performed a metaanalysis of oxytocin augmentation therapy in patients with schizophrenia who received antipsychotic agents. Seven randomized controlled trials were included.[152] Oxytocin was superior to placebo for decreasing the positive and negative syndrome; however, a sensitivity analysis including only oxytocin administration on consecutive days was superior to placebo in negative symptoms only.

Williams and Bürkner in 2017 stated that previous metaanalyses showed inconsistent results (two of them were retracted) leading them to conduct a relatively well-designed updated metaanalysis examining the effect of intranasal oxytocin on the symptomatology in schizophrenia.[153] Eight randomized controlled trials were included demonstrating that oxytocin is not therapeutically effective for treating negative symptoms in schizophrenia.

Bürkner et al. in 2017 performed a Bayesian metaanalysis of the efficacy of intranasal oxytocin to improve areas of social and neurocognition in schizophrenia.[154] Across 12 studies ($N=273$), intranasal oxytocin did not improve social cognition or neurocognition. Moderator analyses revealed that intranasal oxytocin had a significantly larger effect on high-level social cognition (i.e., mentalizing and theory of mind) compared with low-level social cognition (i.e., social cue perception). When restricting the analysis to outcomes for high-level social cognition, there was a significant effect of intranasal oxytocin though the effect was not robust to sensitivity analyses. The authors suggest that their metaanalysis indicates that intranasal oxytocin may have selective effects on high-level social cognition, providing a more focused target for future studies of intranasal oxytocin.

Zheng et al. in 2018 conducted a metaanalysis of 10 randomized controlled trials of subjects with schizophrenia receiving adjunctive intranasal oxytocin.[155] Dose-response effect analyses showed that only high-dose (80 IU/day) intranasal oxytocin had superiority over placebo in improving total psychopathology and positive symptom score. The authors conclude that although their metaanalysis shows no positive effect in general, the higher dose of adjunctive intranasal oxytocin (80 IU/day) appears to be efficacious and safe in improving total psychopathology and positive symptoms in schizophrenia.

To sum up, it seems that although intranasal oxytocin studies in schizophrenia are yet not conclusive, it seems that future well-designed studies should aim at the higher level of social cognition, administer higher doses of oxytocin, and possibly target at certain genetic variants of the oxytocin receptor.

Pharmacological agents affecting the NMDA glutamate receptor as candidate treatments for dysfunctional social cognition

It was reported that *N*-methyl-D-aspartate improved social recognition potency in animal models, while an NMDA antagonist disrupted it.[156, 157]

ASD has been hypothesized to demonstrate abnormal glutamatergic activity, based on both human and animal data.[158, 159]

Thus, several studies researched D-cycloserine (DCS, a partial agonist at the glycine site of the NMDA glutamate receptor) and memantine (an NMDA glutamate receptor antagonist) in ASD.

In this regard, Posey et al.[160] studied DCS in ASD in a small prospective single-blind study reporting that DCS was associated with improvement in social withdrawal, while a group of open-label studies of memantine in ASD has reported improvements in memory[161] and social interaction/withdrawal[161–163] although not all memantine studies have demonstrated improvements in social withdrawal.[164, 165]

Schizophrenia is another disorder reported to have hypoglutamatergic neurotransmission, and thus D-cycloserine was researched in relation to this disorder. In this regard, Lai et al.[166] based upon the hypothesized hypofunction of *N*-methyl-D-aspartate receptor (NMDAR)-mediated signaling pathways in the involvement of cognitive and social impairments in schizophrenia, discussed three NMDAR-related compounds/drugs that could be researched in animal models for their effect on social impairments: D-serine, sarcosine, and D-cycloserine. In this regard, Shimazaki et al.[167] reported that D-serine and sarcosine enhanced social memory in rats.

Finally, Goff et al.[168] reviewed the literature on D-cycloserine (DCS) administration in animal models and in human studies suggesting that preliminary studies in patients with schizophrenia demonstrate that intermittent dosing of DCS may produce persistent improvement of negative symptoms (which include social withdrawal) and enhance learning when combined with cognitive behavioral therapy for delusions or with cognitive remediation.

Psychostimulant and other drugs as candidate treatments for dysfunctional social cognition

It is also of note that some psychostimulant drugs may be researched as candidate treatments for impairments in social cognition including modafinil, a psychostimulant demonstrated to improve via a single dose the emotional recognition of faces (mainly sad ones) in first-episode psychosis,[169] and methylphenidate, reported by Campbell-Meiklejohn et al.[170] to enhance conformity (complying with social norms) following moderate social conflict. Finally, another potential option for intervention in the functional activities of daily living that encompass social cognition is NAP (davunetide, CP201) as detail in Chapter 2 written by Illana Gozes.

Noninvasive brain stimulation techniques

Following preliminary reports on noninvasive brain stimulation techniques and social behavior,[171–175] it seems that transcranial magnetic stimulation (TMS) and transcranial direct-current stimulation (tDCS) may also be used in future studies to manipulate social cognition.

Perspective as to future modes of treatment targeting dysfunctional domains of social cognition

Finally, given the growing understanding that some neuropsychological treatments may alleviate social cognition impairments (this issue is beyond the scope of this chapter; see Green et al., Horan and Green[70, 176])—although many questions are still open in this regard (see a critical review by Horan and Green[176])—it seems that future modes of treatments, hopefully transcending diagnostic boundaries, will include the combination of drug or noninvasive brain stimulation therapies combined with neuropsychological treatments themselves targeting specific impaired domains of social cognition. A step in this direction was taken by Davis et al.[144] who reported beneficial effects of oxytocin-augmented social cognitive skills training in schizophrenia.

References

1. van Os J, Kapur S. Schizophrenia. *Lancet*. 2009;374(9690):635–645. https://doi.org/10.1016/S0140-6736(09)60995-8.
2. Lieberman JA, First MB. Psychotic disorders. Ropper AH, ed *N Engl J Med*. 2018;379(3):270–280. https://doi.org/10.1056/NEJMra1801490.
3. American Psychiatric Association. *Diagnostic and Statistical Manual of Mental Disorders DMS V*; 2013. https://doi.org/10.1176/appi.books.9780890425596.744053.
4. Querfurth HW, LaFerla FM. Alzheimer's disease. *N Engl J Med*. 2010;362(4):329–344. https://doi.org/10.1056/NEJMra0909142.
5. Gross TF. The perception of four basic emotions in human and nonhuman faces by children with autism and other developmental disabilities. *J Abnorm Child Psychol*. 2004;32(5):469–480.
6. Hill EL, Frith U. Understanding autism: insights from mind and brain. Frith U, Hill EL, eds *Philos Trans R Soc Lond Ser B Biol Sci*. 2003;358(1430):281–289. https://doi.org/10.1098/rstb.2002.1209.
7. Schaefer GB, Mendelsohn NJ, Professional Practice and Guidelines Committee. Clinical genetics evaluation in identifying the etiology of autism spectrum disorders: 2013 guideline revisions. *Genet Med*. 2013;15(5):399–407. https://doi.org/10.1038/gim.2013.32.
8. Richards C, Jones C, Groves L, Moss J, Oliver C. Prevalence of autism spectrum disorder phenomenology in genetic disorders: a systematic review and meta-analysis. *Lancet Psychiatry*. 2015;2(10):909–916. https://doi.org/10.1016/S2215-0366(15)00376-4.
9. Henry JD, von Hippel W, Molenberghs P, Lee T, Sachdev PS. Clinical assessment of social cognitive function in neurological disorders. *Nat Rev Neurol*. 2016;12(1):28–39. https://doi.org/10.1038/nrneurol.2015.229.
10. Phillips LH, Scott C, Henry JD, Mowat D, Bell JS. Emotion perception in Alzheimer's disease and mood disorder in old age. *Psychol Aging*. 2010;25(1):38–47. https://doi.org/10.1037/a0017369.

11. Martinez G, Alexandre C, Mam-Lam-Fook C, et al. Phenotypic continuum between autism and schizophrenia: evidence from the movie for the assessment of social cognition (MASC). *Schizophr Res.* 2017;185:161–166. https://doi.org/10.1016/j.schres.2017.01.012.
12. Arioli M, Crespi C, Canessa N. Social cognition through the lens of cognitive and clinical neuroscience. *Biomed Res Int.* 2018;2018:1–18. https://doi.org/10.1155/2018/4283427.
13. Frith CD, Frith U. Implicit and explicit processes in social cognition. *Neuron.* 2008;60(3):503–510. https://doi.org/10.1016/j.neuron.2008.10.032.
14. Porcelli S, Van Der Wee N, van der Werff S, et al. Social brain, social dysfunction and social withdrawal. *Neurosci Biobehav Rev.* 2018; https://doi.org/10.1016/j.neubiorev.2018.09.012.
15. Mundy P. A review of joint attention and social-cognitive brain systems in typical development and autism spectrum disorder. *Eur J Neurosci.* 2018;47(6):497–514. https://doi.org/10.1111/ejn.13720.
16. Tudusciuc O, Adolphs R. Social cognitive neuroscience clinical foundations. In: *Social Cognition in Schizophrenia: From Evidence to Treatment*. New York, NY: Oxford University Press; 2013. https://doi.org/10.1093/med:psych/9780199777587.003.0005.
17. Adolphs R. Cognitive neuroscience of human social behaviour. *Nat Rev Neurosci.* 2003;4(3):165–178. https://doi.org/10.1038/nrn1056.
18. Haxby JV, Hoffman EA, Gobbini MI. The distributed human neural system for face perception. *Trends Cogn Sci.* 2000;4(6):223–233.
19. Christidi F, Migliaccio R, Santamaría-García H, Santangelo G, Trojsi F. Social cognition dysfunctions in neurodegenerative diseases: neuroanatomical correlates and clinical implications. *Behav Neurol.* 2018;2018:1849794.
20. Fortier J, Besnard J, Allain P. Theory of mind, empathy and emotion perception in cortical and subcortical neurodegenerative diseases. *Rev Neurol (Paris).* 2018;174(4):237–246. https://doi.org/10.1016/j.neurol.2017.07.013.
21. Realmuto S, Zummo L, Cerami C, et al. Social cognition dysfunctions in patients with epilepsy: evidence from patients with temporal lobe and idiopathic generalized epilepsies. *Epilepsy Behav.* 2015;47:98–103. https://doi.org/10.1016/j.yebeh.2015.04.048.
22. Adolphs R. The social brain: neural basis of social knowledge. *Annu Rev Psychol.* 2009;60:693–716. https://doi.org/10.1146/annurev.psych.60.110707.163514.
23. Bertrand M-C, Sutton H, Achim AM, Malla AK, Lepage M. Social cognitive impairments in first episode psychosis. *Schizophr Res.* 2007;95(1–3):124–133. https://doi.org/10.1016/j.schres.2007.05.033.
24. Losh M, Adolphs R, Poe MD, et al. Neuropsychological profile of autism and the broad autism phenotype. *Arch Gen Psychiatry.* 2009;66(5):518–526. https://doi.org/10.1001/archgenpsychiatry.2009.34.
25. Cotter J, Granger K, Backx R, Hobbs M, Looi CY, Barnett JH. Social cognitive dysfunction as a clinical marker: a systematic review of meta-analyses across 30 clinical conditions. *Neurosci Biobehav Rev.* 2018;84:92–99. https://doi.org/10.1016/j.neubiorev.2017.11.014.
26. Sasson NJ, Pinkham AE, Carpenter KLH, Belger A. The benefit of directly comparing autism and schizophrenia for revealing mechanisms of social cognitive impairment. *J Neurodev Disord.* 2011;3(2):87–100. https://doi.org/10.1007/s11689-010-9068-x.
27. Baron-Cohen S, Leslie AM, Frith U. Does the autistic child have a "theory of mind"? *Cognition.* 1985;21(1):37–46.
28. Weigelt S, Koldewyn K, Kanwisher N. Face identity recognition in autism spectrum disorders: a review of behavioral studies. *Neurosci Biobehav Rev.* 2012;36(3):1060–1084. https://doi.org/10.1016/j.neubiorev.2011.12.008.
29. Black MH, Chen NTM, Iyer KK, et al. Mechanisms of facial emotion recognition in autism spectrum disorders: insights from eye tracking and electroencephalography. *Neurosci Biobehav Rev.* 2017;80:488–515. https://doi.org/10.1016/j.neubiorev.2017.06.016.

30. Nomi JS, Uddin LQ. Face processing in autism spectrum disorders: from brain regions to brain networks. *Neuropsychologia*. 2015;71:201–216. https://doi.org/10.1016/j.neuropsychologia.2015.03.029.
31. Collin L, Bindra J, Raju M, Gillberg C, Minnis H. Facial emotion recognition in child psychiatry: a systematic review. *Res Dev Disabil*. 2013;34(5):1505–1520. https://doi.org/10.1016/j.ridd.2013.01.008.
32. Harms MB, Martin A, Wallace GL. Facial emotion recognition in autism spectrum disorders: a review of behavioral and neuroimaging studies. *Neuropsychol Rev*. 2010;20(3):290–322. https://doi.org/10.1007/s11065-010-9138-6.
33. Jones W, Carr K, Klin A. Absence of preferential looking to the eyes of approaching adults predicts level of social disability in 2-year-old toddlers with autism spectrum disorder. *Arch Gen Psychiatry*. 2008;65(8):946–954. https://doi.org/10.1001/archpsyc.65.8.946.
34. Chita-Tegmark M. Social attention in ASD: a review and meta-analysis of eye-tracking studies. *Res Dev Disabil*. 2016;48:79–93. https://doi.org/10.1016/j.ridd.2015.10.011.
35. Bakroon A, Lakshminarayanan V. Do different experimental tasks affect psychophysical measurements of motion perception in autism-spectrum disorder? An analysis. *Clin Optom*. 2018;10:131–143. https://doi.org/10.2147/OPTO.S179336.
36. Jack A, Keifer CM, Pelphrey KA. Cerebellar contributions to biological motion perception in autism and typical development. *Hum Brain Mapp*. 2017;38(4):1914–1932. https://doi.org/10.1002/hbm.23493.
37. Robertson CE, Thomas C, Kravitz DJ, et al. Global motion perception deficits in autism are reflected as early as primary visual cortex. *Brain*. 2014;137(Pt 9):2588–2599. https://doi.org/10.1093/brain/awu189.
38. Koldewyn K, Whitney D, Rivera SM. Neural correlates of coherent and biological motion perception in autism. *Dev Sci*. 2011;14(5):1075–1088. https://doi.org/10.1111/j.1467-7687.2011.01058.x.
39. Mundy P, Sullivan L, Mastergeorge AM. A parallel and distributed-processing model of joint attention, social cognition and autism. *Autism Res*. 2009;2(1):2–21. https://doi.org/10.1002/aur.61.
40. Dawson G, Toth K, Abbott R, et al. Early social attention impairments in autism: social orienting, joint attention, and attention to distress. *Dev Psychol*. 2004;40(2):271–283. https://doi.org/10.1037/0012-1649.40.2.271.
41. Mundy PC. *Autism and Joint Attention: Development, Neuroscience, and Clinical Fundamentals*. Guilford Publications; 2016.
42. Grice SJ, Halit H, Farroni T, Baron-Cohen S, Bolton P, Johnson MH. Neural correlates of eye-gaze detection in young children with autism. *Cortex*. 2005;41(3):342–353.
43. Dalton KM, Nacewicz BM, Johnstone T, et al. Gaze fixation and the neural circuitry of face processing in autism. *Nat Neurosci*. 2005;8(4):519–526. https://doi.org/10.1038/nn1421.
44. Pelphrey K, Adolphs R, Morris JP. Neuroanatomical substrates of social cognition dysfunction in autism. *Ment Retard Dev Disabil Res Rev*. 2004;10(4):259–271. https://doi.org/10.1002/mrdd.20040.
45. Lynch CJ, Uddin LQ, Supekar K, Khouzam A, Phillips J, Menon V. Default mode network in childhood autism: posteromedial cortex heterogeneity and relationship with social deficits. *Biol Psychiatry*. 2013;74(3):212–219. https://doi.org/10.1016/j.biopsych.2012.12.013.
46. Gotts SJ, Simmons WK, Milbury LA, Wallace GL, Cox RW, Martin A. Fractionation of social brain circuits in autism spectrum disorders. *Brain*. 2012;135(Pt 9):2711–2725. https://doi.org/10.1093/brain/aws160.
47. Just MA, Keller TA, Malave VL, Kana RK, Varma S. Autism as a neural systems disorder: a theory of frontal-posterior underconnectivity. *Neurosci Biobehav Rev*. 2012;36(4):1292–1313. https://doi.org/10.1016/j.neubiorev.2012.02.007.

48. Supekar K, Uddin LQ, Khouzam A, et al. Brain hyperconnectivity in children with autism and its links to social deficits. *Cell Rep*. 2013;5(3):738–747. https://doi.org/10.1016/j.celrep.2013.10.001.
49. von dem Hagen EAH, Stoyanova RS, Baron-Cohen S, Calder AJ. Reduced functional connectivity within and between "social" resting state networks in autism spectrum conditions. *Soc Cogn Affect Neurosci*. 2013;8(6):694–701. https://doi.org/10.1093/scan/nss053.
50. McCall C, Singer T. The animal and human neuroendocrinology of social cognition, motivation and behavior. *Nat Neurosci*. 2012;15(5):681–688. https://doi.org/10.1038/nn.3084.
51. Vogeley K. Two social brains: neural mechanisms of intersubjectivity. *Philos Trans R Soc Lond Ser B Biol Sci*. 2017;372(1727). https://doi.org/10.1098/rstb.2016.0245.
52. Fernandez M, Mollinedo-Gajate I, Penagarikano O. Neural circuits for social cognition: implications for autism. *Neuroscience*. 2018;370:148–162. https://doi.org/10.1016/j.neuroscience.2017.07.013.
53. Mendelson JL, Gates JA, Lerner MD. Friendship in school-age boys with autism spectrum disorders: a meta-analytic summary and developmental, process-based model. *Psychol Bull*. 2016;142(6):601–622. https://doi.org/10.1037/bul0000041.
54. Velikonja T, Fett A-K, Velthorst E. Patterns of nonsocial and social cognitive functioning in adults with autism spectrum disorder: a systematic review and meta-analysis. *JAMA Psychiatry*. 2019. https://doi.org/10.1001/jamapsychiatry.2018.3645.
55. Chisholm AK, Anderson VA, Pride NA, Malarbi S, North KN, Payne JM. Social function and autism spectrum disorder in children and adults with neurofibromatosis type 1: a systematic review and meta-analysis. *Neuropsychol Rev*. 2018;28(3):317–340. https://doi.org/10.1007/s11065-018-9380-x.
56. Sugranyes G, Kyriakopoulos M, Corrigall R, Taylor E, Frangou S. Autism spectrum disorders and schizophrenia: meta-analysis of the neural correlates of social cognition. *PLoS ONE*. 2011;6(10). e25322. https://doi.org/10.1371/journal.pone.0025322.
57. Chung YS, Barch D, Strube M. A meta-analysis of mentalizing impairments in adults with schizophrenia and autism spectrum disorder. *Schizophr Bull*. 2014;40(3):602–616. https://doi.org/10.1093/schbul/sbt048.
58. Fernandes JM, Cajão R, Lopes R, Jerónimo R, Barahona-Corrêa JB. Social cognition in schizophrenia and autism spectrum disorders: a systematic review and meta-analysis of direct comparisons. *Front Psychol*. 2018;9:504. https://doi.org/10.3389/fpsyt.2018.00504.
59. Bora E, Pantelis C. Meta-analysis of social cognition in attention-deficit/hyperactivity disorder (ADHD): comparison with healthy controls and autistic spectrum disorder. *Psychol Med*. 2016;46(4):699–716. https://doi.org/10.1017/S0033291715002573.
60. Di Martino A, Ross K, Uddin LQ, Sklar AB, Castellanos FX, Milham MP. Functional brain correlates of social and nonsocial processes in autism spectrum disorders: an activation likelihood estimation meta-analysis. *Biol Psychiatry*. 2009;65(1):63–74. https://doi.org/10.1016/j.biopsych.2008.09.022.
61. van Veluw SJ, Chance SA. Differentiating between self and others: an ALE meta-analysis of fMRI studies of self-recognition and theory of mind. *Brain Imaging Behav*. 2014;8(1):24–38. https://doi.org/10.1007/s11682-013-9266-8.
62. Oberman LM, Pineda JA, Ramachandran VS. The human mirror neuron system: a link between action observation and social skills. *Soc Cogn Affect Neurosci*. 2007;2(1):62–66. https://doi.org/10.1093/scan/nsl022.
63. Yang J, Hofmann J. Action observation and imitation in autism spectrum disorders: an ALE meta-analysis of fMRI studies. *Brain Imaging Behav*. 2016;10(4):960–969. https://doi.org/10.1007/s11682-015-9456-7.
64. Patriquin MA, DeRamus T, Libero LE, Laird A, Kana RK. Neuroanatomical and neurofunctional markers of social cognition in autism spectrum disorder. *Hum Brain Mapp*. 2016;37(11):3957–3978. https://doi.org/10.1002/hbm.23288.

65. Rossi A, Galderisi S, Rocca P, et al. The relationships of personal resources with symptom severity and psychosocial functioning in persons with schizophrenia: results from the Italian network for research on psychoses study. *Eur Arch Psychiatry Clin Neurosci.* 2017;267(4):285–294. https://doi.org/10.1007/s00406-016-0710-9.
66. Green MF, Kern RS, Heaton RK. Longitudinal studies of cognition and functional outcome in schizophrenia: implications for MATRICS. *Schizophr Res.* 2004;72(1):41–51. https://doi.org/10.1016/j.schres.2004.09.009.
67. Green MF, Nuechterlein KH, Gold JM, et al. Approaching a consensus cognitive battery for clinical trials in schizophrenia: the NIMH-MATRICS conference to select cognitive domains and test criteria. *Biol Psychiatry.* 2004;56(5):301–307. https://doi.org/10.1016/j.biopsych.2004.06.023.
68. Carter CS, Barch DM, Buchanan RW, et al. Identifying cognitive mechanisms targeted for treatment development in schizophrenia: an overview of the first meeting of the cognitive neuroscience treatment research to improve cognition in schizophrenia initiative. *Biol Psychiatry.* 2008;64(1):4–10. https://doi.org/10.1016/j.biopsych.2008.03.020.
69. Bora E. Relationship between insight and theory of mind in schizophrenia: a meta-analysis. *Schizophr Res.* 2017;190:11–17. https://doi.org/10.1016/j.schres.2017.03.029.
70. Green MF, Horan WP, Lee J. Social cognition in schizophrenia. *Nat Rev Neurosci.* 2015;16(10):620–631. https://doi.org/10.1038/nrn4005.
71. Pinkham AE. Social cognition in schizophrenia. *J Clin Psychiatry.* 2014;75(Suppl 2):14–19. https://doi.org/10.4088/JCP.13065su1.04.
72. Martin AK, Robinson G, Dzafic I, Reutens D, Mowry B. Theory of mind and the social brain: implications for understanding the genetic basis of schizophrenia. *Genes Brain Behav.* 2014;13(1):104–117. https://doi.org/10.1111/gbb.12066.
73. Bosia M, Riccaboni R, Poletti S. Neurofunctional correlates of theory of mind deficits in schizophrenia. *Curr Top Med Chem.* 2012;12(21):2284–2302.
74. Biedermann F, Frajo-Apor B, Hofer A. Theory of mind and its relevance in schizophrenia. *Curr Opin Psychiatry.* 2012;25(2):71–75. https://doi.org/10.1097/YCO.0b013e3283503624.
75. Sarfati Y. Deficit of the theory-of-mind in schizophrenia: clinical concept and review of experimental arguments. *Can J Psychiatr.* 2000;45(4):363–368. https://doi.org/10.1177/070674370004500405.
76. Mier D, Kirsch P. Social-cognitive deficits in schizophrenia. *Curr Top Behav Neurosci.* 2017;30:397–409. https://doi.org/10.1007/7854_2015_427.
77. Thompson AD, Bartholomeusz C, Yung AR. Social cognition deficits and the "ultra high risk" for psychosis population: a review of literature. *Early Interv Psychiatry.* 2011;5(3):192–202. https://doi.org/10.1111/j.1751-7893.2011.00275.x.
78. Mehta UM, Thirthalli J, Naveen Kumar C, Keshav Kumar J, Keshavan MS, Gangadhar BN. Schizophrenia patients experience substantial social cognition deficits across multiple domains in remission. *Asian J Psychiatr.* 2013;6(4):324–329. https://doi.org/10.1016/j.ajp.2013.02.001.
79. Mazza M, Pollice R, Pacitti F, et al. New evidence in theory of mind deficits in subjects with chronic schizophrenia and first episode: correlation with symptoms, neurocognition and social function. *Riv Psichiatr.* 2012;47(4):327–336. https://doi.org/10.1708/1139.12561.
80. Bora E, Pantelis C. Theory of mind impairments in first-episode psychosis, individuals at ultra-high risk for psychosis and in first-degree relatives of schizophrenia: systematic review and meta-analysis. *Schizophr Res.* 2013;144(1–3):31–36. https://doi.org/10.1016/j.schres.2012.12.013.
81. Lee TY, Bin HS, Shin NY, Kwon JS. Social cognitive functioning in prodromal psychosis: a meta-analysis. *Schizophr Res.* 2015;164(1–3):28–34. https://doi.org/10.1016/j.schres.2015.02.008.

82. Corigliano V, De Carolis A, Trovini G, et al. Neurocognition in schizophrenia: from prodrome to multi-episode illness. *Psychiatry Res*. 2014;220(1–2):129–134. https://doi.org/10.1016/j.psychres.2014.07.067.
83. Javed A, Charles A. The importance of social cognition in improving functional outcomes in schizophrenia. *Front Psychol*. 2018;9:157. https://doi.org/10.3389/fpsyt.2018.00157.
84. Green MF. Impact of cognitive and social cognitive impairment on functional outcomes in patients with schizophrenia. *J Clin Psychiatry*. 2016;77(Suppl 2):8–11. https://doi.org/10.4088/JCP.14074su1c.02.
85. Franck N, Demily C. Improving functional outcome of schizophrenia with cognitive remediation. *Presse Med*. 2015;44(3):292–297. https://doi.org/10.1016/j.lpm.2014.06.031.
86. Fett A-KJ, Viechtbauer W, Dominguez M-G, Penn DL, van Os J, Krabbendam L. The relationship between neurocognition and social cognition with functional outcomes in schizophrenia: a meta-analysis. *Neurosci Biobehav Rev*. 2011;35(3):573–588. https://doi.org/10.1016/j.neubiorev.2010.07.001.
87. Couture SM, Penn DL, Roberts DL. The functional significance of social cognition in schizophrenia: a review. *Schizophr Bull*. 2006;32(Suppl 1):S44–S63. https://doi.org/10.1093/schbul/sbl029.
88. Mancuso C, Siciliano R, Barone E. Curcumin and Alzheimer disease: this marriage is not to be performed. *J Biol Chem*. 2011;286(3):le3. author reply le4, https://doi.org/10.1074/jbc.L110.133520.
89. Mike L, Guimond S, Kelly S, et al. Social cognition in early course of schizophrenia: exploratory factor analysis. *Psychiatry Res*. 2019;272:737–743. https://doi.org/10.1016/j.psychres.2018.12.152.
90. Garcia RR, Aliste F, Soto G. Social cognition in schizophrenia: cognitive and neurobiological aspects. *Rev Colomb Psiquiatr*. 2018;47(3):170–176. https://doi.org/10.1016/j.rcp.2017.03.004.
91. Bicks LK, Koike H, Akbarian S, Morishita H. Prefrontal cortex and social cognition in mouse and man. *Front Psychol*. 2015;6:1805. https://doi.org/10.3389/fpsyg.2015.01805.
92. Patin A, Hurlemann R. Social cognition. *Handb Exp Pharmacol*. 2015;228:271–303. https://doi.org/10.1007/978-3-319-16522-6_10.
93. Fujiwara H, Yassin W, Murai T. Neuroimaging studies of social cognition in schizophrenia. *Psychiatry Clin Neurosci*. 2015;69(5):259–267. https://doi.org/10.1111/pcn.12258.
94. Lavoie M-A, Plana I, Bedard Lacroix J, Godmaire-Duhaime F, Jackson PL, Achim AM. Social cognition in first-degree relatives of people with schizophrenia: a meta-analysis. *Psychiatry Res*. 2013;209(2):129–135. https://doi.org/10.1016/j.psychres.2012.11.037.
95. Kohler CG, Walker JB, Martin EA, Healey KM, Moberg PJ. Facial emotion perception in schizophrenia: a meta-analytic review. *Schizophr Bull*. 2010;36(5):1009–1019. https://doi.org/10.1093/schbul/sbn192.
96. Savla GN, Vella L, Armstrong CC, Penn DL, Twamley EW. Deficits in domains of social cognition in schizophrenia: a meta-analysis of the empirical evidence. *Schizophr Bull*. 2013;39(5):979–992. https://doi.org/10.1093/schbul/sbs080.
97. Song MJ, Choi HI, Jang S-K, Lee S-H, Ikezawa S, Choi K-H. Theory of mind in Koreans with schizophrenia: a meta-analysis. *Psychiatry Res*. 2015;229(1–2):420–425. https://doi.org/10.1016/j.psychres.2015.05.108.
98. Bonfils KA, Lysaker PH, Minor KS, Salyers MP. Affective empathy in schizophrenia: a meta-analysis. *Schizophr Res*. 2016;175(1–3):109–117. https://doi.org/10.1016/j.schres.2016.03.037.
99. Bonfils KA, Lysaker PH, Minor KS, Salyers MP. Empathy in schizophrenia: a meta-analysis of the interpersonal reactivity index. *Psychiatry Res*. 2017;249:293–303. https://doi.org/10.1016/j.psychres.2016.12.033.

100. Bora E, Pantelis C. Social cognition in schizophrenia in comparison to bipolar disorder: a meta-analysis. *Schizophr Res.* 2016;175(1–3):72–78. https://doi.org/10.1016/j.schres.2016.04.018.
101. Kronbichler L, Tschernegg M, Martin AI, Schurz M, Kronbichler M. Abnormal brain activation during theory of mind tasks in schizophrenia: a meta-analysis. *Schizophr Bull.* 2017;43(6):1240–1250. https://doi.org/10.1093/schbul/sbx073.
102. Jani M, Kasparek T. Emotion recognition and theory of mind in schizophrenia: a meta-analysis of neuroimaging studies. *World J Biol Psychiatry.* 2018;19(sup3):S86–S96. https://doi.org/10.1080/15622975.2017.1324176.
103. Lyketsos CG, Lopez O, Jones B, Fitzpatrick AL, Breitner J, DeKosky S. Prevalence of neuropsychiatric symptoms in dementia and mild cognitive impairment: results from the cardiovascular health study. *JAMA.* 2002;288(12):1475–1483.
104. Roberts R, Knopman DS. Classification and epidemiology of MCI. *Clin Geriatr Med.* 2013;29(4):753–772. https://doi.org/10.1016/j.cger.2013.07.003.
105. Lindau M, Almkvist O, Kushi J, et al. First symptoms—frontotemporal dementia versus Alzheimer's disease. *Dement Geriatr Cogn Disord.* 2000;11(5):286–293. https://doi.org/10.1159/000017251.
106. Shinagawa S, Ikeda M, Fukuhara R, Tanabe H. Initial symptoms in frontotemporal dementia and semantic dementia compared with Alzheimer's disease. *Dement Geriatr Cogn Disord.* 2006;21(2):74–80. https://doi.org/10.1159/000090139.
107. Saz P, Lopez-Anton R, Dewey ME, et al. Prevalence and implications of psychopathological non-cognitive symptoms in dementia. *Acta Psychiatr Scand.* 2009;119(2):107–116. https://doi.org/10.1111/j.1600-0447.2008.01280.x.
108. Jost BC, Grossberg GT. The evolution of psychiatric symptoms in Alzheimer's disease: a natural history study. *J Am Geriatr Soc.* 1996;44(9):1078–1081.
109. Kumfor F, Irish M, Leyton C, et al. Tracking the progression of social cognition in neurodegenerative disorders. *J Neurol Neurosurg Psychiatry.* 2014;85(10):1076–1083. https://doi.org/10.1136/jnnp-2013-307098.
110. Gregory C, Lough S, Stone V, et al. Theory of mind in patients with frontal variant frontotemporal dementia and Alzheimer's disease: theoretical and practical implications. *Brain.* 2002;125(4):752–764. https://doi.org/10.1093/brain/awf079.
111. Moreau N, Rauzy S, Bonnefoi B, et al. Different patterns of theory of mind impairment in mild cognitive impairment. *J Alzheimers Dis.* 2015;45(2):581–597. https://doi.org/10.3233/JAD-143021.
112. Elamin M, Pender N, Hardiman O, Abrahams S. Social cognition in neurodegenerative disorders: a systematic review. *J Neurol Neurosurg Psychiatry.* 2012;83(11):1071–1079. https://doi.org/10.1136/jnnp-2012-302817.
113. Bora E, Yener GG. Meta-analysis of social cognition in mild cognitive impairment. *J Geriatr Psychiatry Neurol.* 2017;30(4):206–213. https://doi.org/10.1177/0891988717710337.
114. Bora E, Walterfang M, Velakoulis D. Theory of mind in behavioural-variant frontotemporal dementia and Alzheimer's disease: a meta-analysis. *J Neurol Neurosurg Psychiatry.* 2015;86(7):714–719. https://doi.org/10.1136/jnnp-2014-309445.
115. Dermody N, Wong S, Ahmed R, Piguet O, Hodges JR, Irish M. Uncovering the neural bases of cognitive and affective empathy deficits in Alzheimer's disease and the behavioral-variant of frontotemporal dementia. *J Alzheimers Dis.* 2016;53(3):801–816. https://doi.org/10.3233/JAD-160175.
116. Bickart KC, Dickerson BC, Barrett LF. The amygdala as a hub in brain networks that support social life. *Neuropsychologia.* 2014;63:235–248. https://doi.org/10.1016/j.neuropsychologia.2014.08.013.
117. Adenzato M, Cavallo M, Enrici I. Theory of mind ability in the behavioural variant of frontotemporal dementia: an analysis of the neural, cognitive, and social levels. *Neuropsychologia.* 2010;48(1):2–12. https://doi.org/10.1016/j.neuropsychologia.2009.08.001.

118. Cosentino S, Zahodne LB, Brandt J, et al. Social cognition in Alzheimer's disease: a separate construct contributing to dependence. *Alzheimers Dement*. 2014;10(6):818–826. https://doi.org/10.1016/j.jalz.2013.12.021.
119. Engelhardt DM, Rosen B. Implications of drug treatment for the social rehabilitation of schizophrenic patients. *Schizophr Bull*. 1976;2(3):454–462.
120. Gabay AS, Kempton MJ, Mehta MA. Facial affect processing deficits in schizophrenia: a meta-analysis of antipsychotic treatment effects. *J Psychopharmacol*. 2015;29(2):224–229. https://doi.org/10.1177/0269881114560184.
121. Bielsky IF, Young LJ. Oxytocin, vasopressin, and social recognition in mammals. *Peptides*. 2004;25(9):1565–1574. https://doi.org/10.1016/j.peptides.2004.05.019.
122. Donaldson ZR, Young LJ. Oxytocin, vasopressin, and the neurogenetics of sociality. *Science*. 2008;322(5903):900–904. https://doi.org/10.1126/science.1158668.
123. Meyer-Lindenberg A, Domes G, Kirsch P, Heinrichs M. Oxytocin and vasopressin in the human brain: social neuropeptides for translational medicine. *Nat Rev Neurosci*. 2011;12(9):524–538. https://doi.org/10.1038/nrn3044.
124. Corey EJ, Czakó B, Kürti L. *Molecules and Medicine*. John Wiley & Sons; 2012.
125. Shahrestani S, Kemp AH, Guastella AJ. The impact of a single administration of intranasal oxytocin on the recognition of basic emotions in humans: a meta-analysis. *Neuropsychopharmacology*. 2013;38(10):1929–1936. https://doi.org/10.1038/npp.2013.86.
126. Lim MM, Bielsky IF, Young LJ. Neuropeptides and the social brain: potential rodent models of autism. *Int J Dev Neurosci*. 2005;23(2–3):235–243. https://doi.org/10.1016/j.ijdevneu.2004.05.006.
127. Hollander E, Bartz J, Chaplin W, et al. Oxytocin increases retention of social cognition in autism. *Biol Psychiatry*. 2007;61(4):498–503. https://doi.org/10.1016/j.biopsych.2006.05.030.
128. Andari E, Duhamel J-R, Zalla T, Herbrecht E, Leboyer M, Sirigu A. Promoting social behavior with oxytocin in high-functioning autism spectrum disorders. *Proc Natl Acad Sci U S A*. 2010;107(9):4389–4394. https://doi.org/10.1073/pnas.0910249107.
129. Guastella AJ, Einfeld SL, Gray KM, et al. Intranasal oxytocin improves emotion recognition for youth with autism spectrum disorders. *Biol Psychiatry*. 2010;67(7):692–694. https://doi.org/10.1016/j.biopsych.2009.09.020.
130. Domes G, Kumbier E, Heinrichs M, Herpertz SC. Oxytocin promotes facial emotion recognition and amygdala reactivity in adults with asperger syndrome. *Neuropsychopharmacology*. 2014;39(3):698–706. https://doi.org/10.1038/npp.2013.254.
131. Domes G, Heinrichs M, Kumbier E, Grossmann A, Hauenstein K, Herpertz SC. Effects of intranasal oxytocin on the neural basis of face processing in autism spectrum disorder. *Biol Psychiatry*. 2013;74(3):164–171. https://doi.org/10.1016/j.biopsych.2013.02.007.
132. Veening JG, Olivier B. Intranasal administration of oxytocin: behavioral and clinical effects, a review. *Neurosci Biobehav Rev*. 2013;37(8):1445–1465. https://doi.org/10.1016/j.neubiorev.2013.04.012.
133. Lukas M, Neumann ID. Oxytocin and vasopressin in rodent behaviors related to social dysfunctions in autism spectrum disorders. *Behav Brain Res*. 2013;251:85–94. https://doi.org/10.1016/j.bbr.2012.08.011.
134. Gordon I, Vander Wyk BC, Bennett RH, et al. Oxytocin enhances brain function in children with autism. *Proc Natl Acad Sci U S A*. 2013;110(52):20953–20958. https://doi.org/10.1073/pnas.1312857110.
135. LoParo D, Waldman ID. The oxytocin receptor gene (OXTR) is associated with autism spectrum disorder: a meta-analysis. *Mol Psychiatry*. 2015;20(5):640–646. https://doi.org/10.1038/mp.2014.77.
136. Ooi YP, Weng S-J, Kossowsky J, Gerger H, Sung M. Oxytocin and autism spectrum disorders: a systematic review and Meta-analysis of randomized controlled trials. *Pharmacopsychiatry*. 2017;50(1):5–13. https://doi.org/10.1055/s-0042-109400.

137. Keech B, Crowe S, Hocking DR. Intranasal oxytocin, social cognition and neurodevelopmental disorders: a meta-analysis. *Psychoneuroendocrinology*. 2018;87:9–19. https://doi.org/10.1016/j.psyneuen.2017.09.022.
138. Cai Q, Feng L, Yap KZ. Systematic review and meta-analysis of reported adverse events of long-term intranasal oxytocin treatment for autism spectrum disorder. *Psychiatry Clin Neurosci*. 2018;72(3):140–151. https://doi.org/10.1111/pcn.12627.
139. Shilling PD, Feifel D. Potential of oxytocin in the treatment of schizophrenia. *CNS Drugs*. 2016;30(3):193–208. https://doi.org/10.1007/s40263-016-0315-x.
140. Bukovskaya O, Shmukler A. Oxytocin and social cognitions in schizophrenia: a systematic review. *Psychiatry Q*. 2016;87(3):521–543. https://doi.org/10.1007/s11126-015-9407-x.
141. Pedersen CA, Gibson CM, Rau SW, et al. Intranasal oxytocin reduces psychotic symptoms and improves theory of mind and social perception in schizophrenia. *Schizophr Res*. 2011;132(1):50–53. https://doi.org/10.1016/j.schres.2011.07.027.
142. Fischer-Shofty M, Brüne M, Ebert A, Shefet D, Levkovitz Y, Shamay-Tsoory SG. Improving social perception in schizophrenia: the role of oxytocin. *Schizophr Res*. 2013;146(1):357–362. https://doi.org/10.1016/j.schres.2013.01.006.
143. Woolley JD, Chuang B, Lam O, et al. Oxytocin administration enhances controlled social cognition in patients with schizophrenia. *Psychoneuroendocrinology*. 2014;47:116–125. https://doi.org/10.1016/j.psyneuen.2014.04.024.
144. Davis MC, Green MF, Lee J, et al. Oxytocin-augmented social cognitive skills training in schizophrenia. *Neuropsychopharmacology*. 2014;39(9):2070–2077. https://doi.org/10.1038/npp.2014.68.
145. Dagani J, Sisti D, Abelli M, et al. Do we need oxytocin to treat schizophrenia? A randomized clinical trial. *Schizophr Res*. 2016;172(1–3):158–164. https://doi.org/10.1016/j.schres.2016.02.011.
146. Ettinger U, Hurlemann R, Chan RCK. Oxytocin and schizophrenia spectrum disorders. *Curr Top Behav Neurosci*. 2018;35:515–527. https://doi.org/10.1007/7854_2017_27.
147. Bradley ER, Woolley JD. Oxytocin effects in schizophrenia: reconciling mixed findings and moving forward. *Neurosci Biobehav Rev*. 2017;80:36–56. https://doi.org/10.1016/j.neubiorev.2017.05.007.
148. Davis MC, Lee J, Horan WP, et al. Effects of single dose intranasal oxytocin on social cognition in schizophrenia. *Schizophr Res*. 2013;147(2–3):393–397. https://doi.org/10.1016/j.schres.2013.04.023.
149. Goldman MB, Gomes AM, Carter CS, Lee R. Divergent effects of two different doses of intranasal oxytocin on facial affect discrimination in schizophrenic patients with and without polydipsia. *Psychopharmacology*. 2011;216(1):101–110. https://doi.org/10.1007/s00213-011-2193-8.
150. Davis MC, Horan WP, Nurmi EL, et al. Associations between oxytocin receptor genotypes and social cognitive performance in individuals with schizophrenia. *Schizophr Res*. 2014;159(2–3):353–357. https://doi.org/10.1016/j.schres.2014.09.006.
151. Watanabe Y, Kaneko N, Nunokawa A, Shibuya M, Egawa J, Someya T. Oxytocin receptor (OXTR) gene and risk of schizophrenia: case-control and family-based analyses and meta-analysis in a Japanese population. *Psychiatry Clin Neurosci*. 2012;66(7):622. https://doi.org/10.1111/j.1440-1819.2012.02396.x.
152. Oya K, Matsuda Y, Matsunaga S, Kishi T, Iwata N. Efficacy and safety of oxytocin augmentation therapy for schizophrenia: an updated systematic review and meta-analysis of randomized, placebo-controlled trials. *Eur Arch Psychiatry Clin Neurosci*. 2016;266(5):439–450. https://doi.org/10.1007/s00406-015-0634-9.
153. Williams DR, Burkner P-C. Effects of intranasal oxytocin on symptoms of schizophrenia: a multivariate Bayesian meta-analysis. *Psychoneuroendocrinology*. 2017;75:141–151. https://doi.org/10.1016/j.psyneuen.2016.10.013.

154. Burkner P-C, Williams DR, Simmons TC, Woolley JD. Intranasal oxytocin may improve high-level social cognition in schizophrenia, but not social cognition or neurocognition in general: a multilevel bayesian meta-analysis. *Schizophr Bull.* 2017;43(6):1291–1303. https://doi.org/10.1093/schbul/sbx053.
155. Zheng W, Zhu X-M, Zhang Q-E, et al. Adjunctive intranasal oxytocin for schizophrenia: a meta-analysis of randomized, double-blind, placebo-controlled trials. *Schizophr Res.* 2018. https://doi.org/10.1016/j.schres.2018.12.007.
156. Hlinak Z, Krejci I. N-methyl-D-aspartate improved social recognition potency in rats. *Neurosci Lett.* 2002;330(3):227–230.
157. Gao X-M, Elmer GI, Adams-Huet B, Tamminga CA. Social memory in mice: disruption with an NMDA antagonist and attenuation with antipsychotic drugs. *Pharmacol Biochem Behav.* 2009;92(2):236–242. https://doi.org/10.1016/j.pbb.2008.11.016.
158. Carlsson ML. Hypothesis: is infantile autism a hypoglutamatergic disorder? Relevance of glutamate—serotonin interactions for pharmacotherapy. *J Neural Transm.* 1998;105 (4–5):525–535. https://doi.org/10.1007/s007020050076.
159. Blaylock RL, Strunecka A. Immune-glutamatergic dysfunction as a central mechanism of the autism spectrum disorders. *Curr Med Chem.* 2009;16(2):157–170.
160. Posey DJ, Kem DL, Swiezy NB, Sweeten TL, Wiegand RE, McDougle CJ. A pilot study of D-cycloserine in subjects with autistic disorder. *Am J Psychiatry.* 2004;161(11):2115–2117. https://doi.org/10.1176/appi.ajp.161.11.2115.
161. Owley T, Salt J, Guter S, et al. A prospective, open-label trial of memantine in the treatment of cognitive, behavioral, and memory dysfunction in pervasive developmental disorders. *J Child Adolesc Psychopharmacol.* 2006;16(5):517–524. https://doi.org/10.1089/cap.2006.16.517.
162. Chez MG, Burton Q, Dowling T, Chang M, Khanna P, Kramer C. Memantine as adjunctive therapy in children diagnosed with autistic spectrum disorders: an observation of initial clinical response and maintenance tolerability. *J Child Neurol.* 2007;22(5):574–579. https://doi.org/10.1177/0883073807302611.
163. Erickson CA, Posey DJ, Stigler KA, Mullett J, Katschke AR, McDougle CJ. A retrospective study of memantine in children and adolescents with pervasive developmental disorders. *Psychopharmacology.* 2007;191(1):141–147. https://doi.org/10.1007/s00213-006-0518-9.
164. Ghaleiha A, Asadabadi M, Mohammadi M-R, et al. Memantine as adjunctive treatment to risperidone in children with autistic disorder: a randomized, double-blind, placebo-controlled trial. *Int J Neuropsychopharmacol.* 2013;16(4):783–789. https://doi.org/10.1017/S1461145712000880.
165. Niederhofer H. Glutamate antagonists seem to be slightly effective in psychopharmacologic treatment of autism. *J Clin Psychopharmacol.* 2007;27(3):317–318. https://doi.org/10.1097/01.jcp.0000270082.30500.69.
166. Lai W-S, Chang C-Y, Wong W-R, Pei J-C, Chen Y-S, Hung W-L. Assessing schizophrenia-relevant cognitive and social deficits in mice: a selection of mouse behavioral tasks and potential therapeutic compounds. *Curr Pharm Des.* 2014;20(32):5139–5150.
167. Shimazaki T, Kaku A, Chaki S. D-serine and a glycine transporter-1 inhibitor enhance social memory in rats. *Psychopharmacology.* 2010;209(3):263–270. https://doi.org/10.1007/s00213-010-1794-y.
168. Goff DC. D-cycloserine in schizophrenia: new strategies for improving clinical outcomes by enhancing plasticity. *Curr Neuropharmacol.* 2017;15(1):21–34.
169. Scoriels L, Barnett JH, Murray GK, et al. Effects of modafinil on emotional processing in first episode psychosis. *Biol Psychiatry.* 2011;69(5):457–464. https://doi.org/10.1016/j.biopsych.2010.09.043.
170. Campbell-Meiklejohn DK, Simonsen A, Jensen M, et al. Modulation of social influence by methylphenidate. *Neuropsychopharmacology.* 2012;37(6):1517–1525. https://doi.org/10.1038/npp.2011.337.

171. Marini M, Banaji MR, Pascual-Leone A. Studying implicit social cognition with noninvasive brain stimulation. *Trends Cogn Sci*. 2018;22(11):1050–1066. https://doi.org/10.1016/j.tics.2018.07.014.
172. Mielacher C, Scheele D, Hurlemann R. Experimental and therapeutic neuromodulation of emotion and social cognition with non-invasive brain stimulation. *Nervenarzt*. 2015;86(12):1500–1507. https://doi.org/10.1007/s00115-015-4324-7.
173. Boggio PS, Asthana MK, Costa TL, Valasek CA, Osorio AAC. Promoting social plasticity in developmental disorders with non-invasive brain stimulation techniques. *Front Neurosci*. 2015;9:294. https://doi.org/10.3389/fnins.2015.00294.
174. Mehta UM, Thirthalli J, Basavaraju R, Gangadhar BN. Association of intracortical inhibition with social cognition deficits in schizophrenia: findings from a transcranial magnetic stimulation study. *Schizophr Res*. 2014;158(1–3):146–150. https://doi.org/10.1016/j.schres.2014.06.043.
175. Schuwerk T, Schecklmann M, Langguth B, Dohnel K, Sodian B, Sommer M. Inhibiting the posterior medial prefrontal cortex by rTMS decreases the discrepancy between self and other in theory of mind reasoning. *Behav Brain Res*. 2014;274:312–318. https://doi.org/10.1016/j.bbr.2014.08.031.
176. Horan WP, Green MF. Treatment of social cognition in schizophrenia: current status and future directions. *Schizophr Res*. 2019;203:3–11. https://doi.org/10.1016/j.schres.2017.07.013.

SECTION II

Autism

CHAPTER

3

A contemporary view on the molecular basis of neurodevelopmental disorders

Ilse M. van der Werf, Geert Vandeweyer, R. Frank Kooy

Department of Medical Genetics, University of Antwerp, Antwerp, Belgium

Identifying the genetic cause of disease in neurodevelopmental disorders (NDDs) is important on multiple levels as it teaches us about normal and disturbed cellular processes in the development of the brain and can yield potential druggable targets that can be used in drug development. On a more individual scale, it can be helpful in determining the most appropriate support for patients and can inform parents of patients on recurrence risk for next pregnancies. In the era where rapid technological advances promised to drive the identification of the genetic cause of disease in the vast majority of individuals with NDDs, we are still left with "unsolved cases." Therefore, new research strategies have to be applied in the field of NDD genetics. One such strategy may be to focus on gene networks and the global deregulation of these networks in NDDs, rather than on the individual genes within the networks.

Neurodevelopmental disorders

The term NDDs defines the clinical consequences of a disturbed development of the central nervous system. The symptoms of the disorders include deficits in personal, social, and cognitive functioning and manifest early in development.[1] According to the Diagnostic and Statistical Manual of Mental Disorders, 5th edition (DSM-V), the NDD group comprises attention-deficit hyperactivity disorder (ADHD), autism spectrum disorder (ASD), intellectual disability (ID), communication disorders,

https://doi.org/10.1016/B978-0-12-814037-6.00003-3

specific learning disorders, and motor disorders. An important distinction between neurodevelopmental and psychiatric disorders is that NDDs are characterized by an early-onset cognitive deficit and regularly show a rather stable course of disease, whereas psychiatric disorders often present in a remitting and relapsing pattern and manifest later in development.[2] However, in literature, the group of NDDs is often extended by including genetically and clinically related disorders as epilepsy, schizophrenia, and bipolar disorder. This broader definition of NDDs will be used throughout this chapter. Interestingly the independent occurrence of two or more NDDs in one patient, referred to as comorbidity, is higher than expected if the disorders would be independently distributed.[3] Reported comorbidity rates differ quite extensively, in part due to inconsistent terminology and the lack of matching control groups.[4] A metaanalysis of 31 studies evaluating the prevalence rates of chronic health conditions in children with ID showed high comorbidity rates: of the patients with ID, 22.0% also suffers from epilepsy. An autistic disorder is diagnosed in 10.1% of the ID patients, and 7.1% has a pervasive developmental disorder other than autistic disorder (these two groups would be combined into autistic spectrum disorder under the new criteria in DSM-5). ADHD is present in 9.5% of children with ID, and 17.1% have an anxiety disorder.[5] Also, many patients with NDDs have psychiatric disorders and behavioral problems.[6]

Intellectual disability/intellectual development disorder (ID/IDD)

ID is diagnosed when a patient has deficits in the intellectual functions (e.g., reasoning, problem-solving, planning, and learning) and in adaptive behavior (e.g., communication, social skills, and personal independence). These phenotypes have to emerge during the developmental period for a formal diagnosis.[1,7] The latest guidelines for diagnosis in DSM-V classify the severity based on milestones in personal skills, but more often the severity is determined by the IQ of the patient as described in the previous edition of DSM: profound ID (IQ < 20), severe ID (IQ 20–34), moderate ID (IQ 35–49), and mild ID (IQ 50–69). A further distinction is made between syndromic and nonsyndromic ID, based on the exclusive presence of ID in the patient (nonsyndromic) or when the ID is presented together with additional features such as (facial) dysmorphisms or congenital anomalies (syndromic).[8] For children under the age of 5 that fail to meet the expected developmental milestones and cannot be reliably and systematically assessed for intellectual functioning, the diagnosis (global) developmental delay can be made. Theoretically the prevalence of ID in the population is 2%–3%, but the actual number depends on gender, maternal race, and socioeconomic and educational status.[9] In approximately 65% of the moderate to severe ID patients, a genetic aberration that causes the disease can be identified.[8]

Autism spectrum disorders (ASD)

As defined in DSM-V, ASD is a group of disorders that are characterized by two core domains: deficits in social communication and social interaction and repetitive and restrictive behavior. This includes failure to have normal back-and-forth conversations, deficits in nonverbal communication (e.g., abnormalities in eye contact), stereotyped or repetitive patterns of motor behavior, highly restricted interests, and unusual reactivity to sensory input. The phenotype emerges during the early developmental period but may be masked later in life due to adaptive strategies.[10] The DSM-V, unlike previous versions, no longer makes a distinction between different subtypes of autism (autistic disorder; Asperger's disorder; Rett's disorder; childhood disintegrative disorder; and pervasive developmental disorder, not otherwise specified [PDD-NOS]). The prevalence of ASD is estimated at 1.5% in developed countries, with about a 4:1 ratio of males over females.[11] ASD has a strong genetic component, among other evidence based on a high number of mutations and structural variants associated with the disorder to date and on twin studies showing concordance rates of 70%–90% in monozygotic twins versus 0%–10% in dizygotic twins.[12]

Attention-deficit hyperactivity disorder (ADHD)

Patients with ADHD suffer from inattention with or without hyperactivity/impulsivity, which interferes with social and school or work functioning. Examples of the aberrant behavior displayed by patients with ADHD are failure to pay attention to details, problems with organizing tasks and activities, easy distraction by extraneous stimuli, fidgeting with hands or feet, running around in situations where this is socially inappropriate, and excessive talking. The onset of ADHD has to occur before the age of 12 for a formal diagnosis and often persists into adulthood. Three types of ADHD are distinguished based on the main symptoms present in the child: combined presentation, predominantly inattentive presentation, and predominantly hyperactive-impulsive presentation. The prevalence of ADHD is estimated to lie around 7.2%, with an estimated heritability of approximately 70%.[13, 14]

Other neurodevelopmental disorders

The communication disorders comprise disorders affecting communication, language, and speech with childhood onset, other than ASD. This group contains language disorders, speech sound disorders, childhood-onset fluency disorder (stuttering), and social (pragmatic) communication disorders. Specific learning disorder is accompanied by not only difficulties in learning and using academic skills, such as inaccurate or slow word reading; difficulties with understanding the meaning of what is read; and difficulties with spelling and grammar but also problems with mathematical reasoning and mastering number sense and calculation. The group of motor disorders combines developmental coordination disorder, stereotypic movement disorder, and tic disorders. These disorders are characterized by an impaired acquisition and execution of coordinated motor skills, repetitive and purposeless motor behavior, or vocalizations.

Schizophrenia spectrum

Schizophrenia (SZ), schizotypal (personality) disorder, and psychotic disorders are defined by the presence of delusions, hallucinations, disorganized thinking and speech, disorganized or abnormal motor behavior, and negative symptoms. Individuals suffering from delusions have fixed beliefs that cannot be changed even when convincing evidence is presented. During hallucinations, individuals experience clear and strong perceptions of external stimuli that are not present, while they are being awake. The experienced stimuli mostly include auditory hallucinations such as voices. The disorganized thinking becomes apparent when individuals suddenly switch topics or answers are completely unrelated to the question. Problems in motor behavior present as unpredictable agitation, but catatonic behavior is also seen during which a marked decrease in reactivity to the environment is present. Schizophrenic individuals often present negative symptoms, including reduction in the expression of emotion and a decrease in self-initiated purposeful activities. Schizophrenia affects approximately 1% of the population and has an estimated genetic component of 80%.[15]

Bipolar disorders

Bipolar disorders (BD) and related disorders include bipolar I and bipolar II disorder, cyclothymic disorder, and substance/medication-induced bipolar disorder. Bipolar I disorder is more commonly referred to as manic-depressive disorder by the general public. BD is characterized by manic and depressive episodes. A manic episode is a distinct period lasting for more than 1 week during which an individual presents a noticeable change from usual behavior (e.g., inflated self-esteem, decreased need for sleep, pressure to keep talking, and excessive involvement in activities that have a high potential for painful consequences). A major depressive episode is present for more than 2 weeks and involves at least a depressed mood or a loss of interest or pleasure. Other symptoms of a depressive episode include a significant weight loss or gain, fatigue, feelings of worthlessness, diminished ability to think or concentrate, and recurrent thoughts of death. The population prevalence of bipolar I disorder is approximately 0.6%.[16] Based on twin studies the estimated heritability for bipolar disorder is at least 60%.[17]

Epilepsy

Epilepsy is the common term for a group of disorders characterized by recurrent unprovoked seizures.[18] Seizures can be present in different forms, including absence seizures, complex partial seizures, and generalized tonic-clonic seizures. Seizures originating in one or more localized foci are distinctive for partial epilepsies, whereas generalized epilepsies are characterized by seizures commencing simultaneously in both cerebral hemispheres.[18] For a diagnosis the combination of clinical features, the electroencephalographic (EEG) findings, and the potential presence of neurodevelopmental deficits has to be considered.[18] The population frequency of epilepsy is about 3.8%.[19] In approximately 70% of the cases, a genetic defect underlies the disease.[20] An overlap exists in the clinical phenotypes of the several NDDs. Symptoms that were once believed to be characteristic for one disorder can be present at varying levels of severity in other disorders.[3] Based on this observation in combination with the extensive genetic overlap among these disorders, it can be hypothesized that these disorders are not distinct entities, but rather are different representations along a large spectrum of neurological problems.

Genetic causes of NDDs

NDDs can be the result of environmental factors, genetic aberrations, or a combination of these two. Environmental factors include pre- and postnatal infections, pre- and postnatal exposure to toxic compounds (e.g., fetal alcohol syndrome), premature birth, and trauma.[8] However, as shown in heritability and twin studies, a large proportion of the NDDs can be explained by genetic abnormalities. For example, the genetic component in the etiology of ADHD is estimated to be 80%–90%, whereas estimates for ASD range from 50% to 80%.[21–23] NDDs are both clinically and genetically heterogeneous. On a clinical level the disorders can manifest in many different varieties and severities, causing large differences in the clinical presentation of different patients. Genetically, many different genes can be involved in the emergence of the same disorder, with no single gene explaining a substantial proportion of patients. For ID alone, estimates are that up to 2000 genes may be involved in the pathogenesis.[8] Traditionally, large chromosomal aberrations such as aneuploidies or translocations were detected using karyotyping methods, and upon suspicion of a particular syndrome, a set of defined subchromosomal abnormalities could be identified with fluorescence in situ hybridization (FISH). However, recent technologies have accelerated the unbiased discovery of causative variants. Using microarrays for detection of copy number variation (CNVs) and next-generation sequencing (NGS) techniques for single-nucleotide variants (SNVs), it is now possible to construct a comprehensive genetic profile of the entire genome of patients. In patients with severe NDDs, the underlying genetic aberrations appear to be de novo of origin in the majority of cases, explaining the rather stable prevalence rates despite the reduced fecundity of these patients.[24] In contrast, milder NDD phenotypes may be caused by inherited variants that can spread throughout the population.[25] Combined with the pronounced genetic heterogeneity, this may partially explain the failure of genome-wide association studies (GWAS) to identify a large number of SNPs and associated genes involved in the pathogenesis of NDDs, since these studies rely on the "common disease-common variant" theory.[26] Therefore new experimental strategies had to be applied to explain the "missing heritability" in NDDs.[27] Extensive CNV studies showed a significant enrichment of large CNVs (>250kb) in NDD cohorts when compared with control cohorts.[28–31] NGS approaches revealed many rare single-nucleotide variants (SNVs) and small insertions and deletions (indels) as new monogenic causes of NDDs.[32–35] An overview of the different types of genetic anomalies underlying NDDs and some prime examples per category is given in the succeeding texts.

Aneuploidies

The human genome exists of 23 chromosome pairs—22 autosomal pairs supplemented with 2 sex chromosomes that are stored in the cell nucleus—and a small circular mitochondrial DNA molecule that can be found in each mitochondrion. Errors during meiosis can result in haploid gametes that carry an extra copy of a full chromosome or lack one of the chromosomes completely. Upon fertilization, this chromosomal imbalance leads to aneuploidy (monosomy or trisomy).[36] Approximately 0.3% of the liveborns are aneuploid, with trisomy 13 (Patau syndrome), trisomy 18 (Edwards syndrome), trisomy 21 (Down syndrome), and aneuploidies of the sex chromosomes being the most common anomalies. Among stillbirths (fetal death from 20weeks of gestation to term) the prevalence of aneuploidy rises to 4%, and in spontaneous abortions (fetal death between 6 and 20weeks of gestation) the number even rises to 35%.[36] The extra chromosome leading to a trisomy is most often derived from the mother, and the risk for a trisomy increases with the age of the mother, which can partially be explained by the complexity of the oogenesis process. Approximately 1 in 1000 live births is affected by Down syndrome, making it the most prevalent chromosomal aneuploidy associated with

ID.[37] Patients with Down syndrome usually suffer from mild to moderate ID and have characteristic facial features including small ears and a flat, low nose bridge. Congenital heart defects, vision and hearing disorders, and behavioral problems are the other common features for Down syndrome patients.

Repeat expansions

A special group of mutations involved in the emergence of NDDs are dynamic mutations, caused by an unstable expansion of a polymorphic repeat sequence that can result in disease via multiple mechanisms.[38] Expansion of the CGG repeat in the 5′ UTR of the fragile X mental retardation 1 gene (*FMR1*) causes fragile X syndrome (FXS, OMIM #300624).[39–41] FXS is the most common cause of inherited ID and ASD with a population prevalence of about 1 in 5000 males, accounting for 1.9% of patients with ID.[5, 42] Expansion of the repeat beyond 200 trinucleotide repeat units induces hypermethylation of the CGG repeat and a CpG island in the FMR1 promotor, leading to transcriptional silencing of the gene. This results in the loss of the fragile X mental retardation protein (FMRP), which is the prime cause of disease.[40, 43] Patients with FXS have cognitive deficits ranging from mild learning problems to severe ID.[44] Often hyperactivity, short attention span, anxiety, and social avoidance are presented by the patients. Additional phenotypic hallmarks characteristic for FXS are a prominent forehead, narrow face, protruding ears, and macroorchidism.[45] Repeat expansions at other loci in the genome have also been associated with NDDs. Expansion of the CGG repeat in the promotor of *AFF3* on chromosome 2 causes a delay in motor and language development.[46] CGG repeat expansion in a 5′ intron of *ZNF713* has been associated with ASD.[47]

Structural variation

Structural variation (SV) encompasses the rearrangement of chromosomal segments and can occur in different forms.[48] Balanced SVs do not result in the gain or loss of (large) DNA fragments and include inversions and translocations. SVs resulting in the gain or loss of DNA (unbalanced SVs) are better known as CNVs.

Copy number variation in NDDs

With the introduction of genomic microarrays, new opportunities arose to identify CNVs with much higher resolution when compared with light microscopy.[49] The defined minimal size of CNVs is continuously changing, partially due to increasing sensitivity of microarray platforms that enables detection of smaller CNVs. Throughout this chapter, we use the originally suggested cutoff of 1 kb. Several studies showed a significant overrepresentation of de novo CNVs in patient cohorts with ID, ASD, and SZ.[28–31, 50–52] This overrepresentation becomes more pronounced with increasing CNV size, with 11.3% of ID patients carrying a CNV larger than 1.5 Mb, compared with only 0.6% of controls.[28] Larger CNVs are associated with more severe phenotypes.[28] Furthermore, among children with ID and congenital abnormalities carrying a disease-causing CNV, 10.1% carries a second large CNV.[53] It is estimated that approximately 14%–15% of ID cases are caused by CNVs.[28, 54]

Molecular mechanisms underlying CNVs

Copy number variation can originate through five major molecular mechanisms, NAHR, NHEJ, FoSTeS, MMBIR, and L1-mediated retrotransposition.[55] Nonallelic homologous recombination (NAHR) is a mechanism by which two homologous regions misalign and crossover occurs. This often occurs at loci encompassing segmental duplications (SDs) with high sequence similarity that arose during duplication events in the evolution.[55] NAHR can result in the loss or gain of the sequence flanked by the SDs. Due to the presence of SDs, some regions in the genome are prone to NAHR-mediated rearrangements, resulting in the presence of the exact same or very similar CNVs in different individuals. When double-strand breaks in DNA occur, for example, due to radiation, human cells repair this break using nonhomologous end joining (NHEJ). This process does not require or utilize homologous sequences as a template during the repair process, which induces the presence of small insertions and deletions of few nucleotides on the fusion point.[55] Complex rearrangements may be explained by the fork stalling and template switching (FoSTeS) mechanism. During this process the DNA replication fork stalls after which the lagging strand is released and integrates in another replication fork in close proximity. This can give rise to deletions, duplications, and inversions, based on the directionality and position of both replication forks.[56] The similar microhomology-mediated break-induced replication (MMBIR) mechanism also results in deletions, duplications, and inversions but is caused by a collapsed replication fork rather than a stalled fork.[57, 58] Lastly, CNVs can arise through L1 retrotransposition during which long interspersed element-1 (L1) elements are transcribed and the resulting RNA fragments are reverse transcribed and integrated at another site in the genome.[55]

NHAR-induced CNVs

Recurrent CNVs that arise due to homologous recombination are termed genomic disorders.[59] Several genomic disorders are known to date, some with recognizable phenotypic features (syndromic genomic disorders) and others with more variable clinical outcome. When comparing these two groups, a significant excess of boys is seen in the group of patients with genomic disorders with variable clinical presentations.[53]

The 16p11.2 deletion (OMIM #611913) and reciprocal duplication (OMIM #614671) are examples of phenotypically variable genomic disorders, and each have a population prevalence of approximately 1/2000.[60] The phenotypes reported for 16p11.2 rearrangement carriers are heterogeneous, even among first-degree relatives such as siblings or proband and transmitting parent.[61–64] A particular feature of this specific rearrangement is that it may lead to mirrored phenotypes in deletion versus duplication carriers. The deletion was found to be associated with autism, obesity, and macrocephaly, whereas the reciprocal duplication was found to be associated with schizophrenia, a high risk for being underweight, and microcephaly.[60, 65–68] Duplication carriers have a reduced global cognitive functioning when compared with intrafamilial controls, and a substantial proportion of the carriers have an ASD diagnosis.[69] The majority of deletion carriers show reduced cognitive ability, developmental and speech delay, and/or autism.[64] Furthermore, seizures and congenital anomalies were described recurrently in deletion carriers.[61, 66] The Williams-Beuren and Somerville-Van der Aa syndromes are caused by a deletion or its reciprocal duplication on 7q11.23 respectively. The Williams-Beuren syndrome (OMIM #194050) is a syndromic genomic disorder with a population prevalence of 1/7500, characterized by ID, short stature, characteristic facial features, and cardiac defects.[70, 71] The clinical phenotype of Somerville-Van der Aa syndrome (OMIM #609757) is milder and more variable when compared with Williams-Beuren syndrome and has a population prevalence in the range of 1/13,000–1/20,000.[72] Patients have speech and developmental delay; the cognitive abilities range from normal to moderate ID. Some of the patients have ASD, neonatal hypotonia, epilepsy, and congenital malformations. Characteristic facial features include straight eyebrows, high broad nose, thin upper lip, deep-set eyes, and a prominent forehead.

CNV syndromes due to nonhomology-driven mechanisms

For syndromes caused by microdeletions or microduplications that do not have borders defined by homologous segments, the breakpoints and the encompassed interval differ between patients. Delineation of the critical region for these disorders—being the region of overlap of all detected CNVs giving rise to the phenotype—is useful to determine the gene(s) in the CNV interval that contribute to the disease. This strategy is most applicable to syndromic microdeletion and microduplication disorders, as they give rise to a better defined clinical phenotype that can clearly be recognized as absent or present in patients with smaller CNVs. To aid in this delineation, not only small CNVs within the interval in patients with the core symptoms of the phenotype are important, but also small CNVs in the interval in individuals with a dissimilar phenotype are of significant importance.[73] In some cases the phenotype is almost completely due to the loss or gain of a single gene in the CNV interval. The Koolen-de Vries syndrome, for example, was first described as the result of 17q21.31 microdeletions.[74] Patients with this syndrome have ID, developmental delay, hypotonia, epilepsy, characteristic facial features, and congenital malformations. Later, mutations in one of the genes residing within the 17q21.31 microdeletion interval, *KANLS1*, were identified in patients with apparently identical phenotypes.[75] In fact, no distinction can be made between patients with Koolen-de Vries syndrome patients due to a microdeletion or a mutation in *KANSL1*, showing that haploinsufficiency of this single gene determines the clinical outcome.[76]

Triplications

Most CNVs associated to date with NDDs are deletions and duplications. Occasionally, triplications have also been reported to be causal for NDDs[77–80] Triplications can have a more extreme phenotype than the corresponding duplication, and so aid in determining the core symptoms of the duplication disorder.

Complex structural variants

Complex SVs cannot be assigned to the balanced or unbalanced SV category, often because they are defined by a combination of characteristics of both groups.[48] In total, 16 different subclasses of complex SVs can be distinguished. Large and complex chromosomal abnormalities appear to be relatively common in human as in a cohort of 686 ASD patients, each individual was found to have between 6 and 23 complex SVs.[48] CNV-flanked inversions were found to be the largest group of complex SVs, representing 77.2% of all complex SVs.

Single-gene defects

The next technological revolution in genetics enabled complete sequencing of (selected parts of) the genome, which accelerated the identification of single-gene defect in NDDs over the past decade. Using trio-sequencing approaches, studying both parents and the affected proband and looking for private variants in the child, it was quickly discovered that de novo mutations are an important cause for ID.[24] Subsequently, deleterious de novo mutations were found to cause a significant proportion of the severe ID and ASD cases.[34, 81–83] To date, over 700 genes have been identified as causal for ID when mutated.[25]

SNV categories

Single-nucleotide variants (SNVs) and small insertions and/or deletions (indels) can result in single-gene defects. Base pair substitutions can have minor to major effects based on the exact substitution. In case an SNV does not induce alteration of the amino acid sequence upon transcription and translation, this variant is called synonymous. On the contrary, SNVs that do induce an amino acid alteration are termed nonsynonymous. The nonsynonymous variants can be divided in different categories, based on the effect on the amino acid sequence. A variant that induces an amino acid substitution is called a missense variant. If the variant results in a stop codon, thereby terminating translation prematurely, it is defined as a nonsense variant. SNVs can also disrupt the canonical splice sites in a gene by which it may alter the splicing of the mRNA product. Indels can be distinguished from CNVs by means of size with the typically used cutoff used at 1000bp. However, an indel generally comprises <10 nucleotides.[84] For the functional effect of an indel, the number of nucleotides that are deleted or inserted is very important. Indels that are a multiple of three cause in-frame insertions or deletions, causing the protein to gain or lose a number of amino acids. In other cases a frameshift is induced, meaning that the reading frame for the amino acid sequence is altered, which often induces the incorporation of a number of incorrect amino acids and a premature stop codon and therefore an incompletely synthesized protein. The term mutation is used for variants and indels that have a deleterious effect on protein function.

Mutational effects

In general, frameshift and nonsense mutations are considered as most severely hampering protein function as the part of the protein is not produced. To prevent these types of partial proteins to accumulate in the cell, the nonsense-mediated decay (NMD) process is started in case a stop codon is detected before the last exon-exon boundary in the processed mRNA fragment.[85] NMD causes degradation of the erroneous mRNA fragment, thereby preventing the production of affected proteins, resulting in a loss-of-function mutation. At the protein level, this may be compared with the consequences of a microdeletion, which causes the heterozygous loss of the gene(s) encompassed by the deletion interval. Next to a reduced expression, loss-of-function mutations can occur when, for example, an active site of a protein is affected by the mutation or when a mutation is expected to cause a conformational change that generates a nonfunctional protein. Nonsense and frameshift variants that are located in the last exon of a gene can escape from NMD and still be translated. The resulting incorrect proteins may disturb cellular processes, for example, by being incorporated in a protein complex but not being active or by executing an alternative function due to the altered sequence. When a mutation induces a constitutively active protein or induces an abnormal function, it is designated as a gain-of-function mutation. Dominant negative mutations often result in an inactive protein that interferes with the function of the wildtype protein, such as an inactive channel subunit protein that incorporates in a channel complex and thereby inactivates the complete channel.[86, 87]

Distinguishing variants from mutations

Determining whether a certain variant is tolerated or has the potential to induce disease is not always straightforward. Multiple factors have to be taken into account to predict whether a variant can be a disease-causing mutation. At first the type of variant is important, with nonsynonymous variants generally being considered more relevant than synonymous variants. The nonsynonymous variants with more severe effects (e.g., nonsense, frameshift, and splice site variants) are of particular interest. Next the predicted mutational effect of a variant has to be considered, for example, a reduced expression of the protein or a malfunctioning protein due to an altered active site or disrupted binding site. While evaluating the effect the type of protein is also important. For example, proteins encoded by genes for which haploinsufficiency is tolerated may not be severely affected by variants altering the expression level of the gene, but still can be hampered by missense mutations in an active site or binding domains. Therefore it is useful to determine whether a variant is located in a particular protein domain, for example, using the UniProt database.[88] This may assist in defining whether a missense variant potentially disturbs, for example, transcription factor binding to target DNA, binding of proteins in a protein complex or enzymatic activity. To aid in the assessment of the effect of a variant, several prediction tools have been developed. Initially the prediction tools mainly evaluated a single feature, for example, by looking at the conservation of the mutated base pair or by determining the distance in the chemical and physical characteristics of the substituted amino acids.[89, 90] Subsequently, prediction tools were developed that took multiple features into account, such as PolyPhen, SIFT, and MutationTaster.[91–93] Most recently, prediction scores for all possible base substitutions were calculated with algorithms that take multiple prediction tools into account and are trained with datasets containing known pathogenic mutations and neutral variants (e.g., CADD and REVEL scores), which improves the predictive value even further.[94, 95] Next to assessing the predicted effect of variants, it is useful to search databases containing variants reported in patients (e.g., LOVD databases) or control cohorts (e.g., ExAC and gnomAD).[96, 97] The presence of a variant of interest in a relatively high percentage of controls may indicate that the variant is not disease causing. On the other hand the absence of a particular variant or depletion of detrimental variants (nonsense, frameshift, and splice site variants) in a certain gene in a large set of controls may support a hypothesis on the involvement of the variant or gene in disease. Furthermore a literature search may be helpful in the assessment of variants, for example, when the variant has been described as causal for a similar phenotype already (e.g., ClinVar and denovo-db) or when a mouse model of the affected gene displays concordant phenotypes (Mouse Genome Informatics).[98–101] Combining the aforementioned information and predictions will aid in determining whether a variant may be causing the observed phenotype in an individual.

Gene identification strategies

The initial studies applying CNV analysis and later whole exome sequencing (WES) to identify the genetic cause for rare syndromic forms of ID included patients with very similar phenotypes. The lists of genes affected by rare variants and CNVs were compared among different patients and often only one or a few genes were commonly affected in all patients, thereby revealing the genetic cause of disease. Using this method the disease-causing genes for multiple syndromes were identified among which Kabuki syndrome, Schinzel-Giedion syndrome, and Bohring-Opitz syndrome.[102–104] This strategy is also referred to as the phenotype-first approach as highly similar patients were preselected to identify a common underlying genetic defect. Today, exome sequencing is also performed more routinely on other ID cases, revealing genes that are mutated in phenotypically different patients. The syndromes that were identified in this way are referred to as genotype first-based syndromes, clustering patients based on their genetic aberration rather than the phenotypic outcome. Examples of genes that have been discovered through exome sequencing of extensive ID and ASD cohorts in different studies and combining the results to identify patients with the same gene defects are *ADNP*, *CHD8*, *CTNNB1*, and *SCN2A*.[32, 105–108] When a mutation in a new candidate gene is identified in a patient, it is critical to find mutations in the same gene in independent patients to support causality of the gene. Targeted sequencing approaches can be applied to specifically investigate the sequence of a selected set of genes, for example, multiple candidate genes in a large patient cohort. Compared with WES, targeted sequencing is more cost-efficient, thereby allowing a larger number of patients to be screened and increasing chances to identify additional causal mutations in a particular gene. Overall the rapid technological developments have accelerated the identification of many genetic aberrations involved in the pathogenesis of NDDs. Whole-genome sequencing can now be applied to identify both SNVs, SVs, and CNVs in one single experiment and has the potential to be applied as a first-tier test in genetic diagnostics of patients with severe ID, with a prospective diagnostic yield of 62%.[82] Massive cohorts are currently sequenced, including thousands of individuals with NDDs, in a search for novel disease-causing genes or to reach statistical significance for genes that were found mutated in one or few patients already but could not be clearly linked to the disease yet.[81] However, it is apparent that excessive numbers of patients need to be sequenced to identify new disease genes that reach statistical significance and, due to burdensome multiple-testing correction, this will become even more difficult.

Convergence to overlapping networks

The CNV and gene identification efforts in ID and ASD cohorts not only revealed many new genetic defects involved in the pathogenesis of both NDDs but also showed an extensive overlap among these genes and CNVs, meaning that these can cause both ID and ASD. Even more interesting, these genetic defects also showed overlap with genetic aberrations in patients with schizophrenia and epilepsy.[31–33, 35, 64, 68, 109–111] These observations strengthen the hypothesis that NDDs are part of a continuous phenotypic spectrum rather than separate disease entities, with common genetic factors underlying them. The NDD genes identified to date are significantly enriched not only in a subset of protein-protein interaction (PPI) networks, for instance, involved in chromatin remodeling and Wnt signaling, but also in neuron-specific networks responsible for synaptic function and neuronal development.[112, 113] From these observations, two hypotheses can be generated. First, if several genes within a certain PPI network are involved in NDD pathogenesis, it may be expected that the other components when disrupted also cause an NDD phenotype. Second, disturbance of the entire PPI network by a mutation may underlie the emergence of NDDs rather than disruption of the separate genes. The genetic background of an individual may determine the phenotypic presentation of a certain causal mutation, for example, by the presence of additional variants or less detrimental mutations in the same PPI network that exaggerate the effect of the primary mutation.

Gene networks in neurodevelopmental disorders

As it becomes more apparent that common genetic networks underlie the NDD phenotypic spectrum, it is important to stratify the different types of networks and the implication of the differences between these types on research and clinic. Gene or protein interactome networks can be defined in multiple ways and with different layers of complexity.[114] The most straightforward type of networks is one-dimensional and only reflects direct interaction. In this category, three networks can be distinguished: metabolic, gene-regulatory, and protein-protein interaction networks.[114] Metabolic networks are the classical biochemical pathways, where nodes represent catalytic enzymes and the edges represent the metabolites. In gene-regulatory networks the nodes reflect either a transcription factor or gene-regulatory elements in the genome, and the edges reflect physical interaction between the transcription factor and its target regulatory elements. Protein-protein interaction networks, or PPI networks, are composed of proteins (nodes) that have a physical interaction with each other, represented by the edges. This interaction not only can be temporary, for example, when a kinase phosphorylates a protein to (in)activate it, but also can represent a constitutive binding, for example, when a protein is part of a protein complex that consists of multiple units. To fully capture biological processes, the networks have to be extended with multiple layers of additional information to create multidimensional networks. Examples of these more extended types of networks are coexpression networks in which genes showing similar expression profiles are linked. These networks can further be extended with, for example, known functionality of proteins or information resulting from pathway analyses that infer statistical correlations based on gene expression profiles or genotype data.[114, 115] The studies involving networks in NDDs can be divided into at least two groups. The first group starts from genes affected by known disease-causing genetic anomalies and searches for pathways or networks that are overrepresented to identify a link between the different disorders. The second group commences unbiased with genome-wide datasets to identify unknown interactions that can potentially be relevant in the pathogenesis of NDDs.

From disease to networks

Large microarray and sequencing studies identified novel candidate and causal genes for ASD and ID and applied pathway analyses on these genes to reveal a clustering in distinct cellular processes. In an autism cohort, for example, the genes with likely gene-disruptive mutations were significantly enriched for FMRP targets, the protein involved in fragile X syndrome pathogenesis.[33] In another autism cohort a PPI network could be assembled using the genes that were found mutated in the cohort. This network was significantly more interconnected than could be expected by chance and contained members with known function in chromatin remodeling, ubiquitination, and neuronal development.[116] Andrews et al.[117] exploited a highly phenotyped patient cohort with de novo CNVs. First, 408 patient-phenotype groups were built, by grouping the genes affected by CNVs in patients sharing specific Human Phenotype Ontology (HPO) phenotypes. For each of these patient-phenotype groups, multiple functional genomics analyses were performed to identify potentially disrupted biological processes underlying the specific phenotype. Strikingly, 37% of the patient-phenotype groups included genes that significantly clustered within PPI networks, thereby showing that these genes work together in specific molecular pathways. Another interesting finding is that CNVs encompassing genes with significant coexpression patterns in brain are more likely to result in similar overall phenotypes, when compared with CNVs including genes that do not show coordinated expression in the brain.

To identify molecular processes underlying ID, Kochinke et al.[118] constructed a catalog of all known ID genes with associated phenotypes and a clinical classification system. The ID genes are significantly enriched not only for disease-relevant Gene-Ontology terms, including nervous system development and synapse, but also for Gene-Ontology terms reflecting specific cellular processes as DNA repair, ion transport, and Hedgehog signaling. Besides enrichment in biologically relevant processes, the ID genes are significantly enriched for coexpression patterns in the brain, and approximately half of the ID genes physically interacts with another ID gene. The ID gene catalog was supplemented with data from a systematic RNAi screen in *Drosophila melanogaster*, showing that specific Drosophila phenotypes (e.g., "early walker" and "wing veins missing") are highly associated with particular clinical ID classes, which emphasizes the role of ID genes in development.

From networks to disease

On the other end, more fundamental studies identified networks that turned out to be very valid when overlaying them on lists with known and newly identified disease genes. Moen et al.[119] identified all proteins interacting with three highly expressed transcription factors in mouse neural stem cells, building a transcription factor interaction network. The network was significantly enriched for known ID genes and for genes constrained from coding variation in the population, suggestive for a negative selection pressure against mutations in these genes. The burden of mutations in this network correlates with IQ levels; the percentage of loss-of-function mutations in network genes increases when the average IQ in a cohort drops. Corominas et al.[120] built a network based on alternatively spliced, brain-specific, isoforms of ASD risk factors. For in total 168 autism candidate genes, 422 isoforms were detected of which 60% was found to be novel. The spliceform interaction network built with this dataset comprised over 500 physical interactions of which only half would be found in the network when only reference transcripts would be considered. This emphasizes that it is very important to use relevant resources when studying networks in NDDs; for autism, it is more appropriate to use the interactions as defined in the spliceform network derived from brain than public databases containing only reference transcripts that may not be expressed in brain. As exemplified by the different network studies described earlier, both the candidate genes and the unbiased whole-genome approaches not only can result in relevant results but also have specific drawbacks. Whole-genome approaches are data driven without any prior interpretation, with the main advantage of having the ability to detect formerly unknown interactions. However, these analyses often suffer from computational limits and, perhaps more importantly, very burdensome multiple-testing corrections that may result in the loss of relevant interactions.[115] Candidate network and pathway analyses are based on prior knowledge and are hypothesis driven, reducing the chance to identify novel relationships. Nevertheless, we believe that it is of particular interest to study specific networks involved in NDDs as this allows for very extensive manual curation of the included component data and still gives the opportunity to identify novel disease-causing genes.

References

1. APA. *American Psychiatric Association: Diagnostic and Statistical Manual of Mental Disorders DSM-V, Fifth edition, Text Revision*. Washington DC: American Psychiatric Association; 2013.
2. Thapar A, Cooper M, Rutter M. Neurodevelopmental disorders. *Lancet Psychiatry*. 2017;4:339–346.
3. Munir KM. The co-occurrence of mental disorders in children and adolescents with intellectual disability/intellectual developmental disorder. *Curr Opin Psychiatry*. 2016;29:95–102.
4. van Ool JS, Snoeijen-Schouwenaars FM, Schelhaas HJ, Tan IY, Aldenkamp AP, Hendriksen JG. A systematic review of neuropsychiatric comorbidities in patients with both epilepsy and intellectual disability. *Epilepsy Behav*. 2016;60:130–137.
5. Oeseburg B, Dijkstra GJ, Groothoff JW, Reijneveld SA, Jansen DE. Prevalence of chronic health conditions in children with intellectual disability: a systematic literature review. *Intellect Dev Disabil*. 2011;49:59–85.
6. King BH. Psychiatric comorbidities in neurodevelopmental disorders. *Curr Opin Neurol*. 2016;29:113–117.
7. Chiurazzi P. Mental retardation: is naming the real issue? *Am J Med Genet A*. 2011;155A:974–975.
8. van Bokhoven H. Genetic and epigenetic networks in intellectual disabilities. *Annu Rev Genet*. 2011;45:81–104.
9. Leonard H, Wen X. The epidemiology of mental retardation: challenges and opportunities in the new millennium. *Ment Retard Dev Disabil Res Rev*. 2002;8:117–134.
10. Doernberg E, Hollander E. Neurodevelopmental disorders (ASD and ADHD): DSM-5, ICD-10, and ICD-11. *CNS Spectr*. 2016;21:295–299.
11. Lyall K, Croen L, Daniels J, et al. The changing epidemiology of autism spectrum disorders. *Annu Rev Public Health*. 2016;38:81–102.
12. Abrahams BS, Geschwind DH. Advances in autism genetics: on the threshold of a new neurobiology. *Nat Rev Genet*. 2008;9:341–355.
13. Nikolas MA, Burt SA. Genetic and environmental influences on ADHD symptom dimensions of inattention and hyperactivity: a meta-analysis. *J Abnorm Psychol*. 2010;119:1–17.
14. Thomas R, Sanders S, Doust J, Beller E, Glasziou P. Prevalence of attention-deficit/hyperactivity disorder: a systematic review and meta-analysis. *Pediatrics*. 2015;135:e994–1001.
15. Giegling I, Hosak L, Mossner R, et al. Genetics of schizophrenia: a consensus paper of the WFSBP Task Force on Genetics. *World J Biol Psychiatry*. 2017;18:492–505.
16. Merikangas KR, Jin R, He JP, et al. Prevalence and correlates of bipolar spectrum disorder in the world mental health survey initiative. *Arch Gen Psychiatry*. 2011;68:241–251.
17. Gershon ES, Alliey-Rodriguez N, Liu C. After GWAS: searching for genetic risk for schizophrenia and bipolar disorder. *Am J Psychiatry*. 2011;168:253–256.
18. Chang BS, Lowenstein DH. Epilepsy. *N Engl J Med*. 2003;349:1257–1266.

19. Hesdorffer DC, Logroscino G, Benn EK, Katri N, Cascino G, Hauser WA. Estimating risk for developing epilepsy: a population-based study in Rochester, Minnesota. *Neurology*. 2011;76:23–27.
20. Hildebrand MS, Dahl HH, Damiano JA, Smith RJ, Scheffer IE, Berkovic SF. Recent advances in the molecular genetics of epilepsy. *J Med Genet*. 2013;50:271–279.
21. Langner I, Garbe E, Banaschewski T, Mikolajczyk RT. Twin and sibling studies using health insurance data: the example of attention deficit/hyperactivity disorder (ADHD). *PLoS ONE*. 2013;8.e62177.
22. Lichtenstein P, Carlstrom E, Rastam M, Gillberg C, Anckarsater H. The genetics of autism spectrum disorders and related neuropsychiatric disorders in childhood. *Am J Psychiatry*. 2010;167:1357–1363.
23. Sandin S, Lichtenstein P, Kuja-Halkola R, Larsson H, Hultman CM, Reichenberg A. The familial risk of autism. *JAMA*. 2014;311:1770–1777.
24. Vissers LELM, de Ligt J, Gilissen C, et al. A de novo paradigm for mental retardation. *Nat Genet*. 2010;42:1109–1112.
25. Vissers LE, Gilissen C, Veltman JA. Genetic studies in intellectual disability and related disorders. *Nat Rev Genet*. 2016;17:9–18.
26. Manolio TA, Collins FS, Cox NJ, et al. Finding the missing heritability of complex diseases. *Nature*. 2009;461:747–753.
27. Kiser DP, Rivero O, Lesch KP. Annual research review: the (epi)genetics of neurodevelopmental disorders in the era of whole-genome sequencing--unveiling the dark matter. *J Child Psychol Psychiatry*. 2015;56:278–295.
28. Cooper GM, Coe BP, Girirajan S, et al. A copy number variation morbidity map of developmental delay. *Nat Genet*. 2011;43:838–846.
29. de Vries BBA, Pfundt R, Leisink M, et al. Diagnostic genome profiling in mental retardation. *Am J Hum Genet*. 2005;77:606–616.
30. Sebat J, Lakshmi B, Malhotra D, et al. Strong association of de novo copy number mutations with autism. *Science*. 2007;316:445–449.
31. Stefansson H, Rujescu D, Cichon S, et al. Large recurrent microdeletions associated with schizophrenia. *Nature*. 2008;455:232–236.
32. de Ligt J, Willemsen MH, van Bon BW, et al. Diagnostic exome sequencing in persons with severe intellectual disability. *N Engl J Med*. 2012;367:1921–1929.
33. Iossifov I, Ronemus M, Levy D, et al. De novo gene disruptions in children on the autistic spectrum. *Neuron*. 2012;74:285–299.
34. O'Roak BJ, Deriziotis P, Lee C, et al. Exome sequencing in sporadic autism spectrum disorders identifies severe de novo mutations. *Nat Genet*. 2011;43:585–589.
35. Rauch A, Wieczorek D, Graf E, et al. Range of genetic mutations associated with severe non-syndromic sporadic intellectual disability: an exome sequencing study. *Lancet*. 2012;380:1674–1682.
36. Hassold T, Hunt P. To err (meiotically) is human: the genesis of human aneuploidy. *Nat Rev Genet*. 2001;2:280–291.
37. Weijerman ME, de Winter JP. Clinical practice. The care of children with Down syndrome. *Eur J Pediatr*. 2010;169:1445–1452.
38. Nelson DL, Orr HT, Warren ST. The unstable repeats—three evolving faces of neurological disease. *Neuron*. 2013;77:825–843.
39. Heitz D, Rousseau F, Devys D, et al. Isolation of sequences that span the fragile X and identification of a fragile X-related CpG island. *Science*. 1991;251:1236–1239.
40. Pieretti M, Zhang F, Fu Y-H, et al. Absence of expression of the FMR-1 gene in fragile X syndrome. *Cell*. 1991;66:817–822.
41. Verkerk AJMH, Pieretti M, Sutcliffe JS, et al. Identification of a gene (FMR-1) containing a CGG repeat coincident with a breakpoint cluster region exhibiting length variation in fragile X syndrome. *Cell*. 1991;65:905–914.

42. Coffee B, Keith K, Albizua I, et al. Incidence of fragile X syndrome by newborn screening for methylated FMR1 DNA. *Am J Hum Genet*. 2009;85:503–514.
43. Sutcliffe JS, Nelson DL, Zhang F, et al. DNA methylation represses FMR-1 transcription in fragile x syndrome. *Hum Mol Genet*. 1992;1:397–400.
44. Schneider A, Hagerman RJ, Hessl D. Fragile X syndrome—from genes to cognition. *Dev Disabil Res Rev*. 2009;15:333–342.
45. Kidd SA, Lachiewicz A, Barbouth D, et al. Fragile X syndrome: a review of associated medical problems. *Pediatrics*. 2014;134:995–1005.
46. Metsu S, Rooms L, Rainger J, et al. FRA2A is a CGG repeat expansion associated with silencing of *AFF3*. *PLoS Genet*. 2014;10:e1004242.
47. Metsu S, Rainger JK, Debacker K, et al. A CGG-repeat expansion mutation in ZNF713 causes FRA7A: association with autistic spectrum disorder in two families. *Hum Mutat*. 2014;35:1295–1300.
48. Collins RL, Brand H, Redin CE, et al. Defining the diverse spectrum of inversions, complex structural variation, and chromothripsis in the morbid human genome. *Genome Biol*. 2017;18:36.
49. Pinkel D, Segraves R, Sudar D, et al. High resolution analysis of DNA copy number variation using comparative genomic hybridization to microarrays. *Nat Genet*. 1998;20:207–211.
50. Marshall CR, Noor A, Vincent JB, et al. Structural variation of chromosomes in autism spectrum disorder. *Am J Hum Genet*. 2008;82:477–488.
51. Shaw-Smith C, Redon R, Rickman L, et al. Microarray based comparative genomic hybridisation (array-CGH) detects submicroscopic chromosomal deletions and duplications in patients with learning disability/mental retardation and dysmorphic features. *J Med Genet*. 2004;41:241–248.
52. Walsh T, McClellan JM, McCarthy SE, et al. Rare structural variants disrupt multiple genes in neurodevelopmental pathways in schizophrenia. *Science*. 2008;320:539–543.
53. Girirajan S, Rosenfeld JA, Coe BP, et al. Phenotypic heterogeneity of genomic disorders and rare copy-number variants. *N Engl J Med*. 2012;367:1321–1331.
54. Kaminsky EB, Kaul V, Paschall J, et al. An evidence-based approach to establish the functional and clinical significance of copy number variants in intellectual and developmental disabilities. *Genet Med*. 2011;13:777–784.
55. Zhang F, Gu W, Hurles ME, Lupski JR. Copy number variation in human health, disease, and evolution. *Annu Rev Genomics Hum Genet*. 2009;10:451–481.
56. Lee JA, Carvalho CMB, Lupski JR. A DNA replication mechanism for generating nonrecurrent rearrangements associated with genomic disorders. *Cell*. 2007;131:1235–1247.
57. Bauters M, Van Esch H, Friez MJ, et al. Nonrecurrent MECP2 duplications mediated by genomic architecture-driven DNA breaks and break-induced replication repair. *Genome Res*. 2008;18:847–858.
58. Hastings PJ, Ira G, Lupski JR. A microhomology-mediated break-induced replication model for the origin of human copy number variation. *PLoS Genet*. 2009;5.e1000327.
59. Lupski JR. Genomic disorders: structural features of the genome can lead to DNA rearrangements and human disease traits. *Trends Genet*. 1998;14:417–422.
60. Jacquemont S, Reymond A, Zufferey F, et al. Mirror extreme BMI phenotypes associated with gene dosage at the chromosome 16p11.2 locus. *Nature*. 2011;478:97–102.
61. Bijlsma EK, Gijsbers AC, Schuurs-Hoeijmakers JH, et al. Extending the phenotype of recurrent rearrangements of 16p11.2: deletions in mentally retarded patients without autism and in normal individuals. *Eur J Med Genet*. 2009;52:77–87.
62. Fernandez BA, Roberts W, Chung B, et al. Phenotypic spectrum associated with de novo and inherited deletions and duplications at 16p11.2 in individuals ascertained for diagnosis of autism spectrum disorder. *J Med Genet*. 2010;47:195–203.
63. Shen Y, Chen X, Wang L, et al. Intra-family phenotypic heterogeneity of 16p11.2 deletion carriers in a three-generation Chinese family. *Am J Med Genet B Neuropsychiatr Genet*. 2011;156:225–232.

64. Zufferey F, Sherr EH, Beckmann ND, et al. A 600 kb deletion syndrome at 16p11.2 leads to energy imbalance and neuropsychiatric disorders. *J Med Genet*. 2012;49:660–668.
65. McCarthy SE, Makarov V, Kirov G, et al. Microduplications of 16p11.2 are associated with schizophrenia. *Nat Genet*. 2009;41:1223–1227.
66. Shinawi M, Liu P, Kang SH, et al. Recurrent reciprocal 16p11.2 rearrangements associated with global developmental delay, behavioural problems, dysmorphism, epilepsy, and abnormal head size. *J Med Genet*. 2010;47:332–341.
67. Walters RG, Jacquemont S, Valsesia A, et al. A new highly penetrant form of obesity due to deletions on chromosome 16p11.2. *Nature*. 2010;463:671–675.
68. Weiss LA, Shen Y, Korn JM, et al. Association between microdeletion and microduplication at 16p11.2 and autism. *N Engl J Med*. 2008;358:667–675.
69. D'Angelo D, Lebon S, Chen Q, et al. Defining the effect of the 16p11.2 duplication on cognition, behavior, and medical comorbidities. *JAMA Psychiatry*. 2016;73:20–30.
70. Schubert C. The genomic basis of the Williams-Beuren syndrome. *Cell. Mol. Life Sci.* 2009;66:1178–1197.
71. Stromme P, Bjornstad PG, Ramstad K. Prevalence estimation of Williams syndrome. *J Child Neurol*. 2002;17:269–271.
72. Van der Aa N, Rooms L, Vandeweyer G, et al. Fourteen new cases contribute to the characterization of the 7q11.23 microduplication syndrome. *Eur J Med Genet*. 2009;52:94–100.
73. Vandeweyer G, Van der Aa N, Reyniers E, Kooy RF. The contribution of CLIP2 haploinsufficiency to the clinical manifestations of the Williams-Beuren syndrome. *Am J Hum Genet*. 2012;90:1071–1078.
74. Koolen DA, Vissers LELM, Pfundt R, et al. A new chromosome 17q21.31 microdeletion syndrome associated with a common inversion polymorphism. *Nat Genet*. 2006;38:999–1001.
75. Koolen DA, Kramer JM, Neveling K, et al. Mutations in the chromatin modifier gene KANSL1 cause the 17q21.31 microdeletion syndrome. *Nat Genet*. 2012;44:639–641.
76. Koolen DA, Pfundt R, Linda K, et al. The Koolen-de Vries syndrome: a phenotypic comparison of patients with a 17q21.31 microdeletion versus a KANSL1 sequence variant. *Eur J Hum Genet*. 2016;24:652–659.
77. Beunders G, van de Kamp JM, Veenhoven RM, van Hagen JM, Nieuwint AW, Sistermans EA. A triplication of the Williams Beuren Syndrome region in a patient with mental retardation, a severe expressive language delay, behavioural problems and dysmorfisms. *J Med Genet*. 2010;47:271–275.
78. Soler-Alfonso C, Carvalho CM, Ge J, et al. CHRNA7 triplication associated with cognitive impairment and neuropsychiatric phenotypes in a three-generation pedigree. *Eur J Hum Genet*. 2014;22:1071–1076.
79. Van Dijck A, van der Werf IM, Reyniers E, et al. Five patients with a chromosome 1q21.1 triplication show macrocephaly, increased weight and facial similarities. *Eur J Med Genet*. 2015;58:503–508.
80. Yobb TM, Somerville MJ, Willatt L, et al. Microduplication and triplication of 22q11.2: a highly variable syndrome. *Am J Hum Genet*. 2005;76:865–876.
81. Deciphering Developmental Disorders Study. Prevalence and architecture of de novo mutations in developmental disorders. *Nature*. 2017;542:433–438.
82. Gilissen C, Hehir-Kwa JY, Thung DT, et al. Genome sequencing identifies major causes of severe intellectual disability. *Nature*. 2014;511:344–347.
83. Stessman HA, Xiong B, Coe BP, et al. Targeted sequencing identifies 91 neurodevelopmental-disorder risk genes with autism and developmental-disability biases. *Nat Genet*. 2017;49:515–526.
84. Ronemus M, Iossifov I, Levy D, Wigler M. The role of de novo mutations in the genetics of autism spectrum disorders. *Nat Rev Genet*. 2014;15:133–141.

85. Nagy E, Maquat LE. A rule for termination-codon position within intron-containing genes: when nonsense affects RNA abundance. *Trends Biochem Sci*. 1998;23:198–199.
86. Orhan G, Bock M, Schepers D, et al. Dominant-negative effects of KCNQ2 mutations are associated with epileptic encephalopathy. *Ann Neurol*. 2014;75:382–394.
87. Sollis E, Graham SA, Vino A, et al. Identification and functional characterization of de novo FOXP1 variants provides novel insights into the etiology of neurodevelopmental disorder. *Hum Mol Genet*. 2016;25:546–557.
88. UniProt Consortium. UniProt: the universal protein knowledgebase. *Nucleic Acids Res*. 2017;45:D158–D169.
89. Grantham R. Amino acid difference formula to help explain protein evolution. *Science*. 1974;185:862.
90. Pollard KS, Hubisz MJ, Rosenbloom KR, Siepel A. Detection of nonneutral substitution rates on mammalian phylogenies. *Genome Res*. 2010;20:110–121.
91. Adzhubei IA, Schmidt S, Peshkin L, et al. A method and server for predicting damaging missense mutations. *Nat Methods*. 2010;7:248–249.
92. Ng PC, Henikoff S. SIFT: predicting amino acid changes that affect protein function. *Nucleic Acids Res*. 2003;31:3812–3814.
93. Schwarz JM, Rodelsperger C, Schuelke M, Seelow D. MutationTaster evaluates disease-causing potential of sequence alterations. *Nat Methods*. 2010;7:575–576.
94. Ioannidis NM, Rothstein JH, Pejaver V, et al. REVEL: an ensemble method for predicting the pathogenicity of rare missense variants. *Am J Hum Genet*. 2016;99:877–885.
95. Kircher M, Witten DM, Jain P, O'Roak BJ, Cooper GM, Shendure J. A general framework for estimating the relative pathogenicity of human genetic variants. *Nat Genet*. 2014;46:310–315.
96. Fokkema IF, Taschner PE, Schaafsma GC, Celli J, Laros JF, den Dunnen JT. LOVD v.2.0: the next generation in gene variant databases. *Hum Mutat*. 2011;32:557–563.
97. Karczewski KJ, Weisburd B, Thomas B, et al. The ExAC browser: displaying reference data information from over 60 000 exomes. *Nucleic Acids Res*. 2017;45:D840–D845.
98. Blake JA, Eppig JT, Kadin JA, et al. Mouse Genome Database (MGD)-2017: community knowledge resource for the laboratory mouse. *Nucleic Acids Res*. 2017;45:D723–D729.
99. Finger JH, Smith CM, Hayamizu TF, et al. The mouse Gene Expression Database (GXD): 2017 update. *Nucleic Acids Res*. 2017;45:D730–D736.
100. Landrum MJ, Lee JM, Benson M, et al. ClinVar: public archive of interpretations of clinically relevant variants. *Nucleic Acids Res*. 2016;44:D862–D868.
101. Turner TN, Yi Q, Krumm N, et al. denovo-db: a compendium of human de novo variants. *Nucleic Acids Res*. 2017;45:D804–D811.
102. Hoischen A, van Bon BW, Gilissen C, et al. De novo mutations of SETBP1 cause Schinzel-Giedion syndrome. *Nat Genet*. 2010;42:483–485.
103. Hoischen A, van Bon BW, Rodriguez-Santiago B, et al. De novo nonsense mutations in ASXL1 cause Bohring-Opitz syndrome. *Nat Genet*. 2011;43:729–731.
104. Ng SB, Bigham AW, Buckingham KJ, et al. Exome sequencing identifies MLL2 mutations as a cause of Kabuki syndrome. *Nat Genet*. 2010;42:790–793.
105. Bernier R, Golzio C, Xiong B, et al. Disruptive CHD8 mutations define a subtype of autism early in development. *Cell*. 2014;158:263–276.
106. Helsmoortel C, Vulto-van Silfhout AT, Coe BP, et al. A SWI/SNF-related autism syndrome caused by de novo mutations in *ADNP*. *Nat Genet*. 2014;46:380–384.
107. Kuechler A, Willemsen MH, Albrecht B, et al. De novo mutations in beta-catenin (CTNNB1) appear to be a frequent cause of intellectual disability: expanding the mutational and clinical spectrum. *Hum Genet*. 2015;134:97–109.
108. Sanders SJ, Murtha MT, Gupta AR, et al. De novo mutations revealed by whole-exome sequencing are strongly associated with autism. *Nature*. 2012;485:237–241.

109. Ben-Shachar S, Lanpher B, German JR, et al. Microdeletion 15q13.3: a locus with incomplete penetrance for autism, mental retardation, and psychiatric disorders. *J Med Genet*. 2009;46:382–388.
110. Fromer M, Pocklington AJ, Kavanagh DH, et al. De novo mutations in schizophrenia implicate synaptic networks. *Nature*. 2014;506:179–184.
111. O'Roak BJ, Vives L, Fu W, et al. Multiplex targeted sequencing identifies recurrently mutated genes in autism spectrum disorders. *Science*. 2012;338:1619–1622.
112. Krumm N, O'Roak BJ, Shendure J, Eichler EE. A de novo convergence of autism genetics and molecular neuroscience. *Trends Neurosci*. 2014;37:95–105.
113. Pinto D, Delaby E, Merico D, et al. Convergence of genes and cellular pathways dysregulated in autism spectrum disorders. *Am J Hum Genet*. 2014;94:677–694.
114. Vidal M, Cusick ME, Barabasi AL. Interactome networks and human disease. *Cell*. 2011;144:986–998.
115. Ramanan VK, Shen L, Moore JH, Saykin AJ. Pathway analysis of genomic data: concepts, methods, and prospects for future development. *Trends Genet*. 2012;28:323–332.
116. O'Roak BJ, Vives L, Girirajan S, et al. Sporadic autism exomes reveal a highly interconnected protein network of de novo mutations. *Nature*. 2012;485:246–250.
117. Andrews T, Meader S, Vulto-van Silfhout A, et al. Gene networks underlying convergent and pleiotropic phenotypes in a large and systematically-phenotyped cohort with heterogeneous developmental disorders. *PLoS Genet*. 2015;11. e1005012.
118. Kochinke K, Zweier C, Nijhof B, et al. Systematic phenomics analysis deconvolutes genes mutated in intellectual disability into biologically coherent modules. *Am J Hum Genet*. 2016;98:149–164.
119. Moen MJ, Adams HH, Brandsma JH, et al. An interaction network of mental disorder proteins in neural stem cells. *Transl Psychiatry*. 2017;7:e1082.
120. Corominas R, Yang X, Lin GN, et al. Protein interaction network of alternatively spliced isoforms from brain links genetic risk factors for autism. *Nat Commun*. 2014;5:3650.

CHAPTER

4

Autism spectrum disorder: A clinical path to early diagnosis, evaluation, and intervention

Lidia V. Gabis

Director of the Child Development Center at Edmond and Lilly Safra Children's Hospital, Sheba, Tel Hashomer, Israel, Sackler School of Medicine, Tel Aviv University, Tel Aviv, Israel

Diagnosis

Developmental discrepancies in the areas of language and communication serve as "red flags" in the early detection of ASD.[1] Any observed delay in the ability to engage in early interaction (such as using eye contact and responding with a social smile) and significant delay in acquisition of language may prompt a more extensive evaluation. None of the early signs should become a diagnostic element on its own, and other differential diagnoses should be ruled out.

Evaluation of ASD in infancy

A combination of early signs and symptoms may convey a diagnosis of ASD, and the diagnosis is based solely on clinical tools. As such the proposed evaluation should include three parts:

(a) Extensive neurodevelopmental workup
(b) Evaluation of differential diagnosis and comorbidity (such as epilepsy, global delay, sleep, or feeding difficulties)
(c) Assessment of identifiable causes associated with ASD (environmental, genetic, or a combination of both).

In this chapter, we will further elaborate on each of those steps.

https://doi.org/10.1016/B978-0-12-814037-6.00004-5

Recognition of signs and symptoms neurodevelopmental workup

Early diagnosis

Since the diagnosis of ASD is clinically based, the age by which symptoms are assessed and diagnosis is given is influenced by the available diagnostic tools and by the possible presence of a differential diagnosis.[2] From a developmental perspective, a gap of two or more standard deviations below the norm of the expected level of functioning, according to age, is considered significant in determining a delay in one area of development. A standardized score that is two standard deviations or more below the norm in the area of language and communication is required for a diagnosis of ASD[3] and the child meeting DSM-5 criterion A. In addition, the presence of atypical behavior is required (restricted and repeated behaviors, criterion B) in order to complete a diagnosis of ASD.[4]

The American Academy of Pediatrics recommends screening all children for ASD at 18 and 24 months followed by a comprehensive evaluation for children with developmental concerns.[5] However, in recent years, it has been recommended that they be evaluated at a younger age and that the diagnosis be made during the second year of life.[6,7]

Developmental evaluation during first 6 months

A consistent gap of at least two SD (Standard Deviations) in language and communication is much more obvious to observe, measure, and substantiate during early development.[8–10] The infant is born with basic communication abilities that integrate into early interactions with primary caregivers. Communication during the first year of life includes, beyond established eye contact and social smile, the infant's ability to share needs, experiences, and feelings, via nonverbal means such as vocalizations, gestures, movements, and facial expressions.[11] A baby should be able to communicate his needs initially by crying, which is a communicative tool in the first months of life and is already differentiated during this early stage (e.g., a different cry signifying hunger from that representing discomfort or fatigue). The infant should use eye contact to search for caregivers, imitate facial expressions, and use social smile before the age of 2 months. Since the newborn infant is short sighted, preferential gaze at faces and smile should be elicited by looking at the baby from a distance of 20 cm.[12] By 4 months, babies should find other means of communicating their needs, besides crying, like vocalizations and facial expressions. Lack of evolution of "body language" and lack of modulation of eye contact should raise concerns. With the increase in motor control, during the initial 6 months of life, the baby should also find motor methods of expressing needs. As such, motor gaps such as head lag, hypotonia, exaggerated response to sensory stimuli such as noise or touch, or the opposite—lack of response to stimuli—should raise concerns and elicit a more extensive neurological evaluation. Additional "hints" may present as overt motor asymmetries. Minor inconsistent asymmetry is part of normal development, such as asynchronous movements of limbs in infants, but the movements should become more synchronous with time and consistent asymmetry should prompt a more extensive evaluation.

The evaluation at this age should include the following:

- Observation of movement (spontaneous and directed), interaction (with caregiver and with examiner), communication, and response to stimuli (touch, light, sound, etc.)
- Anthropometric measurements (height, weight, and head circumference with marking of trajectory in growth charts).
- Examination of all cranial nerves including vision (tracking of face or object in all directions) and hearing (attention to sounds).
- Evaluation of tone and reflexes including disappearance of primitive reflexes and occurrence of postural reflexes.
- Evaluation of all areas of development—motor, fine motor/visiomotor, language, and communication. A chart such as Denver Developmental Scales or Griffith's Scale can be used.[13]

Neurological abnormalities should prompt an extensive evaluation. Persistence of primitive reflexes, such as Moro and grasp, and abnormalities of tone should prompt a metabolic and imaging workup.

Delays of development without neurological abnormalities should be addressed differently if only one area of delay is present or if delay includes more than one area. In the latter, the diagnosis should be global developmental delay including autism spectrum. Common causes of delays, such as delayed visual maturation and benign hypotonia of infancy, should be considered in the differential diagnosis.[5, 14] It should be emphasized that all four areas of development have a trajectory that commences during the first days of life (and earlier in uterus). As such, all areas should be examined, at all ages and at all phases of development. Example of early milestones of all four areas is shown in Table 4.1.

The additional emphasis is that normal development should be taken "step by step." Atypical development is defined not only by delays in development but also by milestones that occur atypically early and by milestones that occur at a different pace in different areas of development.

Developmental evaluation up to the first year and transition to second year (see Table 4.2)

During the second half of the first year, the motor control evolves along with appearance of a more extensive vocal repertoire such as razzing and

TABLE 4.1 Developmental milestones from birth, in four areas of development

Area	Age—birth	1 Month	2 Months
Gross motor	Flexor position Can move head to sides and asynchronous movement of arms and legs	Some control of head When supine asynchronous gross movements but symmetrical, same amount on both sides Can bring hands to mouth	Moves head to both sides and head lag disappearing Raises head when prone Symmetrical but not simultaneous movements on both sides Occasionally opens both hands
Fine motor/ visiomotor	Eye contact and follows around midline	Follows toward both sides with eyes and starts to turn head	Follows and turns head to both sides to vision and hearing
Language	Awareness to sound, will change expression, open eyes to sound	Sound, some voices, and different cry of pain, hunger, being tired	Voices May start laughing loudly (toward 3 months)
Social	Eye contact May imitate face simple expression	Social smile occasionally Imitates grimacing	Social smile frequently in response to smile (from short distance)

TABLE 4.2 Three "red flags" for ASD at age 3, 6, 12, and 24 months

3 Months	6 Months	1 Year	2 Years
Lack of eye contact coordinated with responsive social smile toward a person	Not asking by eyes and arms to be picked up	Less than two meaningful words/syllables	Less than 50 words
Preference of mobile, colors, objects over human face when placed close (about 20–30 cm)	Unusual interest in specific toys	Lack of consistent response to own name	Lack of symbolic play
Motor asymmetry	Constant irritability	Lack of pointing with an index finger coordinated with eye contact/joint attention	Lack of interest in same age peers

babbling. In addition, even before the transition to the second year, typically developing infants show preference to interaction with humans over objects and interest in reciprocal play with an adult over playing by one's self.[15] The use of language to communicate is measurable around 1 year of age, when an infant should have at least two words in his/her vocabulary. During the second year of life, the infant's amount of vocabulary should increase dramatically with at least 100 single words and two words combinations at age of 2 years.

Following the rapid growth of nonverbal communication during the second year of life, the repertoire of gestures such as pointing, waving, nodding, and clapping should increase spontaneously after 1 year of age. This growth is parallel to the evolution of play: from basic sensorimotor play to functional and then symbolic play.[16] The evolvement of nonverbal communication and play during the first 2 years of life is measurable. Therefore a delay or a deficit in the acquisition of skills in these areas is of concern.

Atypical behaviors

In terms of repetitive behaviors, restricted interests, and sensory abnormalities, the atypical manner may appear before age one.[17] However, with the increase in motor activity, the unusual behaviors become more obvious. For example, hand flapping, walking in circles, lining objects, and a particular interest in spinning objects may be reported by parents early on. But when a child is walking independently and his movements become more purposeful, his conduct would appear abnormal if these are the only activities that represent his behavioral repertoire. In addition, unusual early motor developmental patterns are common in infants that are subsequently diagnosed with ASD and may be an early sign of atypical behaviors. Such as asymmetric movements at the age that symmetry is expected, milestones that would appear earlier than expected, such as early rolling from belly to back, unusual motor interests such as holding a metal object instead of the usual transitional nappies and prolonged interest in mechanical objects or parts of objects such as spinning wheels.

Parental concern and age of diagnosis

Communication delays and the repetitive and restricted behaviors are atypical before age one, but the main "red flags" that bring parents to a diagnostic center emerge following the child's first birthday and toward his second year. Reports include the lack of eye contact, gaze preference to objects rather than faces, less cooperation in early interactions involving song and imitation play, deficit in the development of primary gestural movements (such as raising hands asking to be picked up—which is typical of 6 months), and the lack of fear of strangers that should be present at 9 months. In retrospect, many parents report observing their child not responding to joint attention, exhibiting lack of "pointing with the eyes" and their child not making any effort to communicate his/her needs. Those concerns, however, are very difficult to measure and report in babies. First time parents usually perceive this lack of social communication skills as part of normal development and do not consider these deficits abnormal until another sibling is born or the preschool teacher raises concerns.

After concerns are being raised by parents, it may take a few months up to 1 year until a definite diagnosis is made. The timing of final diagnosis is dependent on symptom severity, gender, ethnic affiliation, access to medical care, and parental socioeconomic status.[18] The average age for which a diagnosis of ASD is given ranges in Europe and the United States between 33 and 50 months[3, 19] despite diagnostic tools being available for younger children. In countries where mandatory surveillance by primary medical care providers or a national "healthy baby" network are in place, the "red flags" can be recognized by professionals before parents raise specific concerns, and the diagnosis can be given earlier. In Israel, there is a national free of cost developmental surveillance network attached to immunizations schedule (Tipat Halav[a]).

In order to detect earlier signs of autism, recent studies have focused on young infants at risk (infants who have older siblings diagnosed with ASD). Studies have utilized various social visual engagement procedures in order to characterize the early communicative and behavioral patterns of those infants who are later diagnosed with ASD.[2, 20–27] Sibling studies have shown that primary social communication skills (i.e., eye contact, social smile, and ability to engage in interaction) may appear to be typical during the first few months of life but tend to diminish during the second half of the first year. However, in many subsequently diagnosed children, differences of early interaction are reported in retrospect by the parents from the first few weeks of life (see in the succeeding text).

Additional early differences and studies in progress

Other alarming early signs are overlooked and analyzed in retrospect such as differences in early physiological bodily functions. Initially, these signs may be ignored and regarded as individual differences within the normal development. Early regulatory difficulties are common during infancy, for example, baby colic. But combined with deficits in communication, they should warrant a more thorough evaluation. Early regulatory

[a] Tipat Halav is a center that provides health and medical services in the field of health promotion and prevention for pregnant women, infants, and children up to age 6 years. Tipat Halav centers are distributed throughout Israel and are operated by the Ministry of Health (Israeli Ministry of health website).

concerning signs are observed in basic areas of functioning: sleep, feeding, and temperament[28–32] (see Table 4.2).

Sleep disturbances are prevalent in ASD and also in specific related disorders such as fragile X and Angelman's syndrome,[33] and abnormal sleep cycle is very common in premature infants, especially after a long NICU hospitalization.[34] In addition, it was shown that insomnia may be an early sign of subsequent autism.[35] Sleep difficulties, insomnia, and abnormal sleep patterns have been described retrospectively in children with subsequent ASD.

General movements are a distinct movement pattern carried out spontaneously without external stimulation and seen in fetuses of 9 weeks gestational age until 21 weeks post term. General movements are helpful in the early diagnosis of an impaired central nervous system and the specific prediction of cerebral palsy[36] and also reflect impairments of brain areas involved in cognitive development.[37]

Measurements of GM and sleep are particularly important to assess in infants born prematurely.

An emerging early "red flag" studied is delayed in empathic responses in infants. Empathy—the ability to feel what another is feeling and to feel concerned for the other as a result—is an important socioemotional capacity. Children diagnosed with ASD suffer from an impaired empathic ability and show lower emotional responses to others in distress compared with typically developing children.[38, 39] Children with ASD also show deficits in the cognitive aspects of empathy, such as theory of mind or emotion recognition.[39] To date, no study has examined empathic responding during the first year of life as a predictor of subsequent ASD diagnosis. However, recent advancements in the field of early empathy development have opened the door to much earlier assessments of empathy as a prodromal marker of ASD.[40, 41]

Related social behaviors relevant to both empathy and ASD are examined in infants, specifically regulatory abilities and preferences for objects versus people (social interest) at 6 months. Children and adults with autism indicate abnormal attention patterns, such as a relative preference for objects over social stimuli.[42, 43] ASD has also been associated with lower regulation/effortful control and increased negative affectivity.[44, 45] These factors are being examined together with empathic abilities at 6 months in a study currently performed at the Keshet Center in Israel.

About the "red flags"

Each of the mentioned symptoms and signs can be observed in typically developing infants and in other developmental disorders. As such, when all three "red flags" appear together, the meaning is that a more extensive evaluation should be performed with a particular focus on communication and interaction patterns. The specificity of these signs is quite vague, but when taken together, their sensitivity is high and should prompt not only an initial evaluation but also a subsequent developmental follow-up.

As such, early high-risk signs of ASD can be recognized during the first year of life. Significant and measurable gaps between different areas of functioning can be evaluated during the second year, and a definite diagnosis of ASD can be given prior to age two.[46]

Etiology

Since the communication pathways are extremely vulnerable, there are many underlying possible causes that may generate the same phenotype.[47,48] Environmental causes such as a viral infection during pregnancy or a genetic cause such as fragile X syndrome may have an immediate impact on the treatment offered and on the outcome. Presently, the probable reasons for the development of ASD can be divided into the following[49]:

- Genetic causes—there are more than 140 single-gene disorders identifies as linked to autism spectrum phenotype and epigenetic processes at the level of RNA or protein. Some of the genetic causes are inherited, and carrier state analysis should be emphasized, specifically—fragile X premutation. Most of the genetic causes identified to date are de novo mutations.
- Specific insult to the central nervous system during vulnerable periods of development.
- Combined etiologies—genetic predisposition and environmental changes causing epigenetic effects and vulnerability of language and communication pathways.

Diagnostic methods

Since the early signs and symptoms may convey a diagnosis of ASD, the diagnostic process is based solely on clinical tools. The main purpose of the evaluation is to answer the additional question of "why"—causes of the disorder, with emphasis on subsequent treatment approach and prevention of a second born infant with autism. A thorough medical evaluation may influence treatment, prognosis, and genetic familial counseling. The proposed medical evaluation should include three parts:

(a) neurodevelopmental evaluation and standardized workup
(b) evaluation of differential diagnosis and oriented tests
(c) etiological evaluation

The pathways leading to normal and synchronized communication are extremely vulnerable, resulting in the same phenotype of ASD being caused by numerous reasons including brain damage during critical fetal brain development and genetic and epigenetic causes. The search for a specific cause can be endless and frustrating for both clinicians and patients. Nevertheless, some of the causes may point to a potentially treatable condition such as hypothyroidism or a creatine deficiency disorder. Other genetic causes may prompt family counseling (such as in the case of fragile X syndrome or other genetic inheritable disorders).

We suggest a stepwise approach (see flowchart)

(a) Extensive neurodevelopmental workup including examination of hearing (consider evoked responses—BAERA) and vision (including eyes exam and fundus), consideration of EEG and neuroimaging (especially if there are additional neurological findings on examination), and specific oriented tests such as examination of the skin with Wood's lamp if warranted (to search for signs of tuberous sclerosis). A behavioral hearing examination should be thorough, and since the child with suspected ASD will probably not cooperate with such test, a brainstem auditory evoked response assay (BAERA) is recommended. Vision should be established as well, since a baby with delayed visual maturation might be delayed not only in making eye contact but also in other milestones.[50]

Brain imaging (MRI) is recommended only when a diagnosis of ASD is accompanied by moderate to severe global delay and additional neurological findings such as dysmorphism, abnormal head size, abnormal motor, or cerebellar examination, for example, the presence of abnormal reflexes or ataxia.[51]

In all other cases of "idiopathic autism" with normal exam, imaging studies are not warranted as part of the clinical workup. The only current indication to perform imaging in a child with ASD and normal exam without global delay or intellectual disability is the rare possibility of finding creatine deficiency syndrome with normal creatine and guanidinoacetate in urine.[52] Most of creatine disorders may cause abnormal levels in urine, so evaluation for this rare but potentially treatable condition should be included in the first tier of tests. If imaging is performed for this specific indication, then it should include MR Spectroscopy as part of the protocol.[53]

The other indication for imaging in a child with ASD without intellectual impairment and otherwise normal exam is an abnormal and asymmetric EEG with focal epilepsy. For those cases, an MRI with coronal cuts should be ordered.

(b) Differential diagnosis and comorbidity—such as epilepsy (especially if there is a clinical course of regression), global delay, or intellectual impairment, either as comorbid with ASD or a rule-out diagnosis. Epilepsy is an important issue to elucidate at all ages,[54] more so in infants. Seizures may be insidious in young children, especially if they have stereotypic behaviors. The pediatric neurology guidelines suggest to consider it, since the EEG in most children with autism may be normal. Since this noninvasive test may reveal an important clue to diagnosis, my personal opinion is that it is always warranted: because of rare cases, such as specific brain waves that may fit a diagnosis of hypsarrhythmia—which is a medical emergency and delay in its treatment may cause regression and intellectual impairment—West syndrome, and specific spikes that may point to a specific diagnosis of Rett syndrome or Angelman's syndrome.[54, 55]
As part of the differential diagnosis, some basic, albeit rare, diagnoses should be ruled out by routine blood testing—basic CBC and SMAC should be performed to search for anemia, hypocholesterolemia, and abnormalities in electrolytes. These results may point to a cause, though more often to a nutritional deficiency or overdose due to poor eating habits of children with ASD. Thyroid testing is advisable, as well as CPK, especially in a child with ASD that is toe walking (rarely it may prompt an evaluation of muscular dystrophy). A more extensive metabolic workup should be considered, especially in countries where the extended neonatal metabolic screen is absent.[56]

(c) Evaluation of the cause of ASD to the extent may change the course of treatment and influence familial genetic counseling. The development of the communication pathways may be altered as a result of the following:
   - **(a)** genetic causes with more than 120 genes linked to autism
   - **(b)** brain insult such as complications of prematurity and intrauterine infections
   - **(c)** a combination of the above

The suggested evaluation should be stepwise and consider family history, additional findings and comorbidity—such as the presence of intellectual disability. The current "state of the art" includes fragile X testing, both in girls and in boys with ASD, and the mCGH to search for multiple genetic syndromes as well as CNV. Whole-exome analysis and whole genome are not yet standard of care but are performed more frequently when all other tests do not reveal an etiology or when the parents are seeking familial planning counseling.[57]

Comprehensive evaluation and planning of treatment (Chart 4.1)

The presence of "red flags" in early development warrants a comprehensive and multidisciplinary evaluation of the infant or young child beyond the neurodevelopmental exam. In the case of ASD, it includes evaluations by health-care professionals (physical, speech and language, and occupational therapists) and a comprehensive psychological evaluation. The diagnostic process includes the use of psychological tools (i.e., intake and clinical observation) and standardized ones (i.e., tests, questionnaires,

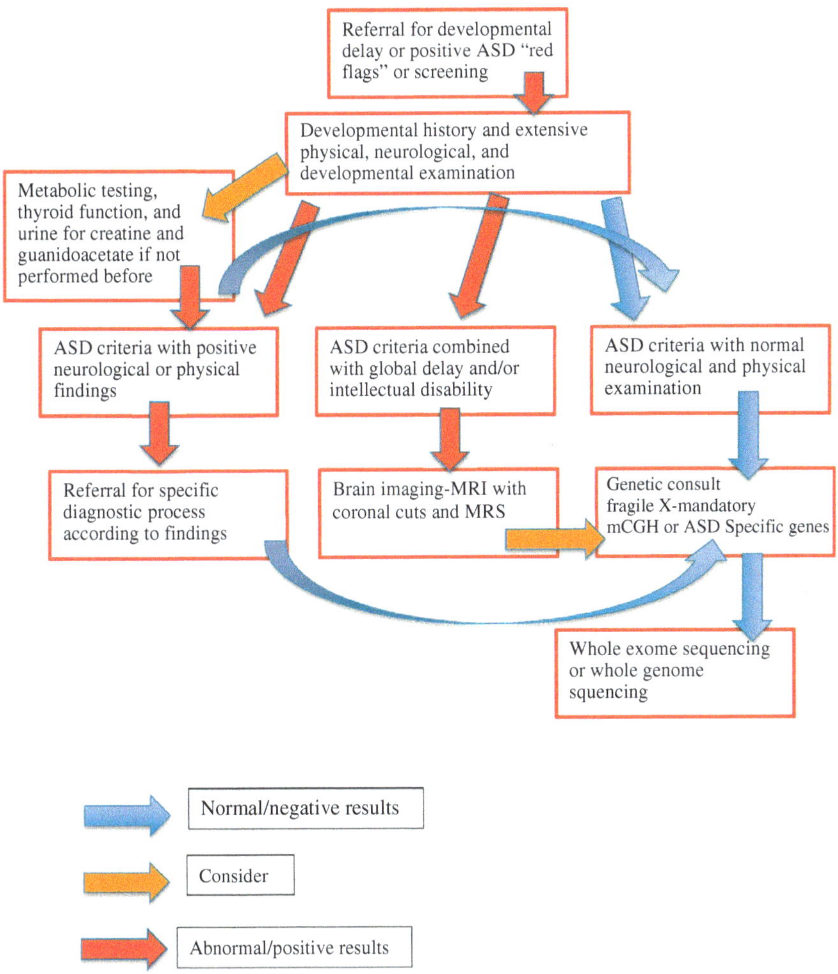

CHART 4.1 Diagnostic process.

and structured observations). In order to fulfill a diagnosis, the child has to comply with DSM-5 criteria for ASD.[1] In addition, baseline assessments by each profession are necessary before commencing treatment. The following are some of the assessment tools being used in the diagnostic process, which are adapted to infants up to 2 years of age:

- **Physical and neurological examination** as described earlier
- **Developmental assessment** using standardized tests—the Mullen Scales of Early Learning[58] or the Bayley Scales of Infant and Toddler Development[59]
- **Specific professional tools**—AIMS (physical therapy),[60] Peabody (gross and fine motor),[61] and Rossetti (language and communication)[62]
- **Structured observations**—the Autism Observation Scale for Infants (AOSI)[63]; Social Attention and Communication Study (SACS)[64, 65]; and Autism Diagnostic Observation Schedule, Toddler Module (ADOS-2)[66]
- **Questionnaires** for parents and caregivers—the Quantitative Checklist for Autism in Toddlers (Q-CHAT),[67] the Modified Checklist for Autism in Toddlers (M-CHAT-R),[68] and the Adaptive Behavior Assessment System (ABAS-2)[69]
- **Structured interviews**—Autism Diagnostic Interview (ADI-R)[70] and Vineland Adaptive Behavior Scales[71]

Professionals who are experts in the field of autism and who have sound developmental knowledge and experience in the assessment of infants and young children should perform the diagnostic evaluation of an infant suspected to have ASD.

Treatment and intervention

When an infant is at risk for ASD or when a final diagnosis of ASD is given, an intensive intervention program should be implemented within the immediate future.

In infancy, the diagnosis of ASD is provisional and dynamic. Early diagnosis, which prompts early intervention, has the potential to change the developmental course of the condition and diminish the developmental gaps significantly.[72] The intervention phase commences already during the evaluation of a child with language and communication delays. The assessment of age appropriate milestones directs the parents and caregivers into increasing communication with the child and decreasing interferences (such as prolonged TV watching). When an infant is found to be at risk for ASD or when a child receives a final diagnosis of ASD, an intervention program should be in place, including intensive 1:1 approach and parental guidance.[73] Several intervention approaches will be discussed; however, the main proven effect refers to the intensity of the treatment more than to the specific method implemented.[74]

Three major intervention approaches are presented; but the hallmark of intervention during infancy is parental guidance and support of ongoing interactions of the infant with the main caregivers. The parent as a partner and as an intervention agent has the potential to markedly improve child's communicative and social skills through consistent and meaningful interactions. This process also empowers the parents and provides a sense of control in view of a difficult diagnosis.

(1) Early Start Denver Model (ESDM): This intervention model is based on the principles of developmental psychology and applied behavior analysis. The intensive teaching is delivered within play-based, relationship-focused routines. The protocol includes a variety of structured strategies for working with very young children in individual (1:1) and group settings and with the aim of promoting development in key domains such as imitation, communication, social, cognitive, and motor skills, adaptive behavior, and play.[75–77] Studies show that children who received ESDM showed significant improvements in IQ, adaptive behavior, and autism diagnosis. These children maintained its rate of growth in adaptive behavior compared with a normative sample of typically developing children. Children who received ESDM also were more likely to experience a change in diagnosis from autism to pervasive developmental disorder, not otherwise specified, than the comparison group.

(2) Applied Behavioral Analysis (ABA): An individualized intervention model that is based on principles of learning theories. Every child's program is individualized to his/her needs and learning style and guided by what is developmentally expected. Teaching occurs either in an isolated, specifically designated environment or within the naturalistic environment. Imitation is the cornerstone for which all skills are built upon.[78] Studies show that the techniques of applied behavior analysis (ABA) are effective in altering the developmental trajectory of some very young children with autism. Research suggests that early, intensive treatment using the methods of ABA enables a significant number of children to enter the educational mainstream and achieve normal intellectual functioning.

(3) Developmental, Individual Difference, Relationship-based model (DIR): Referred to also as the floortime model. An interdisciplinary framework that promotes parents' ability to recognize and respect the emotional experience and expressions of their child, as shown in their actions, ideas, and intentions and to interact in a way that helps their child use his or her natural emotions with a greater sense of purpose. DIR intervention with young children and their parents is based on play aiming at promoting language, attention, mental health, attachment, development, sensory processing, and motor skills.[79, 80] Studies show that children who were treated with DIR made significant changes in emotional functioning, communication, and daily living skills. Moreover, the mothers perceived positive changes in their parent-child interactions.[81]

(4) The Keshet model: Integrated DIR and ESDM models combined with intensive parental guidance and inclusion in mainstream education. Developed and implemented since 2007 at the Weinberg Child Development Center, Sheba Medical Center, Israel.[82] Children receive 10h per week of age-adapted developmentally based treatments for a period of 1 year. Parents are present and involved in all treatment and are considered members of the intervention team. A multidisciplinary team that includes members from all the developmental medical professions provides guidance to the educational teams.

Language interventions and parental guidance: Children with autism may not understand how to initiate and respond to joint attention with another person.[83] In addition to their overall difficulties in social communication,[84] which include difficulty responding to other people's intentions as expressed through language and nonverbal communication (including gestures), organizing and inhibiting their responses may exhibit repetitive verbal responses and echolalia and engage in repetitive behaviors and circumscribed interests. Language development in children with autism can be delayed, and performance is heterogeneous.[85] Impaired language and communication among young children with ASD are among the main targets for intervention. Since parental mediation may support generalization, parent training and guidance in regard to communication with their children is an integral part of therapy.[75] Exposure to content words by parents (CDS) of children with ASD may have an important role in language development.[86] Yet, children with ASD have been found to attend to CDS less so than typical children, limiting their opportunity to learn to interpret meaning resulting from its unique characteristics.[87] Furthermore, joint attention has been found to be related to lexical development in this population.[88, 89] Parent linguistic input that describes and orients the children's focus of attention (i.e., follow-in talking) is associated with their language ability.[90] Hence, parents who provide verbal interpretations that enable their child with ASD to process verbal information can compensate for the joint attention deficit that characterizes many children with ASD. A detailed analysis of CDS may provide useful tools for parental guidance and intervention.

Pharmacological treatment: To date, there are no treatments that target the core features of ASD. In addition, during infancy, we prefer to refrain from pharmacotherapy, unless there is a significant comorbidity that arrests the child's progress. Epilepsy, severe sleep impairment, and other very extreme behaviors are usually treated, but in most children, the cornerstone is intensive intervention rather than medications.

Since it is unethical to withhold intervention, the effect of early adapted intervention on children diagnosed with ASD can only be based on intervention studies.[55] It is plausible that some children with motor or language delays will close the developmental gap naturally or that infants with communication delays may have a natural course of improvement. Nevertheless, since interventions through interaction and play may promote even normal development, it is not advisable to withhold therapy and wait for the natural course to evolve.

The significance of early diagnosis

There is still a significant gap between "state-of-the-art" research on Autism and common practice as they are related to age of diagnosis. This gap varies in its magnitude between countries, among communities, and in relation to socioeconomic status. In addition, there may be a delay of a few years between detection of early signs and the age of definitive diagnosis of ASD.[91] As discussed in the section on intervention, there is an established direct connection between early diagnosis, early intervention, and subsequent outcomes. Yet, the path is not linear.

Early diagnosis prompts early intervention. The main message that we, as health-care providers, convey to parents is that early diagnosis warrants early professional services that are designed to promote the child's communicative, behavioral, and social functioning and assist him/her in acquiring better adaptive skills. The earlier the intervention, the more effective it is in improving functioning.[92] However, at times, early diagnosis of ASD is made based on obvious and significant developmental deficits that are associated with more severe autism. In such cases, early diagnosis does not always assure good prognosis.

A favorite quote in the area of autism is, "If you've met one person with autism, you've met one person with autism" (Dr. Stephen Shore). In spite of the shared features of deficits in social communication and social interaction and the presence of restrictive/repetitive behaviors as listed in DSM-5, there are significant differences across individuals in terms of symptom severity, comorbidities, and treatment response. More recent knowledge aims for more personalized approach to autism according to risk factors (genetic and environmental), biomarkers, phenotypes, and response to a specific treatment.[93]

Summary of Challenges

Liability to ASD is now understood to be due in part to rare genetic variants and environmental causes such as prenatal viruses, insults, and prematurity, the latter a common, recently revealed major contributor to risk. However, despite many advances in our understanding of the biology and genetics of ASD, diagnoses continue to be based on behavioral evidence. In approximately 15% of cases, a genetic finding may support a genetic diagnosis,[94] but these (and other) more objective biomarkers for ASD have not been consistently studied. The current diagnostic model requires significant resources, and assessment results are mainly subjective. We hypothesize that the failure of some behavioral and pharmacological ASD trials are mainly due to mixing biologically different ASD populations and the focus on subjective measures for end points (which are influenced strongly by placebo response). It is plausible that some of the drugs or behavioral interventions could have had significant benefits in a biologically more homogenous subpopulation of ASD. If subgroups could be defined by more objective biomarkers (and end points could also be more objective), we can predict a greater success for ASD interventions.

Another major challenge is the delay in diagnosis. Findings indicate that a stable diagnosis can be made by the age of 2[95] and substantial evidence indicates that optimal outcomes are most assured in children who receive early intervention.[96] However, many children do not receive services until the age of 5.[97] As a result, it is critical to explore methods that may be used to augment ASD risk recognition in an objective, rapid, and reliable manner and as early as possible. Since in many families a genetic predisposition might be recognized after the birth of the first child with ASD, genetic counseling may be warranted before the second child is born. Additionally, if the etiology is known, a procedure such a preimplantation diagnosis may be employed.[98] Use of the Early Start Denver Model (ESDM) of treatment at 6 months of age (as employed at Keshet Center in concert with developmental interventions) leads to amelioration of symptoms in babies identified as having higher genetic risk for autism (so-called baby sibs).

Symptoms and impairments in ASD differ in intensity and can be grouped according to etiology. Personalized medicine (also known as precision medicine) will enable risk assessment, diagnosis, prevention, and therapy specifically tailored to the unique characteristics of the individual, thus enhancing the quality of life and public health (NHLBI strategic planning).

Since development of the newborn starts much earlier, during the prenatal phase, identification of intrauterine markers of atypical development or ASD risk is an additional research fascinating challenge. An antenatal test or marker for autism would make both early monitoring and early intervention possible.

Acknowledgment

The author would like to thank: Dr. Ronit Roth Hannania, Romi Oren-Schwartz and Leora Allen for editing this chapter.

References

1. American Psychiatric Association. *DSM-5*. https://symbiose.uqo.ca/apps/LoginSigparb/LoginPourRessources.aspx?url=http://www.psychiatryonline.org/%5Cnhttp://sbiproxy.uqac.ca/login?url=http://dsm.psychiatryonline.org/book.aspx?bookid=556; 2013.
2. Messinger DS, Young GS, Webb SJ, et al. Early sex differences are not autism-specific: a baby siblings research consortium (BSRC) study. *Mol Autism*. 2015;6(1):https://doi.org/10.1186/s13229-015-0027-y.
3. Salomone E, Charman T, Mcconachie H, Warreyn P. Child's verbal ability and gender are associated with age at diagnosis in a sample of young children with ASD in Europe. *Child Care Health Dev*. 2016;42(1):141–145. https://doi.org/10.1111/cch.12261.
4. Carrington SJ, Kent RG, Maljaars J, et al. DSM-5 autism spectrum disorder: in search of essential behaviours for diagnosis. *Res Autism Spectr Disord*. 2014;8(6):701–715. https://doi.org/10.1016/j.rasd.2014.03.017.
5. Johnson CP, Myers SM. Identification and evaluation of children with autism spectrum disorders. *Pediatrics*. 2007;120(5):1183–1215. https://doi.org/10.1542/peds.2007-2361.
6. Chawarska K, Shic F, Macari S, et al. 18-month predictors of later outcomes in younger siblings of children with autism spectrum disorder: a baby siblings research consortium study. *J Am Acad Child Adolesc Psychiatry*. 2014;53(12):1317–1327. e1, https://doi.org/10.1016/j.jaac.2014.09.015.
7. Christensen DL, Bilder DA, Zahorodny W, et al. Prevalence and characteristics of autism spectrum disorder among 4-year-old children in the autism and developmental disabilities monitoring network. *J Dev Behav Pediatr*. 2016;37(1):1–8. https://doi.org/10.1097/DBP.0000000000000235.
8. Lazenby DWC, Sideridis GD, Huntington N, et al. Language differences at 12 months in infants who develop autism spectrum disorder. *J Autism Dev Disord*. 2016;46(3):899–909. https://doi.org/10.1007/s10803-015-2632-1.
9. Parladé MV, Iverson JM. The development of coordinated communication in infants at heightened risk for autism spectrum disorder. *J Autism Dev Disord*. 2015;45(7):2218–2234. https://doi.org/10.1007/s10803-015-2391-z.
10. Veness C, Prior M, Bavin E, Eadie P, Cini E, Reilly S. Early indicators of autism spectrum disorders at 12 and 24 months of age: a prospective, longitudinal comparative study. *Autism*. 2012;16(2):163–177. https://doi.org/10.1177/1362361311399936.
11. Cocquyt M, Mommaerts MY, Dewart H, Zink I. Measuring pragmatic skills: early detection of infants at risk for communication problems. *Int J Lang Commun Disord*. 2015;50(5):646–658. https://doi.org/10.1111/1460-6984.12167.
12. Dobson V, Teller DY. Visual acuity in human infants: a review and comparison of behavioral and electrophysiological studies. *Vis Res*. 1978. https://doi.org/10.1016/0042-6989(78)90001-9.
13. Frankenburg WK, Dodds JB. The denver developmental screening test. *J Pediatr*. 1967;71(2):181–191. https://doi.org/10.1016/S0022-3476(67)80070-2.

14. Goodman R, Ashby L. Delayed visual maturation and autism. *Dev Med Child Neurol.* 1990;32(9):814–819. https://doi.org/10.1111/j.1469-8749.1990.tb08487.x.
15. Libertus K, Needham A. Face preference in infancy and its relation to motor activity. *Int J Behav Dev.* 2014;38(6):529–538. https://doi.org/10.1177/0165025414535122.
16. Watson LR, Crais ER, Baranek GT, Dykstra JR, Wilson KP. Communicative gesture use in infants with and without autism: a retrospective home video study. *Am J Speech Lang Pathol.* 2013;22(1):25–39. https://doi.org/10.1044/1058-0360(2012/11-0145.
17. Elison JT, Wolff JJ, Reznick JS, et al. Repetitive behavior in 12-month-olds later classified with autism spectrum disorder. *J Am Acad Child Adolesc Psychiatry.* 2014;53(11):1216–1224. https://doi.org/10.1016/j.jaac.2014.08.004.
18. Emerson ND, Morrell HER, Neece C. Predictors of age of diagnosis for children with autism spectrum disorder: the role of a consistent source of medical care, race, and condition severity. *J Autism Dev Disord.* 2016;46(1):127–138. https://doi.org/10.1007/s10803-015-2555-x.
19. Mazurek MO, Handen BL, Wodka EL, Nowinski L, Butter E, Engelhardt CR. Age at first autism spectrum disorder diagnosis: the role of birth cohort, demographic factors, and clinical features. *J Dev Behav Pediatr.* 2014;35(9):561–569. https://doi.org/10.1097/DBP.0000000000000097.
20. Bedford R, Elsabbagh M, Gliga T, et al. Precursors to social and communication difficulties in infants at-risk for autism: gaze following and attentional engagement. *J Autism Dev Disord.* 2012;42(10):2208–2218. https://doi.org/10.1007/s10803-012-1450-y.
21. Chawarska K, Klin A, Volkmar F. Automatic attention cueing through eye movement in 2-year-old children with autism. *Child Dev.* 2003;74(4):1108–1122. https://doi.org/10.1111/1467-8624.00595.
22. Constantino JN, Zhang Y, Frazier T, Abbacchi AM, Law P. Sibling recurrence and the genetic epidemiology of autism. *Am J Psychiatry.* 2010;167(11):1349–1356. https://doi.org/10.1176/appi.ajp.2010.09101470. [pii:appi.ajp.2010.09101470].
23. Jones EJH, Gliga T, Bedford R, Charman T, Johnson MH. Developmental pathways to autism: a review of prospective studies of infants at risk. *Neurosci Biobehav Rev.* 2014;39:1–33. https://doi.org/10.1016/j.neubiorev.2013.12.001.
24. Sandin S, Lichtenstein P, Kuja-Halkola R, Larsson H, Hultman CM, Reichenberg A. The familial risk of autism. *JAMA.* 2014;311(17):1770–1777. https://doi.org/10.1001/jama.2014.4144.
25. Thorup E, Nyström P, Gredebäck G, Bölte S, Falck-Ytter T. The EASE team. Altered gaze following during live interaction in infants at risk for autism: an eye tracking study. *Mol Autism.* 2016;7(1):12. https://doi.org/10.1186/s13229-016-0069-9.
26. Yirmiya N, Gamliel I, Pilowsky T, Feldman R, Baron-Cohen S, Sigman M. The development of siblings of children with autism at 4 and 14 months: social engagement, communication, and cognition. *J Child Psychol Psychiatry Allied Discip.* 2006;47(5):511–523. https://doi.org/10.1111/j.1469-7610.2005.01528.x.
27. Yoder P, Stone WL, Walden T, Malesa E. Predicting social impairment and ASD diagnosis in younger siblings of children with autism spectrum disorder. *J Autism Dev Disord.* 2009;39(10):1381–1391. https://doi.org/10.1007/s10803-009-0753-0.
28. Brisson J, Warreyn P, Serres J, Foussier S, Adrien-Louis J. Motor anticipation failure in infants with autism: a retrospective analysis of feeding situations. *Autism.* 2012;16(4):420–429. https://doi.org/10.1177/1362361311423385.
29. Johnson CR, Turner K, Stewart PA, et al. Relationships between feeding problems, behavioral characteristics and nutritional quality in children with ASD. *J Autism Dev Disord.* 2014;44(9):2175–2184. https://doi.org/10.1007/s10803-014-2095-9.
30. Kozlowski AM, Matson JL, Belva B, Rieske R. Feeding and sleep difficulties in toddlers with autism spectrum disorders. *Res Autism Spectr Disord.* 2012;6(1):385–390. https://doi.org/10.1016/j.rasd.2011.06.012.

31. Lucas RF, Cutler A. Dysregulated breastfeeding behaviors in children later diagnosed with autism. *J Perinat Educ.* 2015;24(3):171–180. https://doi.org/10.1891/1058-1243.24.3.171.
32. Williams LW, Matson JL, Beighley JS, Rieske RD, Adams HL. Comorbid symptoms in toddlers diagnosed with autism spectrum disorder with the DSM-IV-TR and the DSM-5 criteria. *Res Autism Spectr Disord.* 2014;8(3):186–192. https://doi.org/10.1016/j.rasd.2013.11.007.
33. Goldman SE, Bichell TJ, Surdyka K, Malow BA. Sleep in children and adolescents with Angelman syndrome: association with parent sleep and stress. *J Intellect Disabil Res.* 2012;56(6):600–608. https://doi.org/10.1111/j.1365-2788.2011.01499.x.
34. Bart O, Shayevits S, Gabis LV, Morag I. Prediction of participation and sensory modulation of late preterm infants at 12 months: a prospective study. *Res Dev Disabil.* 2011;32(6):2732–2738. https://doi.org/10.1016/j.ridd.2011.05.037.
35. Richdale AL, Schreck KA. Sleep problems in autism spectrum disorders: prevalence, nature, & possible biopsychosocial aetiologies. *Sleep Med Rev.* 2009;13(6):403–411. https://doi.org/10.1016/j.smrv.2009.02.003.
36. Philippi CL, Koenigs M. The neuropsychology of self-reflection in psychiatric illness. *J Psychiatr Res.* 2014;54(1):55–63. https://doi.org/10.1016/j.jpsychires.2014.03.004.
37. Einspieler C, Bos AF, Libertus ME, Marschik PB. The general movement assessment helps us to identify preterm infants at risk for cognitive dysfunction. *Front Psychol.* 2016;7:https://doi.org/10.3389/fpsyg.2016.00406.
38. Hutman T, Rozga A, DeLaurentis AD, Barnwell JM, Sugar CA, Sigman M. Response to distress in infants at risk for autism: a prospective longitudinal study. *J Child Psychol Psychiatry.* 2010;51(9):1010–1020. http://openurl.uquebec.ca:9003/uqam?url_ver=Z39.88-2004&amp%5Cnrft_val_fmt=info:ofi/fmt:kev:mtx:journal&amp%5Cn-genre=article&amp%5Cnsid=ProQ:ProQ:ericshell&amp%5Cnatitle=Response+to+Distress+in+Infants+at+Risk+for+Autism:+A+Prospective+Longitudinal+Study.
39. Yirmiya N, Sigman MD, Kasari C, Mundy P. Empathy and cognition in high-functioning children with autism. *Child Dev.* 1992;63(1):150–160. https://doi.org/10.1111/j.1467-8624.1992.tb03603.x.
40. Davidov M, Zahn-Waxler C, Roth-Hanania R, Knafo A. Concern for others in the first year of life: theory, evidence, and avenues for research. *Child Dev Perspect.* 2013;7(2): 126–131. https://doi.org/10.1111/cdep.12028.
41. Roth-Hanania R, Davidov M, Zahn-Waxler C. Empathy development from 8 to 16 months: early signs of concern for others. *Infant Behav Dev.* 2011;34(3):447–458. https://doi.org/10.1016/j.infbeh.2011.04.007.
42. Maestro S, Muratori F, Cavallaro MC, et al. How young children treat objects and people: an empirical study of the first year of life in autism. *Child Psychiatry Hum Dev.* 2005;35(4):383–396. https://doi.org/10.1007/s10578-005-2695-x.
43. Sasson NJ, Touchstone EW. Visual attention to competing social and object images by preschool children with autism spectrum disorder. *J Autism Dev Disord.* 2014;44(3):584–592. https://doi.org/10.1007/s10803-013-1910-z.
44. Garon N, Bryson SE, Zwaigenbaum L, et al. Temperament and its relationship to autistic symptoms in a high-risk infant sib cohort. *J Abnorm Child Psychol.* 2009;37(1):59–78. https://doi.org/10.1007/s10802-008-9258-0.
45. Samyn V, Roeyers H, Bijttebier P. Effortful control in typically developing boys and in boys with ADHD or autism spectrum disorder. *Res Dev Disabil.* 2011;32(2):483–490. https://doi.org/10.1016/j.ridd.2010.12.038.
46. Zwaigenbaum L, Bryson S, Lord C, et al. Clinical assessment and management of toddlers with suspected autism spectrum disorder: insights from studies of high-risk infants. *Pediatrics.* 2009;123(5):1383–1391. https://doi.org/10.1542/peds.2008-1606. [pii:123/5/1383].
47. Packer A. Neocortical neurogenesis and the etiology of autism spectrum disorder. *Neurosci Biobehav Rev.* 2016;64:185–195. https://doi.org/10.1016/j.neubiorev.2016.03.002.

48. Trottier G, Srivastava L, Walker CD. Etiology of infantile autism: a review of recent advances in genetic and neurobiological research. *J Psychiatry Neurosci*. 1999;24(2):103–115. https://doi.org/10.1038/nm.4102.
49. Sandin S, Lichtenstein P, Kuja-Halkola R, Hultman C, Larsson H, Reichenberg A. The heritability of autism spectrum disorder. *JAMA*. 2017;318(12):1182–1184. https://doi.org/10.1001/jama.2017.12141.
50. Volkmar F, Siegel M, Woodbury-Smith M, King B, McCracken J, State M. Practice parameter for the assessment and treatment of children and adolescents with autism spectrum disorder. *J Am Acad Child Adolesc Psychiatry*. 2014;53(8):931. https://doi.org/10.1016/j.jaac.2014.06.001.
51. Filipek PA, Accardo PJ, Ashwal S, et al. Practice parameter: screening and diagnosis of autism: report of the Quality Standards Subcommittee of the American Academy of Neurology and the Child Neurology Society. *Neurology*. 2000;55(4):468–479. http://www.ncbi.nlm.nih.gov/pubmed/10953176.
52. Braissant O, Henry H, Béard E, Uldry J. Creatine deficiency syndromes and the importance of creatine synthesis in the brain. *Amino Acids*. 2011;40(5):1315–1324. https://doi.org/10.1007/s00726-011-0852-z.
53. Ardon O, Procter M, Mao R, et al. Creatine transporter deficiency: novel mutations and functional studies. *Mol Genet Metab Rep*. 2016;8:20–23. https://doi.org/10.1016/j.ymgmr.2016.06.005.
54. Gabis L, Pomeroy J, Andriola MR. Autism and epilepsy: cause, consequence, comorbidity, or coincidence? *Epilepsy Behav*. 2005;https://doi.org/10.1016/j.yebeh.2005.08.008.
55. Jeste SS, Tuchman R. Autism spectrum disorder and epilepsy: two sides of the same coin? *J Child Neurol*. 2015;30(14):1963–1971. https://doi.org/10.1177/0883073815601501.
56. Schiff M, Benoist JF, Aïssaoui S, et al. Should metabolic diseases be systematically screened in nonsyndromic autism spectrum disorders? *PLoS ONE*. 2011;6(7):https://doi.org/10.1371/journal.pone.0021932.
57. Ouhtit A. Towards understanding the genetics of autism. *Front Biosci*. 2016;8(3):776. https://doi.org/10.2741/776.
58. Mullen EM. *Mullen Scales of Early Learning AGS Ed*. Circle Pines: American Guidance Service Inc; 1995.
59. Bayley N. *Bayley Scales of Infant Development*. 2nd ed. Psychol Corp San Antonio; 1993.https://doi.org/10.1093/cercor/bhv065.
60. Rush Jr. J. Abnormal involuntary movement scale (AIMS). *Am Psychiatr Assoc*. 2000;166–168. https://doi.org/10.1017/CBO9781107415324.004.
61. Eigsti IM. Entry: peabody picture vocabulary test. In: *Encyclopedia of Autism Spectrum Disorders*. 2013. https://doi.org/10.1253/circj.CJ-13-0534.
62. Howard JS, Sparkman CR, Cohen HG, Green G, Stanislaw H. A comparison of intensive behavior analytic and eclectic treatments for young children with autism. *Res Dev Disabil*. 2005;26(4):359–383. https://doi.org/10.1016/j.ridd.2004.09.005.
63. Bryson SE, Zwaigenbaum L, McDermott C, Rombough V, Brian J. The autism observation scale for infants: scale development and reliability data. *J Autism Dev Disord*. 2008;38(4):731–738. https://doi.org/10.1007/s10803-007-0440-y.
64. Barbaro J, Dissanayake C. Prospective identification of autism spectrum disorders in infancy and toddlerhood using developmental surveillance: the social attention and communication study. *J Dev Behav Pediatr*. 2010;31(5):376–385. https://doi.org/10.1097/DBP.0b013e3181df7f3c.
65. Barbaro J, Ridgway L, Dissanayake C. Developmental surveillance of infants and toddlers by maternal and child health nurses in an australian community-based setting: promoting the early identification of autism spectrum disorders. *J Pediatr Nurs*. 2011;26(4):334–347. https://doi.org/10.1016/j.pedn.2010.04.007.

66. Lord C, Rutter M, DiLavore PC, Risi S, Gotham K, Bishop S. *Autism Diagnostic Observation Schedule—Second Edition (ADOS-2)*. Los Angeles West Psychol Serv; 2012.
67. Allison C, Baron-Cohen S, Wheelwright S, et al. The Q-CHAT (Quantitative CHecklist for Autism in Toddlers): a normally distributed quantitative measure of autistic traits at 18-24 months of age: preliminary report. *J Autism Dev Disord*. 2008;38(8):1414–1425. https://doi.org/10.1007/s10803-007-0509-7.
68. Robins DL, Ph D, Fein D, Barton ML, James A. Modified checklist for autism in toddlers (M-CHAT)*. *J Autism Dev Disord*. 2001;Apr;31(2):131–44.
69. Harrison PL, Oakland T. *Adaptive Behavior Assessment System*. 2nd ed. 2003.
70. Lord C, Rutter M, Le Couteur A. Autism diagnostic interview-revised. *J Autism Dev Disord*. 1994;24:659–685. https://doi.org/10.1007/978-1-4419-1698-3_894.
71. Sparrow S, Balla D, Cicchetti D. The vineland adaptive behavior scales: interview edition, survey. In: *Major Psychological Assessment Instruments*. vol. 2. 1984:199–231. https://doi.org/10.1007/s00421-002-0748-4.
72. Granpeesheh D, Dixon DR, Tarbox J, Kaplan AM, Wilke AE. The effects of age and treatment intensity on behavioral intervention outcomes for children with autism spectrum disorders. *Res Autism Spectr Disord*. 2009;3(4):1014–1022. https://doi.org/10.1016/j.rasd.2009.06.007.
73. Sullivan K, Stone WL, Dawson G. Potential neural mechanisms underlying the effectiveness of early intervention for children with autism spectrum disorder. *Res Dev Disabil*. 2014;35(11):2921–2932. https://doi.org/10.1016/j.ridd.2014.07.027.
74. Warren Z, McPheeters ML, Sathe N, Foss-Feig JH, Glasser A, Veenstra-VanderWeele J. A systematic review of early intensive intervention for autism Spectrum disorders. *Pediatrics*. 2011;127(5):e1303–e1311. https://doi.org/10.1542/peds.2011-0426.
75. Dawson G, Rogers S, Munson J, et al. Randomized, controlled trial of an intervention for toddlers with autism: the early start denver model. *Pediatrics*. 2010;125(1):e17–e23. https://doi.org/10.1542/peds.2009-0958.
76. Rogers SJ, Dawson G. *Early Start Denver Model for Young Children With Autism: Promoting Language, Learning, and Engagement*. Guilford Press; 2010.
77. Vivanti G, Dissanayake C, Zierhut C, Rogers SJ. Brief report: predictors of outcomes in the early start denver model delivered in a group setting. *J Autism Dev Disord*. 2013;43(7):1717–1724. https://doi.org/10.1007/s10803-012-1705-7.
78. Harris SL, Delmolino L. Applied behavior analysis: its application in the treatment of autism and related disorders in young children. *Infants Young Child*. 2002;14(3):11–17. https://doi.org/10.1097/00001163-200201000-00006.
79. Greenspan SI. Infant-caregiver interactions and the developmental, individual-difference, relationship-based (dir) model: implications for psychopathology and the psychotherapeutic process. *J Infant Child Adolesc Psychother*. 2007;6(3):211–244. https://doi.org/10.1080/15289160701657787.
80. Wieder S, Greenspan S, Kalmanson B. Autism assessment and intervention: the developmental individual-difference, relationship-based (DIR [R])/floortime [TM] model. *Zero Three*. 2008;28(4):31–37.
81. Liao S, Hwang Y, Chen Y, et al. Home-based DIR/floortime TM intervention program for preschool children with autism spectrum disorders: preliminary findings. *Phys Occup Ther Pediatr*. 2014;2638. https://doi.org/10.3109/01942638.2014.918074.
82. Roth-Hanania R, Gabis LV. *The KESHET Center Model: Pilot Results of an Early Intervention Program for Infants and Toddlers With ASD*. International Society on Early Intervention ISEI; 2016.
83. Mundy P, Block J, Delgado C, Pomares Y, Van Hecke AV, Parlade MV. Individual differences and the development of joint attention in infancy. *Child Dev*. 2007;78(3):938–954. https://doi.org/10.1111/j.1467-8624.2007.01042.x.

84. Stone WL, Ousley OY, Yoder PJ, Hogan KL, Hepburn SL. Nonverbal communication in two- and three-year-old children with autism. *J Autism Dev Disord*. 1997;27(6):677–696. https://doi.org/10.1023/A:1025854816091.
85. Gernsbacher MA, Morson EM, Grace EJ. Language and speech in autism. *Annu Rev Linguist*. 2016;2(1):413–425. https://doi.org/10.1146/annurev-linguistics-030514-124824.
86. Haebig E, McDuffie A, Ellis Weismer S. Brief report: parent verbal responsiveness and language development in toddlers on the autism spectrum. *J Autism Dev Disord*. 2013;43(9):2218–2227. https://doi.org/10.1007/s10803-013-1763-5.
87. Watson LR, Roberts JE, Baranek GT, Mandulak KC, Dalton JC. Behavioral and physiological responses to child-directed speech of children with autism spectrum disorders or typical development. *J Autism Dev Disord*. 2012;42(8):1616–1629. https://doi.org/10.1007/s10803-011-1401-z.
88. McDuffie A, Yoder P, Stone W. Prelinguistic predictors of vocabulary in young children with autism spectrum disorders. *J Speech Lang Hear Res*. 2005;48(5):1080. https://doi.org/10.1044/1092-4388(2005/075.
89. Mundy P, Sigman M, Kasari C. A longitudinal study of joint attention and language development in autistic children. *J Autism Dev Disord*. 1990;20(1):115–128. https://doi.org/10.1007/BF02206861.
90. Hurwitz S, Watson LR. Joint attention revisited: finding strengths among children with autism. *Autism*. 2016;20(5):538–550. https://doi.org/10.1177/1362361315593536.
91. Cornish KM, Kogan CS, Li L, Turk J, Jacquemont S, Hagerman RJ. Lifespan changes in working memory in fragile X premutation males. *Brain Cogn*. 2009;69(3):551–558. https://doi.org/10.1016/j.bandc.2008.11.006.
92. Hume K, Pratt C. *The usage and perceived outcomes of early intervention and early childhood programs for young children with autism spectrum disorder*. 2005;207:195–208.
93. Gabis LV, Pomeroy J. An etiologic classification of autism spectrum disorders. *Isr Med Assoc J*. 2014;16(5):295–298.
94. Devlin B, Scherer SW. Genetic architecture in autism spectrum disorder. *Curr Opin Genet Dev*. 2012;22(3):229–237. https://doi.org/10.1016/j.gde.2012.03.002.
95. Risi S, Lord C, Gotham K, et al. Combining information from multiple sources in the diagnosis of autism spectrum disorders. *J Am Acad Child Adolesc Psychiatry*. 2006;45(9):1094–1103. https://doi.org/10.1097/01.chi.0000227880.42780.0e.
96. Sallows GO, Graupner TD. Intensive behavioral treatment for children with autism: four-year outcome and predictors. *Am J Ment Retard*. 2005;110(6):417. https://doi.org/10.1352/0895-8017(2005)110[417:IBTFCW]2.0.CO; 2.
97. DCD. *CDC Screening and Diagnosis of Autism Spectrum Disorder (ASD)*. Centers for Disease Control and Prevention; 2015. https://www.cdc.gov/ncbddd/autism/screening.html; (Accessed March 10, 2017).
98. Gabis LV, Hochberg O, Leon Attia O, Banet-Levi Y, Topf D, Shefer S. Prolonged time lag to final diagnosis of fragile X syndrome. *J Pediatr*. 2017. https://doi.org/10.1016/j.jpeds.2017.10.008.

CHAPTER

5

Neuroinflammation and neuroprotection in schizophrenia and autism spectrum disorder

Ehud Mekori-Domachevsky[a,b], *Hadar Segal-Gavish*[a,b], *Raz Gross*[b,c]

[a]Division of Child and Adolescent Psychiatry, The Chaim Sheba Medical Center, Tel Hashomer, Israel, [b]Sackler School of Medicine, Tel Aviv University, Ramat-Aviv, Israel, [c]Division of Psychiatry, The Chaim Sheba Medical Center, Tel Hashomer, Israel

Schizophrenia and autism spectrum disorder (ASD) are both severe, complex psychiatric disorders with a heterogeneous clinical presentation. There are some common features to both disorder, such as aberrant social interaction, peculiar thought process, and cognitive deficits, and while ASD is considered a "true" neurodevelopmental disorder, current knowledge also suggests a significant etiological role for early-life antecedents in the evolution of schizophrenia.

Despite decades of research, the multifactorial etiology of both disorders is still mostly unknown. The inflammatory (or neuroinflammatory) hypothesis that suggests that pathological functioning of the immune system plays a central role in schizophrenia and in ASD is not new. Nevertheless, mounting evidence from current advanced laboratory studies has moved the immune system to the center stage of psychiatric or more specifically immunopsychiatric research.

The aim of this chapter is to outline the different aspects of the immune system that are implicated in schizophrenia and ASD from a neuroprotective perspective. The chapter is divided into sections for the sake for organization, but it is obvious that all of the factors and elements are in constant and intricate interactions with each other—leading to the complicated pathogenic and protective processes that underlay both schizophrenia and ASD.

https://doi.org/10.1016/B978-0-12-814037-6.00011-2

PART I
Schizophrenia

Microglia

Microglia are the main immune cells of the brain and account for about 10% of the cells in the CNS. They belong to the monocytic lineage and have a myeloid origin. Microglia are ramified cells with fine and motile process that constantly scavenge the brain's parenchyma. Microglia exert their inflammatory functions through several mechanisms. Among these are proinflammatory cytokines that trigger further inflammatory cascades and oxidative agents that cause physical damage to the target cells. They also use complement to assist their phagocytic abilities (Fig. 5.1).

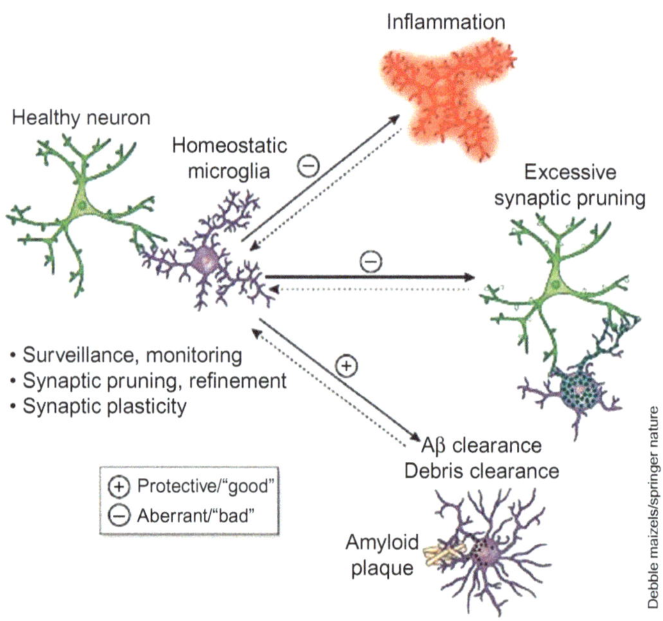

FIG. 5.1 Microglia states in health and disease. Microglia have complex roles that are both beneficial and detrimental to disease pathogenesis, including engulfing or degrading toxic proteins (i.e., amyloid plaques) and promoting neurotoxicity through excessive inflammatory cytokine release. Aberrations in microglia's normal homeostatic functions such as surveillance, synaptic pruning, and plasticity may also contribute to excessive synapse loss and cognitive dysfunction in AD and other diseases. From Salter MW, Stevens B. Microglia emerge as central players in brain disease. *Nat. Med.* 2017;23(9):1018.

Yet, microglia have a much more diverse role in the CNS than only fighting infections. They play a central role in synaptic pruning during development; they facilitate apoptosis of excess neurons as part of normal neurogenesis and regulate synaptic plasticity.[1] During brain development, microglia take part in the normal process for synaptic pruning, a process that has been implicated in the pathophysiology of schizophrenia and ASD. As phagocytic cells, microglia also phagocytize neurons that die as a result of planned cell death, as part of normal development and neurogenesis.[1]

Microglia are affected by environmental factors that alter their function. Factors that cause inflammation and trigger microglia into overactivity during critical periods of development may have consequences later in life. This could explain the well-known association between inflammation during pregnancy and schizophrenia.[1] Neuronal pruning defect, which causes hyper- to hypoconnectivity in the brain, is seen in schizophrenia and ASD.[2]

The hypothesis of microglial overactivation suggests that this process impairs normal from a developmental point of view (e.g., microglia malfunction and therefore the individual's brain develops in a way that makes him in greater risk for having schizophrenia or ASD) and from an acute point of view (e.g., aberrant activation takes part in active psychotic episodes as well as recovery in schizophrenia).[3]

Several studies have examined the effect of antipsychotic medication (the mainstay of treatment for schizophrenia) on microglia. As dopamine acts as a "bridge" between the immune system and the nervous system (dopamine receptors are found in all immune cells),[4] it has been suggested that dopamine has an antiinflammatory or immunoregulatory role.[5] Therefore, it makes sense that antipsychotic medication (with their antidopaminergic properties) will also affect microglia. Studies with minocycline, an antibiotic that suppresses microglial activation, have found a beneficial effect of minocycline in schizophrenia.[6,7]

Cytokines

Cytokines are low-molecular-weight glycoproteins that are secreted by different cells in the body, mainly cells of the immune system. They have a wide array of functions in the innate and adaptive immune system, where they regulate the activation and inhibition of the immune response. Accordingly, cytokines can be divided into proinflammatory (such as interleukin-6, interleukin-1β, and tumor necrosis factor alpha (TNFα)) or antiinflammatory (interleukin-10). Throughout the human body, proinflammatory cytokines are mainly secreted by endothelial cells and monocytes, while in the brain, they are mainly secreted by microglia cells (the primary immune cells of the brain, of monocyte origin). Nevertheless, in the central nervous system (CNS), cytokines also have roles in the regulation of normal brain function. For example, some cytokines (such as IL-1α) encourage neurogenesis, while others (like IL-1β) decreases neurogenesis and increase gliosis.[8]

Under normal conditions, interleukin-6 has several functions in the CNS. In addition, it is also a proinflammatory cytokine that has been implicated in the pathophysiology of SCZ.[9] Studies done in maternal immune activation in mice models have implicated that IL-6 plays a key role in the neurodevelopmental deficits that the offspring exhibited. Furthermore, anti-IL6 antibodies were able to reverse the MIA phenotype.[10] Mouse models of maternal immune activation have abnormal cytokine levels in pregnancy that will impede normal brain development. The source of the excess levels does not affect the outcome (whether it is endogenous or exogenous).[11]

Elevated levels of IL-6 in peripheral blood were found to be in association with psychopathology[12] and cognition.[13] It should be noted that IL-6 has been postulated to increase dopamine levels in a manner that resembles the hyperdopaminergic hypothesis theory of SCZ and to decrease glutamate (another hypothesis of SCZ).[9]

Past studies have examined the effect of antipsychotic medication on cytokine levels in schizophrenia and ASD. Although results are not always consistent, they are usually considered antiinflammatory (Fig. 5.2).

Tocilizumab, a humanized monoclonal antibody against IL-6, is used to treat different autoimmune disorders. Because of the central role of IL-6 in the psychopathology of schizophrenia, it only makes sense to utilize tocilizumab for it also. One small (six patients) preliminary open-label trial demonstrated an improvement in cognition but not in psychopathology.[14] Another larger study RCT sought to find out whether tocilizumab can alleviate residual symptoms in schizophrenia. It did not.[15] The authors have suggested that the medication does not cross the blood-brain barrier, but it also should be suggested whether it should be used at the onset of the disorder.

Oxidative stress

Oxygen metabolites and free radical are an integral product of human metabolism and cellular breathing processes, done by the mitochondria. If their concentration exceeds a certain level or if antioxidant mechanism are insufficient, cell and tissue damage might happen as a consequence of lipid peroxidation and DNA damage. Aberrant activation of the immune system might trigger this process by causing neuroinflammation by activation of microglia and lymphocytes. This process is thought to play a role in the pathophysiology of schizophrenia by damaging neuronal and glial functioning by affecting the integrity of DNA, proteins, and lipids.[16]

Biomarker	Associated clinical features	
	ASD	Schizophrenia
Immune function		
IL-6	NF	↑ in drug-naïve FEP patients compared to healthy controls; ↓ after 10-week treatment with risperidone; ↓ in patients with depressive symptoms (but not in those without); clozapine treatment appears to ↑ IL-6; treatment with typical anti-psychotics ↓ IL-6 and sIL-6R; ↓ after anti-psychotic treatment in relapsed patients ↑ IL-6R in patients compared to controls
TNF-α	NF	↑ in drug-naïve FEP patients compared to healthy controls; ↓ after 10-week treatment with risperidone; ↓ in patients with depressive symptoms (but not in those without)
		Anti-psychotic treatment reported to have no effect or ↓ levels of TNF-α; clozapine reported to ↑ TNF-α
sTNF-R	NF	Clozapine treatment appears to increase sTNF-R
IL-1β	NF	↑ in drug-naïve FEP patients compared to healthy controls; after risperidone treatment levels returned to baseline at 6 months; ↓ levels after anti-psychotic treatment
IL-1RA	Levels did not change after 8-week treatment with risperidone despite clinical improvement	↑ drug-naïve FEP patients; ↓ after 6-week treatment with risperidone or olanzapine anti-psychotic treatment reported to have no effect on IL-1RA; clozapine treatment ↑ IL-1RA; ↑ sIL-1RA with anti-psychotic treatment
IL-12	NF	Possibly ↑ with anti-psychotic treatment
IFN-Υ	Levels did not change after 8-week treatment with risperidone despite clinical improvement	↓ after anti-psychotic treatment
TGF-β	NF	↑ in un-medicated FEP and schizophrenia relapse patients; further increased after treatment with anti-psychotics in FEP; unaffected by anti-psychotic treatment
Chemokines	↓ Eotaxin and MCP-1 after 8-week treatment with risperidone; MCP-1 levels did not change after 8-week treatment with risperidone despite clinical improvement in another study	↓ CC16 in patients compared to controls; increase after treatment with clozapine
	↑ EGF in children with ASD; levels did not change after 8-week treatment with risperidone despite clinical improvement	↑ S100B in drug-naïve and medicated patients compared to controls and also in drug-naïve compared to medicated patients; ↓ S100B with haloperidol and clozapine
IL-2	NF	Unaffected by anti-psychotic treatment; ↓ by first and second generation anti-psychotics
IL-2R	NF	Increased in younger patients; treatment with clozapine increases sIL-2R levels
IL-10	↑ drug-naïve FEP patients; ↓ after treatment with risperidone or olanzapine; changes in IL-10 correlated with improvements in negative, general and total symptom scores; another study reported no effect of anti-psychotics on IL-10	
IL-4		↓ after 10-week treatment with risperidone; another study reported no effect with anti-psychotics; ↓ after anti-psychotic treatment in FEP and relapse patents
IL-15		↑ drug-naïve in FEP
IL-13, IL-17, IL-1	↓ IL-13 in children with ASD compared to controls; levels of IL-13, IL-17, and IL-1 did not change after 8-week treatment with risperidone despite clinical improvement	IL-13 possibly ↓ by anti-psychotic treatment
IL-27		↓ after anti-psychotic treatment in FEP
Oxidative stress		
SOD and NO levels	NF	↑ in patients with schizophrenia compared to controls Risperidone and haloperidol ↓ superoxide dismutase levels (but not nitric oxide levels) ↓ SOD levels at baseline predicted greater symptom improvement during treatment and greater change in SOD was correlated with greater symptom improvement
PON1 activity, TRAP, and LOOH levels	NF	↓ PON1 activity and ↑ TRAP in FEP ↑ PON1 activity and ↓ LOOH levels after 11 weeks of risperidone treatment

NF, no studies found.

FIG. 5.2 Biomarkers and anti-psychotic treatment in ASD and schizophrenia. Prata J, Santos SG, Almeida MI, Coelho R, Barbosa MA. Bridging Autism Spectrum Disorders and Schizophrenia through inflammation and biomarkers—pre-clinical and clinical investigations. *J Neuroinflammation.* 2017;14(1):179.

Several mechanisms were suggested as possible to cause the cognitive impairment observed in schizophrenia: (1) cell death induced by oxidative stress in the frontal and prefrontal cortex; (2) a reduction of BDNF (that was linked to catalase activity, important in protecting cells for oxidative stress); (3) damaged caused to white matter by ROS (white matter is very susceptible to lipid peroxidation); and (4) NMDA receptor hypofunction, due to the redox-sensitive nature of these proteins.[17]

Glutathione (GSH) is a key player in protecting the body from free radical that induces damage. It mitigates thiol oxidation of cysteine residues, producing oxidized glutathione (GSSG), that in turn is reduced back to GSH by GSH reductase. The intracellular ratio of GSH/GSSG is typically >100:1, serving as an important redox buffer system. But in the conditions of oxidative stress, this ration may drop to 10:1 or even 1:1. The reduced GSH serves as a substrate for enzymes that scavenge reactive oxygen species (ROS), inactivate electrophilic species, and restore reduced cysteine-thiol moieties on proteins, with the concomitant production of GSSG.[16, 18]

Several lines of evidence have connected oxidative stress and schizophrenia. Postmortem studies done on the brains of patients with schizophrenia have demonstrated lower GSH levels in several regions—caudate nucleus and prefrontal cortex—and also in plasma and erythrocytes. These findings were supported by imaging studies that showed an inverse correlation between GSH levels in the posterior medial frontal

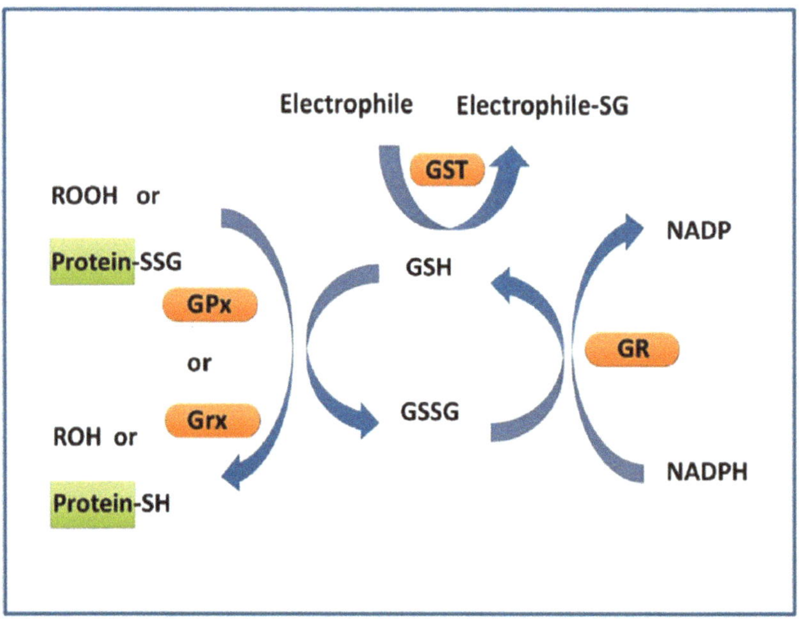

FIG. 5.3 Source: From Johnson WM, Wilson-Delfosse AL, Mieyal J. Dysregulation of glutathione homeostasis in neurodegenerative disease. *Nutrients*. 2012;4(10):1399–1440.

cortex and negative symptoms in schizophrenia. Also, low GSH levels were correlated with higher PANSS scores and worse functioning in patients with schizophrenia (Fig. 5.3).[16]

N-Acetylcysteine is a precursor of GSH, and it has been postulated that it might have a beneficial effect in schizophrenia. Indeed, oral NAC has been shown to increase plasma concentrations of GSH, and also, NAC itself has an antioxidant effect. It stands to reason that an increase in GSH will have a neuroprotective as it will encourage neurogenesis and improve defense from intracellular ROS damage. Moreover, GSH was shown to facilitate DNA repair activity, potentially meaning restoration of already damaged cells and improving network connectivity and cognition.[17]

A recent metaanalysis that included 447 patients with schizophrenia in seven randomized control trials found that NAC supplementation was associated with an improvement in total, general, and negative subscales of PANSS (an effect not observed in the placebo groups). There were no adverse effects reported in the studies included in this metaanalysis. The beneficial effect was most prominent in studies that were relatively short term (<24 weeks of follow-up), with lower doses of NAC (<2 g a day) and with patients with <10 years of duration on illness.[19]

BDNF

Brain-derived neurotrophic factor (BDNF) is part of the neurotrophin family. These proteins regulate growth, maintenance, and apoptosis on neurons in the central nervous system and peripherally. Although not the only neurotrophin, BDNF is believed to play a role in the pathophysiology of different neurocognitive disorders due to its role in neuronal maturation, differentiation, and synaptic plasticity.

BDNF's complex mechanism of action will not be discussed here in detail. Overall, it binds two main receptors: the high-affinity TrKB receptor and the low-affinity p75 neurotrophin receptor. After binding to its' receptors, several downstream signaling cascades are activated, including the mitogen-activated protein kinase (MAPK), phospholipase C-γ (PLC-γ), and phosphoinositide-3-kinase (PI3K) pathways (see Fig. 5.4).[20]

Synaptic plasticity occurs via altering the strength or the efficacy of existing synapses, creating new synapses, or pruning others—all according to synaptic activity. Long-term potentiation (LTP) and long-term depression (LTD) are the most investigated forms of synaptic plasticity. The NMDA complex (which has been repeatedly implicated in the pathophysiology of schizophrenia) plays a key role in LTP by elevating Ca^{2+} influx, thus activating several mechanisms—one of them is the secretion of BDNF.[21] The release of BDNF may also involve neuronal nitric oxide synthase-nitric oxide (nNOS-NO) signaling.[22] Ample lines of evidence

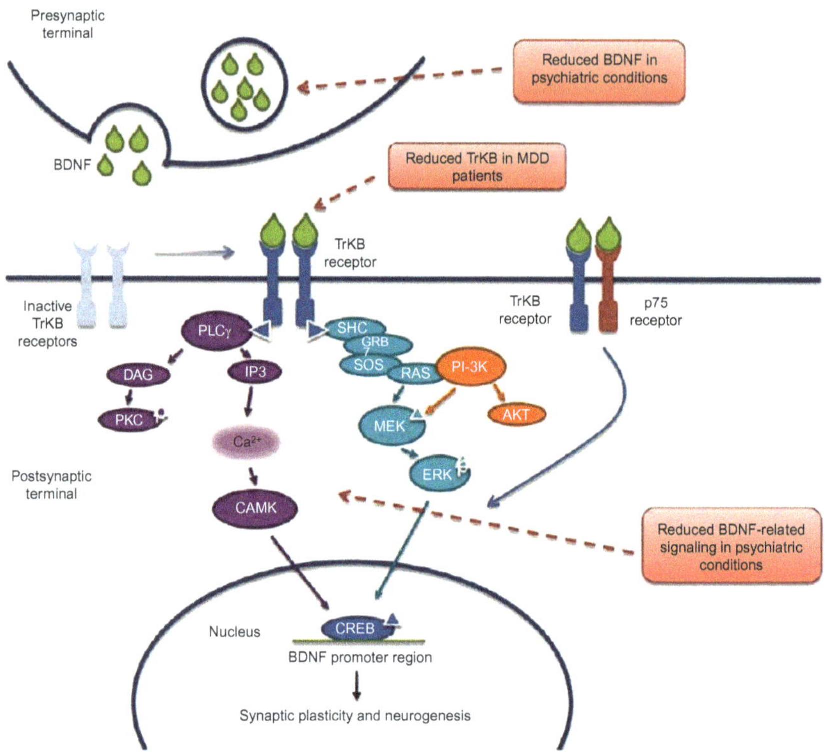

FIG. 5.4 Schematic representation of the BDNF-TrkB signaling pathways and their alterations in pathological conditions. When secreted BDNF binds to its high-affinity TrkB receptors, there is an activation of several downstream signal transduction cascades, including the MAPK, PLC-γ, and PI3K pathways, leading to the prevention of programmed cell death and neuronal differentiation (see text). Reduced BDNF levels and impaired BDNF signaling have been reported in several psychiatric diseases, including depression and schizophrenia (see text). From Begni V, Riva MA, Cattaneo A. Cellular and molecular mechanisms of the brain-derived neurotrophic factor in physiological and pathological conditions. *Clin Sci.* 2016;131:123–138.

have proven an inversed relationship between inflammatory activity and BDNF. In animal models, infusion of different kinds of proinflammatory protein (such as LPS or interleukin-1β) induced a decrease in BDNF levels.

A possible mechanism of schizophrenia is that the neurodevelopmental abnormalities in this disorder have similar features of normal aging[23] among those changes in neuronal plasticity.[24] BDNF plays a central role in neuronal plasticity; this is implicated as central to this pathophysiological process.

In clinical studies done in patients with schizophrenia, BDNF was suggested as a key player. Although there is heterogeneity among different studies, most have suggested an overall defect of BDNF expression in schizophrenia. Initial studies have shown an increase in BDNF levels, while later studies have shown a decrease among psychotic patients.[20]

Treatment with antipsychotic drugs can upregulate BDNF in the hippocampus and prefrontal cortex in animal models and normalize aberrant behaviors.[20]

Antipsychotic drugs are the mainstay of treatment in schizophrenia, and thus, their neuroprotective properties may be due to their effect on BDNF. Preclinical studies have suggested that they may increase BDNF levels. Indeed, some studies have shown an increase in patients with schizophrenia treated with olanzapine. However, other studies with olanzapine and other antipsychotics have not demonstrated such effects.

References

1. Salter MW, Stevens B. Microglia emerge as central players in brain disease. *Nat Med.* 2017;23(9):1018.
2. Cowan M, Petri Jr WA. Microglia: immune regulators of neurodevelopment. *Front Immunol.* 2018;9:2576.
3. Monji A, Kato TA, Mizoguchi Y, et al. Neuroinflammation in schizophrenia especially focused on the role of microglia. *Prog Neuro-Psychopharmacol Biol Psychiatry.* 2013;42:115–121.
4. Sarkar C, Basu B, Chakroborty D, Dasgupta PS, Basu S. The immunoregulatory role of dopamine: an update. *Brain Behav Immun.* 2010;24(4):525–528.
5. Prata J, Santos SG, Almeida MI, Coelho R, Barbosa MA. Bridging autism spectrum disorders and schizophrenia through inflammation and biomarkers-pre-clinical and clinical investigations. *J Neuroinflammation.* 2017;14(1):179.
6. Tikka T, Fiebich BL, Goldsteins G, Keinänen R, Koistinaho J. Minocycline, a tetracycline derivative, is neuroprotective against excitotoxicity by inhibiting activation and proliferation of microglia. *J Neurosci.* 2001;21(8):2580–2588.
7. Chaves C, Marque CR, Maia-de-Oliveira JP, et al. Effects of minocycline add-on treatment on brain morphometry and cerebral perfusion in recent-onset schizophrenia. *Schizophr Res.* 2015;161(2–3):439–445.
8. Borsini A, Zunszain PA, Thuret S, Pariante CM. The role of inflammatory cytokines as key modulators of neurogenesis. *Trends Neurosci.* 2015;38(3):145–157.
9. Girgis RR, Kumar SS, Brown AS. The cytokine model of schizophrenia: emerging therapeutic strategies. *Biol Psychiatry.* 2014;75(4):292–299.
10. Smith SE, Li J, Garbett K, Mirnics K, Patterson PH. Maternal immune activation alters fetal brain development through interleukin-6. *J Neurosci.* 2007;27(40):10695–10702.
11. Burd I, Balakrishnan B, Kannan S. Models of fetal brain injury, intrauterine inflammation, and preterm birth. *Am J Reprod Immunol.* 2012;67(4):287–294.
12. Miller BJ, Buckley P, Seabolt W, Mellor A, Kirkpatrick B. Meta-analysis of cytokine alterations in schizophrenia: clinical status and antipsychotic effects. *Biol Psychiatry.* 2011;70(7):663–671.
13. Miller B, Mellor A, Buckley P. Interleukin-6 and cognition in non-affective psychosis. *Schizophr Bull.* 2013;39:S242–S243.
14. Miller BJ, Dias JK, Lemos HP, Buckley PF. An open-label, pilot trial of adjunctive tocilizumab in schizophrenia. *J Clin Psychiatry.* 2016;77(2):275–276.
15. Girgis RR, Ciarleglio A, Choo T, et al. A randomized, double-blind, placebo-controlled clinical trial of tocilizumab, an interleukin-6 receptor antibody, for residual symptoms in schizophrenia. *Neuropsychopharmacology.* 2017;43:1317.

16. Calabrese V, Giordano J, Crupi R, et al. Hormesis, cellular stress response and neuroinflammation in schizophrenia: early onset versus late onset state. *J Neurosci Res*. 2017;95(5):1182–1193.
17. Yolland CO, Phillipou A, Castle DJ, et al. Improvement of cognitive function in schizophrenia with N-acetylcysteine: a theoretical review. *Nutr Neurosci*. 2018;1–10.
18. Johnson WM, Wilson-Delfosse AL, Mieyal J. Dysregulation of glutathione homeostasis in neurodegenerative diseases. *Nutrients*. 2012;4(10):1399–1440.
19. Ghaderi A, Bussu A, Tsang C, Jafarnejad S. Effect of N-acetyl cysteine (NAC) supplementation on positive and negative syndrome scale in schizophrenia: a systematic review and meta-analysis of randomised controlled trials. *Eur J Clin Pharmacol*. 2018;1–13.
20. Begni V, Riva MA, Cattaneo A. Cellular and molecular mechanisms of the brain-derived neurotrophic factor in physiological and pathological conditions. *Clin Sci*. 2017;131(2):123–138.
21. Jourdi H, Hsu Y-T, Zhou M, Qin Q, Bi X, Baudry M. Positive AMPA receptor modulation rapidly stimulates BDNF release and increases dendritic mRNA translation. *J Neurosci*. 2009;29(27):8688–8697.
22. Patterson SL, Abel T, Deuel TA, Martin KC, Rose JC, Kandel ER. Recombinant BDNF rescues deficits in basal synaptic transmission and hippocampal LTP in BDNF knockout mice. *Neuron*. 1996;16(6):1137–1145.
23. Kirkpatrick B, Messias E, Harvey PD, Fernandez-Egea E, Bowie CR. Is schizophrenia a syndrome of accelerated aging? *Schizophr Bull*. 2007;34(6):1024–1032.
24. Islam F, Mulsant BH, Voineskos AN, Rajji TK. Brain-derived neurotrophic factor expression in individuals with schizophrenia and healthy aging: testing the accelerated aging hypothesis of schizophrenia. *Curr Psychiatry Rep*. 2017;19(7):36.

PART II

Autism spectrum disorder (ASD)

Microglial dysfunction in ASD

New insights into the role of microglia in the pathophysiology of autism and other neurodevelopmental disorders have been accumulating in recent years. Evidence from human and animal studies suggests that microglia are abnormal in ASD.

The importance of microglia in the brain dysfunction of ASD is primarily supported by postmortem studies. Numerous studies utilizing postmortem autistic brain tissue samples have showed increased microglial cell density, upregulation of microglial cell gene expression, and an activated microglial morphology, predominantly in the prefrontal cortex and the cerebellum.[1–5] Furthermore, transcriptomic analysis of postmortem autistic brains demonstrated upregulation of genes related to the activation of microglia and astrocytes and genes belonging to immune and inflammatory categories.[6, 7] Interestingly, immune-glial transcriptome changes observed in ASD brains had a less pronounced genetic component and thus were speculated to be either secondary to abnormal ongoing plasticity in the ASD brain or a response to environmental insults.[6]

Consistent with postmortem findings, an in vivo brain imaging study using positron emission tomography (PET) analysis with a radiotracer that specifically binds activated microglia has revealed marked activation of microglia in several brain regions in young adults with ASD, the most prominent of them being the cerebellum.[8]

Animal models of ASD also contribute to our growing body of knowledge concerning the involvement of microglia in ASD pathobiology. The inbred BTBR T+tf/J (BTBR) mouse strain has been widely used as an animal model for idiopathic autism.[9] In line with human postmortem findings, BTBR mice exhibit increased proportion of MHC class II-expressing microglia, suggesting an elevated state of activation of microglia.[10]

Rett syndrome, a neurodevelopmental disorder used to be classified under ASD due to its autism-like clinical features, is usually caused by mutations in the gene encoding methyl-CpG-binding protein 2 (Mecp2).[11] A study by Schafer et al. aiming to elucidate the contribution of microglia to Rett syndrome in *Mecp2* null mice showed that while microglia-specific loss or gain of *Mecp2* expression is insufficient to induce synapse loss and phenotypic abnormalities, microglia secondarily contribute by dismantling synaptic circuits in complete *Mecp2* null mice. These results suggest that microglia aberrantly respond to the loss of *Mecp2* in neurons or other CNS cell types.[12] This observation is further supported by the results of another research demonstrating that the restoration of *Mecp2* expression in microglia through either bone marrow transplantation or genetics does not rescue neurological deficits in preclinical models of Rett syndrome.[13]

Mounting evidence suggests a relationship between maternal immune activation (MIA) and neurodevelopmental disorders such as ASD. Indeed, MIA generated by the exposure of pregnant rodents to viral or bacterial components, such as poly(I:C) or LPS, respectively, elicits behaviors in the offspring that are reminiscent of ASD, such as aberrant social interaction, elevated repetitive behaviors, and altered ultrasonic vocalizations.[14, 15] Studies evaluating microglia derived from MIA animal models have provided conflicting results. Some studies exhibited elevated microglial activation in the brain of descendants of poly(I:C)-exposed pregnant rodents,[16, 17] while others failed to replicate these findings.[18, 19] Nevertheless, there have been consistent reports of increased microglia-derived cytokines in the brains of MIA rodent offspring.[20, 21] Cytokines dysregulation will be discussed further later.

As already stated, microglia have a major role in pruning synapses during neurodevelopment. Intriguingly, mice lacking the chemokine receptor Cx3cr1 expressed by microglia exhibit a transient reduction in microglia numbers in the developing brain and a resultant deficit in synaptic pruning.[22] A following study demonstrated that deficient synaptic pruning in the Cx3cr1 knockout mice is associated with decreased functional brain connectivity and alterations in social and repetitive behaviors, phenotypes that are related to autism.[23] These results suggest that microglia-mediated synaptic pruning during development may contribute to ASD and other neurodevelopmental disorders.

Cytokine alterations in ASD

Accumulating evidence points to aberrations in cytokine profiles and signaling in ASD. A metaanalysis performed on studies comparing peripheral blood cytokine concentrations in ASD and healthy controls found elevations in proinflammatory cytokines, such as IFN-γ, IL-1β, and IL-6, and reduced concentrations of the antiinflammatory cytokine, such as TGF-β1 in autistic patients. Moreover, levels of eotaxin, IL-8, and macrophage chemoattractant protein (MCP)-1, chemokines associated with recruitment of inflammatory cells from the circulation, were increased in participants with ASD.[24]

Some studies have reported alteration of cytokines not only in the periphery but also in the brain of individuals with ASD. Protein arrays of postmortem brain tissue from ASD patients[1] demonstrated elevations of two proinflammatory chemokines, MCP-1, and thymus- and activation-regulated chemokine (TARC), in several brain regions. Interestingly, the antiinflammatory cytokine TGF-β1 was also increased in the brain regions studied, in contrast to the periphery in which TGF-β1 levels were found to be decreased based on several reports.[25–27] A subsequent study, aimed at further characterizing cytokine profiles in the brain tissue of ASD patients using flow cytometry method, showed that the proinflammatory cytokines TNF-α, IL-6 and GM-CSF, Th1 cytokine IFN-γ, and chemokine IL-8 were increased in the frontal cortices of autistic subjects in comparison with the age-matched controls.[28] Additional work investigated specifically IL-6 in postmortem brain tissue from ASD patients, due to recent evidence indicating its crucial role within the CNS in terms of synaptic function.[29] This study by Malik et al. exhibited that IL-6 is elevated in the cerebellum of autistic subjects. Moreover, the researchers demonstrated that in vitro IL-6 overexpression in cerebellar granule cells leads to impairment in adhesion and migration and the formation of granule cell excitatory synapses while having no effect on inhibitory synapses and thereby causing an imbalance of excitatory and inhibitory circuits.[30] Notably, mice overexpressing IL-6 in the brain develops autistic-like behaviors and abnormal dendritic spine morphology and synaptic function,[31] thus providing additional support for the role of IL-6 in the pathophysiology of ASD.

Ashwood et al. investigated correlations between circulating antiinflammatory cytokine TGF-β1 levels and measurements of behaviors and found that lower TGF-β1 levels were associated with lower adaptive behaviors and worse behavioral symptoms.[25] A more comprehensive work showed associations between plasma levels of several cytokines and severity of certain core measures of ASD; increased IL-4 was associated with greater impairments in nonverbal communication, while aberrant behaviors (e.g., increased lethargy, stereotypy, and hyperactivity) were associated with increased levels of IL-1β, IL-6, IL-8, and IL-12p40.[32]

Findings from animal models provide further evidence for cytokine dysregulation in ASD. In line with human findings, BTBR mice have elevated expression of proinflammatory cytokines in the brain, including IL-1β,[10] IL-6, and TNF-α.[33] Furthermore, as mentioned previously, poly(I:C)-induced maternal immune activation (MIA) mouse model shows chronic, region-specific changes in brain and serum cytokines in offspring throughout development,[20] similar to ASD patients. Importantly, a mechanistic study utilizing poly(I:C)-injected pregnant mice showed that T-helper 17 (T_H17) cells and the effector cytokine IL-17a are required in mothers for MIA-induced ASD-like behavioral abnormalities in offspring and that treatment with IL-17a-blocking antibody in pregnant dams ameliorated MIA-associated behavioral deficits.[34] IL-17a has also been implicated from human studies; increased levels of IL-17a have been reported in the serum of children with ASD and correlated with the severity of behavioral symptoms.[35] Based on emerging clinical evidence showing a link between use of valproic acid (VPA) in pregnancy and risk of autism, rodent models of prenatal exposure to VPA displaying autistic-like behaviors have been developed.[36] A recent review showed that several alterations in cytokines in the brain of VPA model indeed reflect changes observed in patients with ASD, such as increased levels of IL-1β, IL-6, and TNF-α.[37]

Studying cytokine dysregulation in individuals with ASD may facilitate the identification of subpopulations within ASD that share similar traits, in addition to providing biological markers of response to treatments, and eventually lead to specialized therapy.

Immune-modulating drugs for ASD

Currently, no pharmacotherapy has been proven to effectively target core symptoms of ASD. Indeed, a growing body of evidence implicates immune dysregulation in ASD, and as the field moves forward, data from preliminary clinical trials investigating antiinflammatory and immune-modulatory treatments in ASD patients are accumulating.

Corticosteroids are antiinflammatory and immunosuppressive agents that inhibit proinflammatory cytokine synthesis, among other mechanisms.[38] Interestingly, ASD patients treated with corticosteroids due to other indication experienced an improvement in autistic features. A case report of a child with ASD who regressed at 18 months and had a comorbid autoimmune lymphoproliferative syndrome (ALPS) described improvement in language and communication skills with chronic oral prednisolone treatment.[39] A similar earlier case report depicted amelioration of language abilities and behavior following corticosteroid treatment.[40] One retrospective study compared 20 steroid-treated patients with regressive autism to 24 nontreated ASD patients and found that steroid treatment was associated with improvement in language and behavior scores.[41] In the two trials of administration of adrenocorticotropic hormone (ACTH) synthetic analog[42, 43] to children with autism, aimed to induce antiinflammatory/immunosuppressive effects indirectly through the stimulation of corticosteroids synthesis, increased social interaction and decreased stereotypic behavior were observed.

Intravenous immunoglobulin (IVIG) is a human blood product consists of pooled polyclonal IgG from thousands of healthy donors. IVIG therapy exerts a beneficial immunomodulatory effect in inflammatory and autoimmune diseases; however, understanding of the mechanisms of action is still quite limited.[44] IVIG has been used to treat children with ASD in a double-blind crossover trial, in which their behaviors were compared with baseline performance, 6 weeks following the administration of IVIG or placebo. Parent and teacher ratings indicated improvements in eye contact and speech and a reduction in hyperactivity in the IVIG treatment group. However, clinician ratings did not show significant differences between placebo and IVIG treatment.[45] Another larger-scale placebo-controlled trial evaluated the effect of oral human immunoglobulin therapy in children with autism and persistent GI symptoms after 12 weeks of treatment. No significant differences were observed between treatment groups in terms of reduction in gastrointestinal symptoms nor in the severity of ASD symptoms.[46]

The efficacy of celecoxib, a cyclooxygenase-2 inhibitor with antiinflammatory effects, as adjunctive therapy to risperidone for the treatment of ASD has been examined in a 10-week randomized double-blind placebo-controlled study. Treatment of risperidone plus celecoxib resulted in significant improvements in irritability, social withdrawal, and stereotypic behavior when compared with risperidone plus placebo. Moreover, complete response was achieved by 20% of patients in the risperidone plus placebo group compared with 55% in the risperidone plus celecoxib group.[47] A study with a parallel design was conducted with pentoxifylline, an immune-modulating compound known to inhibit proinflammatory cytokines, as an add-on to risperidone. The group that received risperidone plus pentoxifylline exhibited significantly greater improvement in symptoms of irritability, social withdrawal, stereotypic behavior, hyperactivity, and inappropriate speech.[48]

Pioglitazone is an agonist of peroxisome proliferator-activated receptor (PPAR)-γ and a member of the thiazolidinedione class of medication. It is commonly used as treatment for type II diabetes mellitus but is also known to have antiinflammatory effects.[48] A 16-week prospective cohort, single-blind, single-arm, 2-week placebo run-in, dose-finding study of pioglitazone performed recently showed improvement in core ASD symptoms, social withdrawal, and repetitive behaviors, as well as associated symptoms—irritability, hyperactivity, and anxiety. Notably, pioglitazone treatment significantly decreased the proinflammatory cytokine IL-6 and increased the antiinflammatory cytokine IL-10.[49]

The US Food and Drug Administration (FDA) has approved two second-generation antipsychotic drugs, risperidone and aripiprazole, for the treatment of ASD-related irritability.[50] In vitro and in vivo studies have attempted to elucidate possible antiinflammatory effects of risperidone and aripiprazole. In addition to their known effects on dopamine D2 receptors, both medications were found to decrease production of proinflammatory cytokines and inhibit microglial activation.[51–56] Clinical studies provided conflicting results regarding the effects of risperidone on cytokines levels; Choi et al. found a risperidone-induced decrease in two chemokines, eotaxin and MCP-1, previously recognized as abnormally upregulated in children with ASD,[57] while another study showed that improvement in irritability symptoms in ASD patients following risperidone treatment was not associated with changes in cytokine levels.[58]

Brain-derived neurotropic factor (BDNF) dysregulation in ASD

Converging lines of evidence link BDNF to the pathophysiology of ASD in human patients. Several metaanalyses consistently reported that peripheral BDNF levels are higher in autistic subjects compared with controls,[59–62] in contrast to reduced levels of peripheral BDNF in schizophrenia.[63] Studies in rats have demonstrated positive correlations between whole-blood and hippocampal BDNF levels[64] and between serum and cortical BDNF levels.[65] In mice, blood BDNF levels were undetectable.[64, 66] In humans, a clear evidence for periphery-brain correlation is still absent, and thus, further research is warranted. There are few reports on BDNF expression in the brain of ASD patients from postmortem studies. Consistent with studies on peripheral BDNF levels, two studies found increased levels of BDNF protein in the basal forebrain[67] and fusiform gyrus[68] in postmortem brain tissues of ASD subjects. However, there is also a report indicating decreased levels of BDNF in the cerebral cortex of autistic subjects in comparison with controls.[69]

Two of the metaanalyses mentioned earlier reported elevated levels of peripheral BDNF levels specifically in children with ASD.[60, 62] Findings from animal models, such as the ASD mouse model of prenatal exposure to VPA[70] and the BTBR mouse model of idiopathic autism,[71] suggest increased BDNF levels in the fetal brain. Furthermore, at 5 weeks of age, zinc transporter 3 (ZnT3) null mice that exhibit autistic-like behaviors were shown to have increased levels of BDNF and TrkB proteins in both the hippocampus and neocortex.[72] It has been hypothesized that hyperactivity of BDNF during neurodevelopment may account for early postnatal neocortical overgrowth that is often observed in children with ASD.[73] Evidence of excess number of neurons[73] and synapses[74] in ASD subjects is consistent with the known functions of BDNF as a promoter of neuronal growth and synapse formation.[75]

Further results from animal studies showing conflicting trends in BDNF levels add to the complexity. For example, mice with a mutant Ca^{2+}-dependent activator protein for secretion 2 (Caps2), which results in reduced axonal secretion of BDNF, exhibit autism-related behaviors.[76] Moreover, reduced levels of BDNF mRNA,[77, 78] BDNF protein, and BDNF receptor tyrosine kinase B (TrkB) protein[79] were found in the hippocampus of BTBR mice. Notably, upregulation of BDNF in the hippocampus of ASD animal models, using different methods including microRNA 124a silencing[80] and mesenchymal stem cell transplantation,[81] ameliorated autism-like behaviors.

Findings concerning different isoforms of BDNF are adding another level of complexity. BDNF is initially synthesized as a precursor, pro-BDNF, which is later processed either into a truncated form or into mature BDNF.[82] A study that showed increased BDNF levels in the fusiform gyrus of individuals with ASD using ELISA identified pro-BDNF as the isoform responsible for the observed increase, as Western blotting demonstrated an elevation in pro-BDNF and a reduction in truncated BDNF.[68] In addition, ZnT3 knockout mice that display autism-related behaviors exhibited increased levels of mature BDNF in both the hippocampus and neocortex, while levels of pro-BDNF were elevated in the cortex but not in the hippocampus.[72] These results suggest altered proteolytic processing of pro-BDNF in ASD.

To conclude, further studies are required to determine the spatiotemporal expression pattern of BDNF and its isoforms for providing better understanding of the role of BDNF in the pathophysiology of ASD.

Oxidative stress in ASD

A role for oxidative stress in the pathogenesis of ASD has long been hypothesized. Deth et al. described a "redox/methylation hypothesis" suggesting that in genetically susceptible subjects, oxidative stress initiated by environmental insults leads to impaired methylation that interferes with epigenetic events that are fundamental to normal neurodevelopment and synchronization of neural networks.[83]

A metaanalysis examining evidence from studies on oxidative stress-related blood biomarkers in ASD showed that autistic subjects exhibit decreased blood levels of reduced glutathione, glutathione peroxidase, methionine, and cysteine and increased concentrations of oxidized glutathione as compared with controls, suggesting a role for glutathione metabolism, the transmethylation cycle, and the transsulfuration pathway in the pathophysiology of ASD. Furthermore, in regard to blood levels of vitamins, the authors reported a lack of association for folic acid and vitamin B12 and a rare association for vitamins A, B6, C, D, and E.[84]

Evidence from human postmortem studies further supports the involvement of oxidative stress in ASD pathophysiology. One postmortem study investigating the levels of mitochondrial electron transport chain (ETC) complexes in brain samples of autism patients and age-matched controls found decreased expression of ETC complexes in the cerebellum and the frontal and temporal regions of the brain in children with ASD, which could result in aberrant energy metabolism and oxidative stress.[85] Consistent with this report, additional study demonstrated mitochondrial defects within the temporal cortex of children with ASD.[86] Another postmortem study has indicated increased levels of 3-nitrotyrosine (3-NT), a marker of oxidative stress, in the cerebellum of ASD subjects.[87]

Animal studies provide additional evidence for oxidative stress in ASD. BTBR mice exhibit oxidative DNA lesions and an altered pattern of

DNA methylation in the cerebellum,[88] as well as increased expression of ROS-generating enzymes, increased lipid and protein oxidation products, decreased glutathione levels, and dysregulated enzymatic antioxidant network in the cerebellum and blood.[89] Mice exposed to either methylmercury (MeHg) or valproic acid (VPA) in early postnatal life display ASD-related behaviors. The antioxidant protection of vitamin E and its derivative, shown to ameliorate neurobehavioral deficits induced by both MeHg and VPA, suggests that oxidative stress could be a common pathophysiological mechanism in these ASD mouse models.[90]

References

1. Vargas DL, Nascimbene C, Krishnan C, Zimmerman AW, Pardo CA. Neuroglial activation and neuroinflammation in the brain of patients with autism. *Ann Neurol.* 2005;57:67–81.
2. Morgan JT, et al. Microglial activation and increased microglial density observed in the dorsolateral prefrontal cortex in autism. *Biol Psychiatry*. 2010;68:368–376.
3. Edmonson C, Ziats MN, Rennert OM. Altered glial marker expression in autistic post-mortem prefrontal cortex and cerebellum. *Mol Autism*. 2014;5:3.
4. Tetreault NA, et al. Microglia in the cerebral cortex in autism. *J Autism Dev Disord.* 2012;42:2569–2584.
5. Morgan JT, et al. Abnormal microglial–neuronal spatial organization in the dorsolateral prefrontal cortex in autism. *Brain Res*. 2012;1456:72–81.
6. Voineagu I, et al. Transcriptomic analysis of autistic brain reveals convergent molecular pathology. *Nature*. 2011;474:380–386.
7. Gupta S, et al. Transcriptome analysis reveals dysregulation of innate immune response genes and neuronal activity-dependent genes in autism. *Nat Commun*. 2014;5:5748.
8. Suzuki K, et al. Microglial activation in young adults with autism spectrum disorder. *JAMA Psychiatry*. 2013;70:49.
9. Meyza KZ, Blanchard DC. The BTBR mouse model of idiopathic autism—current view on mechanisms. *Neurosci Biobehav Rev*. 2017;76:99–110.
10. Heo Y, Zhang Y, Gao D, Miller VM, Lawrence DA. Aberrant immune responses in a mouse with behavioral disorders. *PLoS ONE*. 2011;6:e20912.
11. Amir RE, et al. Rett syndrome is caused by mutations in X-linked MECP2, encoding methyl- CpG-binding protein 2. *Nat Genet*. 1999;23:185–188.
12. Heller CT, et al. Microglia contribute to circuit defects in Mecp2 null mice independent of microglia-specific loss of Mecp2 expression. *elife*. 2016;5.
13. Reichardt HM, et al. Wild-type microglia do not reverse pathology in mouse models of Rett syndrome. *Nature*. 2015;521:E1–E4.
14. Careaga M, Murai T, Bauman MD. Maternal immune activation and autism spectrum disorder: from rodents to nonhuman and human primates. *Biol Psychiatry*. 2017;81:391–401.
15. Estes ML, McAllister AK. Maternal immune activation: implications for neuropsychiatric disorders. *Science*. 2016;353:772–777.
16. Zhu F, Zheng Y, Liu Y, Zhang X, Zhao J. Minocycline alleviates behavioral deficits and inhibits microglial activation in the offspring of pregnant mice after administration of polyriboinosinic–polyribocytidilic acid. *Psychiatry Res*. 2014;219:680–686.
17. Juckel G, et al. Microglial activation in a neuroinflammational animal model of schizophrenia—a pilot study. *Schizophr Res*. 2011;131:96–100.

18. Missault S, et al. The risk for behavioural deficits is determined by the maternal immune response to prenatal immune challenge in a neurodevelopmental model. *Brain Behav Immun*. 2014;42:138–146.
19. Giovanoli S, et al. Late prenatal immune activation causes hippocampal deficits in the absence of persistent inflammation across aging. *J Neuroinflammation*. 2015;12:221.
20. Garay PA, Hsiao EY, Patterson PH, McAllister AK. Maternal immune activation causes age- and region-specific changes in brain cytokines in offspring throughout development. *Brain Behav Immun*. 2013;31:54–68.
21. Mattei D, et al. Minocycline rescues decrease in neurogenesis, increase in microglia cytokines and deficits in sensorimotor gating in an animal model of schizophrenia. *Brain Behav Immun*. 2014;38:175–184.
22. Paolicelli RC, et al. Synaptic pruning by microglia is necessary for normal brain development. *Science*. 2011;333:1456–1458.
23. Zhan Y, et al. Deficient neuron-microglia signaling results in impaired functional brain connectivity and social behavior. *Nat Neurosci*. 2014;17:400–406.
24. Masi A, et al. Cytokine aberrations in autism spectrum disorder: a systematic review and meta-analysis. *Mol Psychiatry*. 2015;20:440–446.
25. Ashwood P, et al. Decreased transforming growth factor beta1 in autism: a potential link between immune dysregulation and impairment in clinical behavioral outcomes. *J Neuroimmunol*. 2008;204:149–153.
26. Okada K, et al. Decreased serum levels of transforming growth factor-β1 in patients with autism. *Prog Neuropsychopharmacol Biol Psychiatry*. 2007;31:187–190.
27. El Gohary TM, El Aziz NA, Darweesh M, Sadaa ES. Plasma level of transforming growth factor β 1 in children with autism spectrum disorder. *Egypt J Ear Nose Throat Allied Sci*. 2015;16:69–73.
28. Li X, et al. Elevated immune response in the brain of autistic patients. *J Neuroimmunol*. 2009;207:111–116.
29. Gruol DL. IL-6 regulation of synaptic function in the CNS. *Neuropharmacology*. 2015;96:42–54.
30. Malik M, et al. IL-6 is increased in the cerebellum of autistic brain and alters neural cell adhesion, migration and synaptic formation. *J Neuroinflammation*. 2011;8:52.
31. Wei H, et al. Brain IL-6 elevation causes neuronal circuitry imbalances and mediates autism-like behaviors. *Biochim Biophys Acta Mol basis Dis*. 2012;1822:831–842.
32. Ashwood P, et al. Elevated plasma cytokines in autism spectrum disorders provide evidence of immune dysfunction and are associated with impaired behavioral outcome. *Brain Behav Immun*. 2011;25:40–45.
33. Schwartzer JJ, et al. Maternal immune activation and strain specific interactions in the development of autism-like behaviors in mice. *Transl Psychiatry*. 2013;3:e240.
34. Choi GB, et al. The maternal interleukin-17a pathway in mice promotes autism-like phenotypes in offspring. *Science*. 2016;351:933–939.
35. AL-Ayadhi LY, Mostafa GA. Elevated serum levels of interleukin-17A in children with autism. *J Neuroinflammation*. 2012;9:595.
36. Roullet FI, Lai JKY, Foster JA. In utero exposure to valproic acid and autism—a current review of clinical and animal studies. *Neurotoxicol Teratol*. 2013;36:47–56.
37. Deckmann I, Schwingel GB, Fontes-Dutra M, Bambini-Junior V, Gottfried C. Neuroimmune alterations in autism: a translational analysis focusing on the animal model of autism induced by prenatal exposure to valproic acid. *Neuroimmunomodulation*. 2018;25:285–299.
38. Schweingruber N, Reichardt SD, Lühder F, Reichardt HM. Mechanisms of glucocorticoids in the control of neuroinflammation. *J Neuroendocrinol*. 2012;24:174–182.
39. Shenoy S, Arnold S, Chatila T. Response to steroid therapy in autism secondary to autoimmune lymphoproliferative syndrome. *J Pediatr*. 2000;136:682–687.

40. Stefanatos GA, Grover W, Geller E. Case Study: corticosteroid treatment of language regression in pervasive developmental disorder. *J Am Acad Child Adolesc Psychiatry*. 1995;34:1107–1111.
41. Duffy FH, et al. Corticosteroid therapy in regressive autism: a retrospective study of effects on the Frequency Modulated Auditory Evoked Response (FMAER), language, and behavior. *BMC Neurol*. 2014;14:70.
42. Buitelaar JK, van Engeland H, van Ree JM, de Wied D. Behavioral effects of Org 2766, a synthetic analog of the adrenocorticotrophic hormone (4–9), in 14 outpatient autistic children. *J Autism Dev Disord*. 1990;20:467–478.
43. Buitelaar JK, et al. The adrenocorticotrophic hormone (4–9) analog ORG 2766 benefits autistic children: report on a second controlled clinical trial. *J Am Acad Child Adolesc Psychiatry*. 1992;31:1149–1156.
44. Wong PH, White KM. Impact of immunoglobulin therapy in pediatric disease: a review of immune mechanisms. *Clin Rev Allergy Immunol*. 2016;51:303–314.
45. Niederhofer H, Staffen W, Mair A. Immunoglobulins as an alternative strategy of psychopharmacological treatment of children with autistic disorder. *Neuropsychopharmacology*. 2003;28:1014–1015.
46. Handen BL, et al. A double-blind, placebo-controlled trial of oral human immunoglobulin for gastrointestinal dysfunction in children with autistic disorder. *J Autism Dev Disord*. 2009;39:796–805.
47. Asadabadi M, et al. Celecoxib as adjunctive treatment to risperidone in children with autistic disorder: a randomized, double-blind, placebo-controlled trial. *Psychopharmacology*. 2013;225:51–59.
48. Akhondzadeh S, et al. Double-blind placebo-controlled trial of pentoxifylline added to risperidone: effects on aberrant behavior in children with autism. *Prog Neuropsychopharmacol Biol Psychiatry*. 2010;34:32–36.
49. Capano L, et al. A pilot dose finding study of pioglitazone in autistic children. *Mol Autism*. 2018;9:59.
50. LeClerc S, Easley D. Pharmacological therapies for autism spectrum disorder: a review. *P T*. 2015;40:389–397.
51. Obuchowicz E, Bielecka-Wajdman AM, Paul-Samojedny M, Nowacka M. Different influence of antipsychotics on the balance between pro- and anti-inflammatory cytokines depends on glia activation: an in vitro study. *Cytokine*. 2017;94:37–44.
52. Maeda K, Sugino H, Futamura T, Marunaka Y, Mitsumoto Y. Atypical antipsychotics suppress production of proinflammatory cytokines and up-regulate interleukin-10 in lipopolysaccharide-treated mice. *Prog Neuropsychopharmacol Biol Psychiatry*. 2008;33:303–307.
53. MacDowell KS, et al. Risperidone normalizes increased inflammatory parameters and restores anti-inflammatory pathways in a model of neuroinflammation. *Int J Neuropsychopharmacol*. 2013;16:121–135.
54. Suzuki SO, et al. Inhibitory effects of aripiprazole on interferon--induced microglial activation via intracellular Ca^{2+} regulation in vitro. *J Neurochem*. 2008;106:815–825.
55. Sato-Kasai M, et al. Aripiprazole inhibits polyI:C-induced microglial activation possibly via TRPM7. *Schizophr Res*. 2016;178:35–43.
56. Kato T, Monji A, Hashioka S, Kanba S. Risperidone significantly inhibits interferon-γ-induced microglial activation in vitro. *Schizophr Res*. 2007;92:108–115.
57. Choi JE, et al. Change in plasma cytokine levels during risperidone treatment in children with autism. *J Child Adolesc Psychopharmacol*. 2014;24:586–589.
58. Tobiasova Z, et al. Risperidone-related improvement of irritability in children with autism is not associated with changes in serum of epidermal growth factor and interleukin-13. *J Child Adolesc Psychopharmacol*. 2011;21:555–564.
59. Saghazadeh A, Rezaei N. Brain-derived neurotrophic factor levels in autism: a systematic review and meta-analysis. *J Autism Dev Disord*. 2017;47:1018–1029.

60. Zheng Z, et al. Peripheral brain-derived neurotrophic factor in autism spectrum disorder: a systematic review and meta-analysis. *Sci Rep*. 2016;6:31241.
61. Armeanu R, Mokkonen M, Crespi B. Meta-analysis of BDNF levels in autism. *Cell Mol Neurobiol*. 2017;37:949–954.
62. Qin X-Y, et al. Association of peripheral blood levels of brain-derived neurotrophic factor with autism spectrum disorder in children. *JAMA Pediatr*. 2016;170:1079.
63. Steiner J, et al. Peripheral brain-derived neurotrophic factor in schizophrenia and the role of antipsychotics: meta-analysis and implications. *Mol Psychiatry*. 2014;20:1108–1119.
64. Klein AB, et al. Blood BDNF concentrations reflect brain-tissue BDNF levels across species. *Int J Neuropsychopharmacol*. 2011;14:347–353.
65. Karege F, Schwald M, Cisse M. Postnatal developmental profile of brain-derived neurotrophic factor in rat brain and platelets. *Neurosci Lett*. 2002;328:261–264.
66. Radka SF, Holst PA, Fritsche M, Altar CA. Presence of brain-derived neurotrophic factor in brain and human and rat but not mouse serum detected by a sensitive and specific immunoassay. *Brain Res*. 1996;709:122–301.
67. Perry EK, et al. Cholinergic activity in autism: abnormalities in the cerebral cortex and basal forebrain. *Am J Psychiatry*. 2001;158:1058–1066.
68. Garcia KLP, et al. Altered balance of proteolytic isoforms of pro-brain-derived neurotrophic factor in autism. *J Neuropathol Exp Neurol*. 2012;71:289–297.
69. Sheikh AM, et al. BDNF-Akt-Bcl2 antiapoptotic signaling pathway is compromised in the brain of autistic subjects. *J Neurosci Res*. 2010;88:2641–2647.
70. Almeida LEF, Roby CD, Krueger BK. Increased BDNF expression in fetal brain in the valproic acid model of autism. *Mol Cell Neurosci*. 2014;59:57–62.
71. Hwang SR, et al. Altered expression levels of neurodevelopmental proteins in fetal brains of BTBR T+tf/J Mice with autism-like behavioral characteristics. *J Toxicol Environ Health A*. 2015;78:516–523.
72. Yoo MH, Kim TY, Yoon YH, Koh JY. Autism phenotypes in ZnT3 null mice: involvement of zinc dyshomeostasis, MMP-9 activation and BDNF upregulation. *Sci Rep*. 2016;6:28548.
73. Courchesne E, et al. Neuron number and size in prefrontal cortex of children with autism. *JAMA*. 2011;306:2001–2010.
74. Tang G, et al. Loss of mTOR-dependent macroautophagy causes autistic-like synaptic pruning deficits. *Neuron*. 2014;83:1131–1143.
75. Park H, Poo MM. Neurotrophin regulation of neural circuit development and function. *Nat Rev Neurosci*. 2013;14:7–23.
76. Sadakata T, et al. Reduced axonal localization of a Caps2 splice variant impairs axonal release of BDNF and causes autistic-like behavior in mice. *Proc Natl Acad Sci*. 2012;109:21104–21109.
77. Stephenson DT, et al. Histopathologic characterization of the BTBR mouse model of autistic-like behavior reveals selective changes in neurodevelopmental proteins and adult hippocampal neurogenesis. *Mol Autism*. 2011;2:7.
78. Daimon CM, et al. Hippocampal transcriptomic and proteomic alterations in the BTBR mouse model of autism spectrum disorder. *Front Physiol*. 2015;6:324.
79. Scattoni ML, Martire A, Cartocci G, Ferrante A, Ricceri L. Reduced social interaction, behavioural flexibility and BDNF signalling in the BTBR T+tf/J strain, a mouse model of autism. *Behav Brain Res*. 2013;251:35–40.
80. Bahi A. Hippocampal BDNF overexpression or microR124a silencing reduces anxiety- and autism-like behaviors in rats. *Behav Brain Res*. 2017;326:281–290.
81. Segal-Gavish H, et al. Mesenchymal stem cell transplantation promotes neurogenesis and ameliorates autism related behaviors in BTBR mice. *Autism Res*. 2016;9:17–32.
82. Mowla SJ, et al. Biosynthesis and post-translational processing of the precursor to brain-derived neurotrophic factor. *J Biol Chem*. 2001;276:12660–12666.

83. Deth R, Muratore C, Benzecry J, Power-Charnitsky VA, Waly M. How environmental and genetic factors combine to cause autism: a redox/methylation hypothesis. *Neurotoxicology*. 2008;29:190–201.
84. Frustaci A, et al. Oxidative stress-related biomarkers in autism: systematic review and meta-analyses. *Free Radic Biol Med*. 2012;52:2128–2141.
85. Chauhan A, et al. Brain region-specific deficit in mitochondrial electron transport chain complexes in children with autism. *J Neurochem*. 2011;117:209–220.
86. Tang G, et al. Mitochondrial abnormalities in temporal lobe of autistic brain. *Neurobiol Dis*. 2013;54:349–361.
87. Sajdel-Sulkowska EM, Xu M, Koibuchi N. Increase in cerebellar neurotrophin-3 and oxidative stress markers in autism. *Cerebellum*. 2009;8:366–372.
88. Shpyleva S, et al. Cerebellar oxidative DNA damage and altered DNA methylation in the BTBR T + tf/J mouse model of autism and similarities with human post mortem cerebellum. *PLoS ONE*. 2014;9:e113712.
89. Nadeem A, et al. Increased oxidative stress in the cerebellum and peripheral immune cells leads to exaggerated autism-like repetitive behavior due to deficiency of antioxidant response in BTBR T + tf/J mice. *Prog Neuropsychopharmacol Biol Psychiatry*. 2019;89:245–253.
90. Ming X, Cheh MA, Yochum CL, Halladay AK, Wagner GC. Evidence of oxidative stress in autism derived from animal models. *Am J Biochem Biotechnol*. 2008;4:218–225.

SECTION III

Neuropsychiatric disorders

CHAPTER

6

Neuroprotective roles of neurotrophic factors in depression

Christina Gross[a,b], *Kim B. Seroogy*[c]

[a]Division of Neurology, Cincinnati Children's Hospital Medical Center, Cincinnati, OH, United States, [b]Department of Pediatrics, University of Cincinnati College of Medicine, Cincinnati, OH, United States, [c]Department of Neurology, University of Cincinnati College of Medicine, Cincinnati, OH, United States

Introduction

Mood disorders comprise a subset of neuropsychiatric diseases characterized by pathological disturbances in an individual's mood, which are often long-lasting and may change abruptly. The most common mood disorders are depression, bipolar disorder, and schizoaffective disorder. Mounting evidence suggests that impairments in neurotrophic factor expression or signaling contribute to these disorders and may be potential treatment targets. Defects at the cellular level associated with depression, such as reduced neurogenesis, elevated risk for neuroinflammation, and increased vulnerability to neurotoxicity, can all be explained by deficiencies in signaling through neurotrophic factors. Insufficient neurotrophic stimuli in mood disorders are not directly linked to neurodegeneration; however, the underlying mechanisms and resulting potential therapeutic targets are shared with neurodegenerative diseases such as Parkinson's and Alzheimer's diseases. Notably, many neurodegenerative diseases are associated with depression suggesting overlapping disease etiologies and neuropathological substrates. In this chapter, we will provide a concise overview of impaired neurotrophic factor signaling in depression and

https://doi.org/10.1016/B978-0-12-814037-6.00007-0

comment on how impaired neuroprotection may contribute to symptoms of depression in neurodegenerative diseases associated with depressive disorders. We will discuss some of the ongoing efforts to target these defects in neurotrophic factor signaling for therapeutic treatment and explain what is needed to fully exploit altered neurotrophic factor signaling as a treatment target in depression and in neurodegenerative disorders with comorbid depression.

The need for novel treatment strategies in depression

Depression is a mood disorder associated with consistent feelings of sadness, worthlessness, and guilt, as well as the loss of interest and pleasure. Other symptoms include insomnia or hypersomnia, changes in appetite and weight, fatigue, inability to focus, and indecisiveness. Together, these symptoms have a huge negative impact on the quality of life of affected individuals and impair their relationships with family and friends as well as their productivity in the work force.[1] The most severe form of depression, major depressive disorder (MDD), is defined as experiencing at least five of these symptoms during a 2-week period (The Diagnostic and Statistical Manual of Mental Disorders, 5th edition [DSM-5]).[2] Individuals with MDD are at risk for suicide and often need lifelong treatment. Current therapeutic treatments for depression include antidepressant medications (e.g., selective serotonin reuptake inhibitors, SSRIs, and most recently a form of the fast-acting NMDA receptor antagonist, ketamine), deep brain stimulation, and psychotherapy.[3] Up to 50% of all individuals with depression, especially those falling under the category of MDD, are treatment resistant.[4] Treatment resistance can even evolve over time, which poses a constant challenge to affected individuals and their clinical providers to identify alternative treatments.[5]

The etiology of depression is multifaceted. Genetic and environmental factors, such as chronic stress and early-life adverse events, have been identified as disease contributors.[6–8] Depressive disorders are also frequently associated with neurodegenerative disorders, for example Alzheimer's disease,[9] frontotemporal dementia,[10] Parkinson's disease,[11] amyotrophic lateral sclerosis,[12] and multiple sclerosis,[13] worsening the associated conditions and often complicating the treatment of affected individuals. The high incidence of depression in neurodegenerative disorders suggests shared underlying pathomechanisms of the two diseases, which may offer novel treatment targets. In this chapter, we will discuss current evidence that one of these shared etiologies may be impaired neuroprotection mediated by insufficient neurotrophic factor signaling.

Evidence suggestive of impaired neuroprotection and neurotrophic factors signaling in depression

A role for the loss of neuroprotection and altered neurotrophic factor signaling in depression has been supported by various findings in both humans and animal models. Those findings range from brain imaging, postmortem brain analyses, and quantification of blood serum levels of neurotrophic factors in humans with depression to genetic, molecular, and behavioral analyses in mouse and rat models of the disease. More recently, genetic studies have associated mutations in genes encoding neurotrophic factors with depression. Lastly, some preclinical studies in animal models suggest that enhancing neuroprotection may be a potential therapeutic strategy. The following paragraphs will provide a brief overview of these findings, which together, make a strong argument for the importance of neurotrophic factor-mediated neuroprotection in depression and other mood disorders.

Altered brain volume, connectivity, and cellular atrophy in depression

For decades, mood disorders have been associated with reduced brain activity, impaired synaptic plasticity, and reduced brain volume in structures of the prefrontal-limbic network, which is important for socioemotional development.[14] One of the first reports used magnetic resonance imaging (MRI) to show reduced activity and gray matter volume in the prefrontal cortex of bipolar and unipolar depressive disorders.[15] Since then, similar observations were made in other brain regions, such as the cingulate cortex, amygdala, and hippocampus.[16–19] Early metaanalyses of brain volumetric MRI studies suggested overall reduced brain volume in the frontal brain regions, hippocampus, and subcortical nuclei, such as the thalamus and nucleus accumbens.[20, 21] Although not all studies confirm brain volume reductions and atrophy in the affected regions[22, 23] and the volumetric changes may be restricted to certain subgroups of depression,[24] overall, these studies strongly suggest that reduced volume of brain areas involved in reward and stress responses contributes to depression.

Importantly, these findings were supported by postmortem analyses of cell density and cell soma size in these brain regions that pointed in a similar direction. Several studies reported reduced soma size or reduced numbers of pyramidal and granule neurons, interneurons, and glial cells in the brain. Examples are reduced soma size of hippocampal pyramidal neurons in autopsy tissue from individuals with depression compared with controls[25] and reduced numbers of hippocampal dentate granule cells in brains of individuals with major depression not currently treated with antidepressants.[26] In the prefrontal cortex, glial cell density and neuronal soma size were consistently shown to be reduced in several different studies in individuals with major depression[27,28] or bipolar disorder.[29] Similar results were observed in the anterior cingulate cortex.[30]

In addition to deficiencies at brain morphological and cellular levels, correlated impairments were also observed at the ultrastructural and molecular levels.[14] For example, Kang and colleagues used microarray gene profiling and stereological analyses of in situ hybridization for synapse-related genes in postmortem brains of individuals with MDD and revealed reduced mRNA expression of synaptic genes and reduced numbers of synapses in the dorsolateral prefrontal cortex.[31]

Together, these findings point toward a hypotrophic state of the brain in depression. A limitation of all of the described studies is that they did not assess whether the reduced size and density of neurons leading to reduced brain volume precedes the depressive phenotype or whether a depressed brain effects molecular changes, for example, in neurotrophic factor signaling, that may lead to these structural alterations. Answering this question will be important to determine if neurotrophic treatment strategies could alter disease progression by correcting disease-contributing mechanisms.

Impaired neurogenesis in depression

Adult neurogenesis, the capability of certain areas in the brain to produce new neurons throughout adulthood, has been observed in many different mammals, including humans.[32] Despite a recent report contradicting the widely shared assumption that neurogenesis continues in the aged human brain,[33] there is still considerable evidence that humans, like rodents, are capable of producing new neurons throughout the life span.[34–36] Animal studies have provided strong evidence that adult neurogenesis is impaired in depression. Yet, the question of whether impaired neurogenesis is an underlying cause of depression is still mired in controversy.[37–39] So far, only a few studies support a clear causal relationship between impaired neurogenesis and depression in mice. For example, most behaviors usually tested to establish a depressive-like phenotype in mice, such as the forced swim test,[40] were not affected by ablation of neurogenesis using antimitotic agents or cranial irradiation.[37] Notably, there is increasing evidence that ablating neurogenesis enhances, that is, aggravates, the response to stress,[41,42] but even here, the relationship is not clear.[43,44] Taken together, these studies suggest that neurogenesis is required for resilience to stress and adverse life events, but that impairments in neurogenesis may not per se lead to depressive-like phenotypes. In line with this notion, it was recently proposed that adult neurogenesis is required for cognitive flexibility, deficiency of which may affect resilience to stress and induce depression.[45] Of interest in this context, several studies have found evidence that impairing neurogenesis during antidepressant treatment abolishes at least some of the treatment-induced positive effects,[46–48] but this, again, was contradicted by others[49] (further discussed in the succeeding text).

There are only a few studies in humans addressing impaired neurogenesis in depression. Impairments in cell maturation in the dentate gyrus were suggested by postmortem brain analyses of individuals with MDDs showing reduced numbers of dentate granule cells without a concomitant reduction in neuronal progenitor cells.[26] This suggested potentially impaired neurogenesis, but another postmortem study could not confirm these findings.[50] It is important to note that stress and traumatic events are likewise associated with reduced neurogenesis.[51] Mood disorders are often triggered by stress[52] and associated with increased glucocorticoid receptor signaling,[53] which, in turn, may impair adult neurogenesis. In support of the "neurogenesis hypothesis of depression," one could argue that the previously discussed reductions in brain volume in individuals with MDDs may be indirect evidence of reduced neurogenesis.

In summary, keeping in mind the issues with modeling complex brain disorders like depression in rodents[54] and the recent research suggesting differences in adult neurogenesis between humans and other mammals, a clear assessment of the role of neurogenesis in depression is difficult based on current knowledge and requires more research. Impaired adult neurogenesis in individuals with depression and its requirement for antidepressant efficacy, if confirmed, would support the hypothesis that neurotrophic factor signaling, an important driver of neuronal maturation, is diminished in depression.

Neuroinflammation and neurotoxicity in depression

Inflammatory processes are an essential and necessary part of the body's response to infections, environmental contaminants, and injury that help to fight off pathogens and promote wound healing. In the brain, neuroinflammation as an acute response to viral infections and injury, for example, can have both beneficial and adverse effects.[55–57] Chronically increased inflammation in the brain has often been associated with not only neurodegenerative disorders, such as Alzheimer's disease, Parkinson's disease, and amyotrophic lateral sclerosis,[58] but also mood disorders, such as depression and schizophrenia.[59] Multiple studies have also shown that stress, a major contributor to depressive disorders, leads to increased neuroinflammation.[60, 61]

What role do neurotrophic factors play in neuroinflammation? Neuroinflammation in the brain is characterized by activated microglia and increased signaling through proinflammatory pathways, including chemokines.[62] Chemokines like IL-1β or TNFα can increase glutamatergic signaling in the brain leading to excitotoxicity and neuronal cell death.[63, 64] Elevated serum levels of chemokines, overactive microglia in the brain, and increased neurotoxicity have all been detected in patients with depression.[65–67] Strategies that increase neuroprotection and lead to reduced excitotoxicity and neuronal cell death may overcome the detrimental effects of neuroinflammation. This hypothesis is supported by studies showing that increasing the levels of the neuroprotective neurotrophin brain-derived neurotrophic factor (BDNF) in the brains of animal stroke models reduces neuroinflammation and promotes neuron survival.[68, 69] Notably, BDNF has been shown to be reduced by neuroinflammation.[70] This suggests that a status of chronically increased neuroinflammation, as observed in depression, may contribute to reduced neuroprotection in depression, adding rationale for neurotrophic factor-targeted treatments to reverse or ameliorate neuroinflammation-induced alterations of brain function in depression.

Altered neurotrophic factor signaling in depression

The most direct evidence supporting impaired neuroprotection has been provided by studies that identified impaired levels or activity of neurotrophic factors and their receptors in depression. The family of neurotrophins consists of four main factors, BDNF, nerve growth factor (NGF), neurotrophin-3 (NT-3), and neurotrophin-4/5 (NT-4/5).[71] Neurotrophins are secreted proteins that are synthesized as pre-pro-neurotrophins and are processed to the mature factor to exert their function. They can bind and activate two different classes of transmembrane receptor proteins, the p75 neurotrophin receptor and the tropomyosin receptor kinases (Trk) A, B, and C. In contrast to the high-affinity Trk receptors, the p75 receptor does not have an intrinsic catalytic domain and thus mainly functions via interactions with effector proteins, for example, other membrane receptors, including Trk receptors. p75 was originally thought to mediate signaling for cell death but was recently shown to also play important roles in survival, growth, synaptic plasticity, and immune response regulation through its interaction with other receptors.[72] Trk receptors, as implicated by the name, possess an intrinsic kinase activity and regulate many different intracellular pathways. Phospholipase C-γ (PLC-γ) is directly activated by Trk receptor kinases, which triggers hydrolysis of phosphatidylinositol-(4,5)-bisphosphate to inositol triphosphate (IP3) and diacylglycerol (DAG). These second messengers have multiple effects on intracellular signaling. IP3 triggers the release of calcium from internal stores. DAG activates protein kinase C, which then induces the Ras-MAP kinase pathway.[73] Due to their widespread effects on cellular signaling, neurotrophins are important in neurodevelopment, cell survival, differentiation, synaptic plasticity, and neuroinflammation. Defects in neurotrophin signaling are thus expected to have profound consequences on neuronal circuits and brain functioning.

Following is a summary of findings associating reduced neurotrophin signaling with depression.

Reduced levels of neurotrophic factors in individuals with depression

It has long been postulated that neurotrophic factors, in particular BDNF, are involved in the pathology of depression and the mechanism of action of antidepressants.[74] There is increasing evidence that expression levels of neurotrophic factors are altered in mood disorders. In MDDs, neurotrophic factor serum levels have even been suggested as a biomarker for treatment success. Notably, these effects are not limited to one type of neurotrophic factor but rather affect all four neurotrophin family members. For example, NGF was shown to be reduced in the serum of 94 individuals with MDD compared with 47 healthy controls.[75] BDNF, the most extensively studied neurotrophin in mood disorders and which has a widespread expression in the brain, was shown to be reduced in blood serum and postmortem brain tissue of patients with depression.[76–78] Reduced NGF, BDNF, and glial cell line-derived neurotrophic factor (GDNF) were also detected in adolescent individuals with depression compared with healthy controls.[79] The use of BDNF serum levels as a surrogate for its brain levels is supported by a study analyzing three different mammalian species that found positive correlations between blood and brain BDNF levels.[80] Not all studies found reduced neurotrophin levels in humans, though. NT-3, for example, was *increased* in cerebrospinal fluid of elderly patients with MDD compared with healthy individuals and patients with Alzheimer's disease.[81] Several other studies suggested increased NT-3 serum levels in depression; however, these studies were mostly done in individuals who suffered from bipolar disorder or had additional mood disorders, such as schizophrenia.[82, 83] Studies in the human population, in particular in persons with mood disorders who usually require medication to manage their condition, are notoriously difficult to control and to compare. The aforementioned studies differ in age, gender distribution, and medication usage. Nonetheless, there seems to be a broad consensus that depression in humans is associated with altered neurotrophin expression, and in most cases, reduced levels of neurotrophins were found.

Evidence from genetic studies supporting impaired neurotrophic factor signaling in depression

Depression is thought to be caused by the interplay of mutations in genes that increase susceptibility to depression and environmental factors, such as adverse childhood events. Several studies have shown that interactions between childhood adversity and susceptibility genes lead to more severe changes in brain structure than either of these factors alone.[84] Adverse childhood experience itself can lead to structural abnormalities in the brain.[18] Another factor influenced by early-life experience is epigenetic regulation of gene expression. Several studies show that BDNF is affected by those epigenetic changes.[85] The gene structure of BDNF makes it especially vulnerable to epigenetic regulation: with 11 exons and 9 different promoters, it is highly regulated at the posttranscriptional level, including alternative splice sites and polyadenylation sites, which can influence its subcellular distribution and function.[86] A prominent genetic variation of BDNF is the Val66Met single-nucleotide polymorphism leading to a valine-to-methionine substitution in proBDNF that affects processing and sorting and was originally shown to alter activity-induced secretion of BDNF and memory performance in humans.[87] A metaanalysis reported an association of this polymorphism with anxiety disorders in humans[88] and increased susceptibility to poststroke depression,[89] but a clear association of the polymorphism with MDD could not be confirmed in another study.[90] Apart from gene mutations, epigenetic modifications of the BDNF gene, in particular increased methylation expected to lead to reduced transcription of BDNF mRNA, have been reported in several studies and were associated with brain structural changes such as cortical thinning (reviewed in Ref. 91).

In addition to BDNF, mutations in the gene coding for the neurotrophin receptors TrkB and p75[92, 93] have been associated with depression, but a recent metaanalysis studying polymorphisms in *BDNF*, *NTRK2* (gene coding for TrkB), and NGFR (gene coding for p75) only found a strong correlation for the Val66Met polymorphism of BDNF.[94] The evaluation of mutations and polymorphisms as well as epigenetic alterations in genes coding for neurotrophic factors and their receptors as a contributing factor to depression is hampered by the limited size of sample numbers and the many confounding factors affecting human studies. Larger, well-controlled studies are needed to further assess genetic and epigenetic susceptibilities in neurotrophin signaling in depression.

Stress and stress-induced depression and neurotrophic factor signaling in mouse and rat models

As already alluded to, chronic stress or stress induced by early-life adverse events often contributes to or can even elicit depression. On a molecular level, similar pathways seem to be affected by both conditions, suggesting common underlying pathological mechanisms for stress and depression.[95] In fact, mouse or rat models of depression are frequently generated through the exposure of the animals to various forms of repetitive stress.[96] Although depression in mice and rats is difficult to assess,[97] animal models have the advantage of being highly controlled for genetic background and environmental influences, which should make the data better interpretable and more consistent. As described in more detail in the succeeding text, work in these rodent models showed that both acute stress and depressive phenotype-inducing chronic stress lead to alterations in neurotrophic factor signaling.

The majority of studies using animal models have shown that various forms of stress lead to reduction of BDNF mRNA in the brain (e.g., reviewed in Ref. 98). BDNF in the context of stress has been studied in many different animal models, suggesting a robust and widespread effect. The models encompassed different time frames, such as chronic stress and acute stress, and a variety of different stressors, including immobilization, social defeat, foot shock, and swimming. Many of these animal models consistently showed reduced BDNF levels and corroborate the importance of impaired BDNF signaling as a molecular outcome of stress.[98] However, there are also a few studies showing no change (e.g., Ref. 99) or increased levels of BDNF in rodent models of stress (e.g., Ref. 84). These discrepancies could be due to the type of stressor used or the genetic background or age of the animal model. Nonetheless, these studies in mouse and rat models support a role for altered BDNF signaling in stress and stress-induced depression, and the majority of the work seems to align with findings in depressed humans that indicated reduced neurotrophic factor signaling.

The relevance of altered neurotrophic factor signaling in stress is further supported by a few animal studies that quantified the levels of components of other neurotrophic factor signaling pathways and showed that they are affected similarly. For example, NGF and NT-3 are reduced by stress induced through extended immobilization of rats.[100] While not members of the neurotrophin family, it is interesting to note that vascular endothelial growth factor (VEGF) and its type 2 receptor are also both decreased in the hippocampus by chronic unpredictable stress in rats.[101] The described work in mice and rats overall supports a correlation between a stress-induced depressive phenotype and reduced levels of neurotrophic factors.

To evaluate a potential causal relationship between altered BDNF levels and depression, a variety of mouse models with genetically altered BDNF levels have been generated and assessed for defects in behavioral-, cognitive-, and depressive-like phenotypes.[102] The tested mouse models include heterozygous Bdnf deletion mice, forebrain-specific Bdnf knockout mice with early deletion or adult onset of knockout, inducible forebrain-specific Bdnf knockout mice, mouse models carrying the highly prevalent Val66Met polymorphism, and mice overexpressing BDNF, either whole-body or forebrain-restricted. As comprehensively reviewed by Lindholm and Castrén, these studies did not clearly confirm a causal relationship between low levels of BDNF and anxiety- and depression-related behaviors.[102] Interestingly, Bdnf heterozygous mice appeared to be more sensitive to stress (e.g., reduced maternal care),[103,104] suggesting an interaction between low BDNF levels and environment in inducing anxiety- or depression-associated phenotypes. Similarly, increased sensitivity to stress-induced behavioral alterations was observed in female mice with inducible Bdnf knockout in the forebrain.[105] Two studies showed that mice carrying the Val66Met polymorphism have anxiety- and depression-like behavior[106] that may worsen with exposure to stress.[107] Overexpression of BDNF appeared to be antidepressant in one study[108] but increased anxiety in another.[109] Equally conflicting results were observed in mice with reduced levels of TrkB, the high-affinity receptor for BDNF, which showed hyperactivity in one study and increased anxiety in another,[110,111] whereas Koponen and colleagues found that TrkB overexpression is antidepressant and anxiolytic.[112] These discrepancies could be explained by different mouse models, genetic backgrounds and tests used, and the fact that a complex neuropsychiatric disorder like depression is difficult to model in mice.

Despite these somewhat controversial findings, the multiple studies showing that susceptibility to stress-induced behavioral abnormalities is enhanced in some mouse models with reduced BDNF levels support the hypothesis that depression can be caused by the combined effect of environmental factors, such as adverse life events, and a genetic predisposition to reduced neurotrophic factor signaling.

Evidence for enhanced neuroprotection as an underlying mechanism of current successful therapeutic strategies in depression

Except for research done with genetic mouse models, the studies described earlier showed mainly correlative evidence for altered neurotrophic factor signaling and neuroprotection in depression. Work demonstrating that successful therapeutic strategies in depression may function through neurotrophic and neuroprotective pathways has provided further support for a potential causal relationship between impaired neuroprotection and depression. Below, we briefly discuss the direct and indirect evidence for increased neurotrophic factor signaling and neuroprotection as a common underlying theme of effective treatment of depression.

Several antidepressant treatments have neuroprotective roles in depression

Several studies have shown that antidepressant treatment leads to a neuroprotective effect in the brain. A metaanalysis based on 32 published studies recently confirmed that electroconvulsive shock therapy, which is used to treat drug-resistant MDDs, does not damage the brain, but instead increases the volume in frontolimbic brain areas of depressed individuals.[113] These changes may be induced by neuroplastic events, including neurogenesis. Indeed, several studies have demonstrated that not only multiple antidepressant medications, such as fluoxetine, but also electroconvulsive shock therapy induced neurogenesis in brains of mouse models.[48, 114] Conversely, using x-irradiation to abolish neurogenesis during antidepressant treatment, David and colleagues provided evidence that the generation of new neurons in the adult brain is necessary for antidepressant-mediated improvement for some, but not all, depression-associated behaviors in mice.[46] A few studies also suggested that antidepressants can ameliorate neuroinflammation in mice.[115] Lastly, less conventional antidepressant interventions, such as exercise, which has a stress-reducing and anxiolytic effect in humans, was shown to induce neurogenesis[116] and reduce neuroinflammation[117] in mice. Taken together, whereas most of these studies do not provide proof for a causal relationship, they support the hypothesis that effective antidepressant therapies exert neuroprotective effects in the brain.

Increased BDNF may be a mechanism of action contributing to the effect of antidepressants

One of the first associations between depression and neurotrophic factor signaling stems from the attempt to identify a common mechanism of action underlying the long-term effects of clinically used antidepressants. Based on several studies in rodent models of depression (e.g., Refs. 118–121), Duman et al.[74] postulated that a shared mechanism of antidepressants involves 3′,5′-cyclic adenosine monophosphate (cAMP)-mediated intracellular signaling pathways leading to the transcriptional upregulation of BDNF through cAMP-response element-binding protein (CREB). This hypothesis was, in part, developed based on results of the group showing that electroconvulsive shock therapy and several antidepressant drugs (but not antipsychotic drugs) increased BDNF, TrkB, and CREB mRNA in the hippocampus of rats.[120, 122] Studies in humans provided further correlative evidence for BDNF as a factor involved in antidepressant actions by showing that BDNF levels in serum and brain tissue of individuals suffering from depression are increased following antidepressant treatment. In a metaanalysis of 20 different studies that assessed BDNF levels in blood or brain tissue from patients with or without antidepressant treatment, Brunoni et al. found overall a significant increase in BDNF levels in treated patients compared with untreated.[123]

Several follow-up studies corroborated this hypothesis through experiments in rodent models demonstrating that acutely increasing BDNF, for example, in the hippocampal region through intraparenchymal injections or peripherally via subcutaneous injections, resulted in antidepressant effects.[121, 124] However, a few studies showed that BDNF may have different roles in mediating depressive-like behavior in mouse models depending on which region of the brain is manipulated: in contrast to the hippocampus, BDNF injection into the ventral tegmental area (VTA) induces depressive-like behaviors,[125] whereas reduction of BDNF in the nucleus accumbens, to which VTA neurons project, protects mice from chronic stress-induced depression.[126] Such findings also illustrate the distinctive effects of BDNF/TrkB signaling in closely adjacent nuclei and circuits. Substantial populations of dopaminergic neurons in the VTA and adjoining substantia nigra (SN) express both BDNF and its cognate TrkB receptor.[127, 128] However, whereas infusion of the neurotrophin into the VTA elicits a depressive phenotype, this is not the case in the SN in which BDNF administration affects motor behaviors, elicits multiple neurochemical changes in the nigrostriatal system, and is neuroprotective for injured dopaminergic neurons.[129–133] Thus the neurotrophic actions of locally administered BDNF vary in the origins of the juxtapositioned mesolimbic and mesostriatal pathways, via mechanisms that may include distinct autocrine and/or paracrine functions.

A few studies also showed that impaired expression of BDNF or TrkB in mice blunted the effect of antidepressant drug treatments,[134–137] including the fast-acting antidepressant and NMDA receptor antagonist ketamine.[138] This work further corroborates the relevance of neurotrophic factor signaling in depression.

The potential of neurotrophic factors in disease-modifying therapies for mood disorders and beyond

In conclusion, this limited overview of the literature comprising studies in humans and in animal models illustrates that there is strong correlative and some causal support of the hypothesis that impaired neuroprotection and reduced neurotrophic factor signaling may contribute to the etiology of depression. As discussed throughout this chapter, there are still many open questions to answer and conundrums to solve before a neurotrophin- or BDNF-targeted treatment strategy for depression can be developed. For example, are the neurotrophins considered an exogenous treatment strategy or an endogenous target for depression therapy? If the former, many considerations arise, including site of administration, delivery method, possible blood-brain barrier issues, doses, duration of treatment, etc. If the latter, quite possibly a more parsimonious option, questions of ensuring directed upregulation of the tropic factor to relevant brain regions arise and of monitoring altered neurotrophin levels in the brain or the accuracy of using peripheral blood levels of BDNF as a gauge of central levels. The fact that more recent treatment options for depression such as deep brain stimulation[139] and focused ultrasound[140] elicit increases in BDNF in the hippocampus in animal models offers promising new avenues of targeted neuroprotective therapies for the disease.

An intriguing aspect of this literature is the striking overlap in mechanisms with neurodegenerative disorders, including Alzheimer's disease and Parkinson's disease. These diseases have a high incidence of depression,[141,142] which further confirms the hypothesis that impaired neuroprotection, apart from inducing neuronal death that can cause cognitive or motor problems associated with neurodegenerative disorders, also leads to mood disorders such as depression. Interestingly, reduced serum levels of BDNF were found in individuals with Parkinson's disease and depression compared with those without depression.[143] This suggests that altered BDNF may be used as a biomarker to identify individuals with increased risk of depression and could contribute to the development of depression in this disease. Moreover, this also raises the intriguing possibility that exercise or physical activity, which has been shown to upregulate BDNF and/or TrkB in blood in Parkinson's patients[144] and in the striatum and hippocampus in animal models of Parkinson's disease,[145] could have beneficial effects upon both motor and mood symptoms in the disease.

Overall, the studies described herein point toward impaired neurotrophic factor signaling as part of the disease etiology underlying depression, both in major depressive disorder and when observed as a comorbidity of other diseases associated with the loss of neuroprotection. More work is needed to assess whether neurotrophic factor signaling is impaired in all forms of depression and whether it could be an effective therapeutic target in depressive disorders.

Acknowledgments

The authors apologize to all neuroscientists whose important work could not be cited due to space limitations. CG is supported by NIH grants R01NS092705 and R21HD090333, a 2017 Independent Investigator Award from the Brain and Behavior Research Foundation and the Cincinnati Children's Research Foundation. KBS is supported by a grant from the University of Cincinnati Gardner Neuroscience Institute-Neurobiology Research Center, the Selma Schottenstein Harris Lab for Research in Parkinson's, the Gardner Family Center for Parkinson's Disease and Movement Disorders, the Parkinson's Disease Support Network of Ohio, Kentucky and Indiana, and the Kerman Family Fund.

References

1. Kessler RC, Akiskal HS, Ames M, et al. Prevalence and effects of mood disorders on work performance in a nationally representative sample of U.S. workers. *Am J Psychiatry*. 2006;163(9):1561–1568.
2. American Psychiatric Association. *Diagnostic and Statistical Manual of Mental Disorders (DSM-5®)*. Washington, DC: American Psychiatric Pub; 2013.

3. Kraus C, Kadriu B, Lanzenberger R, Zarate Jr. CA, Kasper S. Prognosis and improved outcomes in major depression: a review. *Transl Psychiatry*. 2019;9(1):127.
4. Akil H, Gordon J, Hen R, et al. Treatment resistant depression: a multi-scale, systems biology approach. *Neurosci Biobehav Rev*. 2018;84:272–288.
5. Thase ME, Schwartz TL. Choosing medications for treatment-resistant depression based on mechanism of action. *J Clin Psychiatry*. 2015;76(6):720–727.
6. Flint J, Kendler KS. The genetics of major depression. *Neuron*. 2014;81(3):484–503.
7. Yang L, Zhao Y, Wang Y, et al. The effects of psychological stress on depression. *Curr Neuropharmacol*. 2015;13(4):494–504.
8. Chen Y, Baram TZ. Toward understanding how early-life stress reprograms cognitive and emotional brain networks. *Neuropsychopharmacology*. 2016;41(1):197–206.
9. Khundakar AA, Thomas AJ. Neuropathology of depression in Alzheimer's disease: current knowledge and the potential for new treatments. *J Alzheimers Dis*. 2015;44(1):27–41.
10. Blass DM, Rabins PV. Depression in frontotemporal dementia. *Psychosomatics*. 2009;50(3):239–247.
11. Gallagher DA, Schrag A. Psychosis, apathy, depression and anxiety in Parkinson's disease. *Neurobiol Dis*. 2012;46(3):581–589.
12. Roos E, Mariosa D, Ingre C, et al. Depression in amyotrophic lateral sclerosis. *Neurology*. 2016;86(24):2271–2277.
13. Feinstein A, Magalhaes S, Richard JF, Audet B, Moore C. The link between multiple sclerosis and depression. *Nat Rev Neurol*. 2014;10(9):507–517.
14. Duman RS, Aghajanian GK. Synaptic dysfunction in depression: potential therapeutic targets. *Science*. 2012;338(6103):68–72.
15. Drevets WC, Price JL, Simpson Jr. JR, et al. Subgenual prefrontal cortex abnormalities in mood disorders. *Nature*. 1997;386(6627):824–827.
16. Sheline YI, Gado MH, Price JL. Amygdala core nuclei volumes are decreased in recurrent major depression. *NeuroReport*. 1998;9(9):2023–2028.
17. Videbech P, Ravnkilde B. Hippocampal volume and depression: a meta-analysis of MRI studies. *Am J Psychiatry*. 2004;161(11):1957–1966.
18. Teicher MH, Samson JA, Anderson CM, Ohashi K. The effects of childhood maltreatment on brain structure, function and connectivity. *Nat Rev Neurosci*. 2016;17(10):652–666.
19. Sheline YI. 3D MRI studies of neuroanatomic changes in unipolar major depression: the role of stress and medical comorbidity. *Biol Psychiatry*. 2000;48(8):791–800.
20. Ancelin ML, Carriere I, Artero S, et al. Lifetime major depression and grey-matter volume. *J Psychiatry Neurosci*. 2019;44(1):45–53.
21. Koolschijn PC, van Haren NE, Lensvelt-Mulders GJ, Hulshoff Pol HE, Kahn RS. Brain volume abnormalities in major depressive disorder: a meta-analysis of magnetic resonance imaging studies. *Hum Brain Mapp*. 2009;30(11):3719–3735.
22. Soares JC, Mann JJ. The anatomy of mood disorders—review of structural neuroimaging studies. *Biol Psychiatry*. 1997;41(1):86–106.
23. Geerlings MI, Gerritsen L. Late-life depression, hippocampal volumes, and hypothalamic-pituitary-adrenal axis regulation: a systematic review and meta-analysis. *Biol Psychiatry*. 2017;82(5):339–350.
24. Bora E, Harrison BJ, Davey CG, Yucel M, Pantelis C. Meta-analysis of volumetric abnormalities in cortico-striatal-pallidal-thalamic circuits in major depressive disorder. *Psychol Med*. 2012;42(4):671–681.
25. Stockmeier CA, Mahajan GJ, Konick LC, et al. Cellular changes in the postmortem hippocampus in major depression. *Biol Psychiatry*. 2004;56(9):640–650.
26. Boldrini M, Santiago AN, Hen R, et al. Hippocampal granule neuron number and dentate gyrus volume in antidepressant-treated and untreated major depression. *Neuropsychopharmacology*. 2013;38(6):1068–1077.
27. Rajkowska G, Miguel-Hidalgo JJ, Wei J, et al. Morphometric evidence for neuronal and glial prefrontal cell pathology in major depression. *Biol Psychiatry*. 1999;45(9):1085–1098.

28. Cotter D, Mackay D, Chana G, Beasley C, Landau S, Everall IP. Reduced neuronal size and glial cell density in area 9 of the dorsolateral prefrontal cortex in subjects with major depressive disorder. *Cereb Cortex*. 2002;12(4):386–394.
29. Rajkowska G, Halaris A, Selemon LD. Reductions in neuronal and glial density characterize the dorsolateral prefrontal cortex in bipolar disorder. *Biol Psychiatry*. 2001;49(9):741–752.
30. Cotter D, Mackay D, Landau S, Kerwin R, Everall I. Reduced glial cell density and neuronal size in the anterior cingulate cortex in major depressive disorder. *Arch Gen Psychiatry*. 2001;58(6):545–553.
31. Kang HJ, Voleti B, Hajszan T, et al. Decreased expression of synapse-related genes and loss of synapses in major depressive disorder. *Nat Med*. 2012;18(9):1413–1417.
32. Christian KM, Song H, Ming G-L. Functions and dysfunctions of adult hippocampal neurogenesis. *Annu Rev Neurosci*. 2014;37(1):243–262.
33. Sorrells SF, Paredes MF, Cebrian-Silla A, et al. Human hippocampal neurogenesis drops sharply in children to undetectable levels in adults. *Nature*. 2018;555(7696):377–381.
34. Kempermann G, Gage FH, Aigner L, et al. Human adult neurogenesis: evidence and remaining questions. *Cell Stem Cell*. 2018;23(1):25–30.
35. Danzer SC. Adult neurogenesis in the human brain: paradise lost? *Epilepsy Curr*. 2018;18(5):329–331.
36. Moreno-Jimenez EP, Flor-Garcia M, Terreros-Roncal J, et al. Adult hippocampal neurogenesis is abundant in neurologically healthy subjects and drops sharply in patients with Alzheimer's disease. *Nat Med*. 2019;25(4):554–560.
37. Petrik D, Lagace DC, Eisch AJ. The neurogenesis hypothesis of affective and anxiety disorders: are we mistaking the scaffolding for the building? *Neuropharmacology*. 2012;62(1):21–34.
38. Kang E, Wen Z, Song H, Christian KM, Ming GL. Adult neurogenesis and psychiatric disorders. *Cold Spring Harb Perspect Biol*. 2016;8(9):a019026.
39. Yun S, Reynolds RP, Masiulis I, Eisch AJ. Re-evaluating the link between neuropsychiatric disorders and dysregulated adult neurogenesis. *Nat Med*. 2016;22(11):1239–1247.
40. Castagne V, Moser P, Roux S, Porsolt RD. Rodent models of depression: forced swim and tail suspension behavioral despair tests in rats and mice. *Curr Protoc Neurosci*. 2011; [Chapter 8:Unit 8.10A.].
41. Egeland M, Guinaudie C, Du Preez A, et al. Depletion of adult neurogenesis using the chemotherapy drug temozolomide in mice induces behavioural and biological changes relevant to depression. *Transl Psychiatry*. 2017;7(4):e1101.
42. Anacker C, Luna VM, Stevens GS, et al. Hippocampal neurogenesis confers stress resilience by inhibiting the ventral dentate gyrus. *Nature*. 2018;559(7712):98–102.
43. Levone BR, Cryan JF, O'Leary OF. Role of adult hippocampal neurogenesis in stress resilience. *Neurobiol Stress*. 2015;1:147–155.
44. Hanson ND, Owens MJ, Boss-Williams KA, Weiss JM, Nemeroff CB. Several stressors fail to reduce adult hippocampal neurogenesis. *Psychoneuroendocrinology*. 2011;36(10):1520–1529.
45. Anacker C, Hen R. Adult hippocampal neurogenesis and cognitive flexibility—linking memory and mood. *Nat Rev Neurosci*. 2017;18(6):335–346.
46. David DJ, Samuels BA, Rainer Q, et al. Neurogenesis-dependent and -independent effects of fluoxetine in an animal model of anxiety/depression. *Neuron*. 2009;62(4):479–493.
47. Santarelli L, Saxe M, Gross C, et al. Requirement of hippocampal neurogenesis for the behavioral effects of antidepressants. *Science*. 2003;301(5634):805–809.
48. Malberg JE, Eisch AJ, Nestler EJ, Duman RS. Chronic antidepressant treatment increases neurogenesis in adult rat hippocampus. *J Neurosci*. 2000;20(24):9104–9110.
49. Surget A, Saxe M, Leman S, et al. Drug-dependent requirement of hippocampal neurogenesis in a model of depression and of antidepressant reversal. *Biol Psychiatry*. 2008;64(4):293–301.

50. Reif A, Fritzen S, Finger M, et al. Neural stem cell proliferation is decreased in schizophrenia, but not in depression. *Mol Psychiatry*. 2006;11(5):514–522.
51. Egeland M, Zunszain PA, Pariante CM. Molecular mechanisms in the regulation of adult neurogenesis during stress. *Nat Rev Neurosci*. 2015;16(4):189–200.
52. Han MH, Nestler EJ. Neural substrates of depression and resilience. *Neurotherapeutics*. 2017;14(3):677–686.
53. Calfa G, Kademian S, Ceschin D, Vega G, Rabinovich GA, Volosin M. Characterization and functional significance of glucocorticoid receptors in patients with major depression: modulation by antidepressant treatment. *Psychoneuroendocrinology*. 2003;28(5):687–701.
54. Nestler EJ, Hyman SE. Animal models of neuropsychiatric disorders. *Nat Neurosci*. 2010;13(10):1161–1169.
55. Jassam YN, Izzy S, Whalen M, McGavern DB, El Khoury J. Neuroimmunology of traumatic brain injury: time for a paradigm shift. *Neuron*. 2017;95(6):1246–1265.
56. Kraft AD, Harry GJ. Features of microglia and neuroinflammation relevant to environmental exposure and neurotoxicity. *Int J Environ Res Public Health*. 2011;8(7):2980–3018.
57. Tohidpour A, Morgun AV, Boitsova EB, et al. Neuroinflammation and infection: molecular mechanisms associated with dysfunction of neurovascular unit. *Front Cell Infect Microbiol*. 2017;7:276.
58. Ransohoff RM. How neuroinflammation contributes to neurodegeneration. *Science*. 2016;353(6301):777–783.
59. Bhattacharya A, Derecki NC, Lovenberg TW, Drevets WC. Role of neuro-immunological factors in the pathophysiology of mood disorders. *Psychopharmacology*. 2016;233(9):1623–1636.
60. Kim Y-K, Won E. The influence of stress on neuroinflammation and alterations in brain structure and function in major depressive disorder. *Behav Brain Res*. 2017;329:6–11.
61. Calcia MA, Bonsall DR, Bloomfield PS, Selvaraj S, Barichello T, Howes OD. Stress and neuroinflammation: a systematic review of the effects of stress on microglia and the implications for mental illness. *Psychopharmacology*. 2016;233(9):1637–1650.
62. Shabab T, Khanabdali R, Moghadamtousi SZ, Kadir HA, Mohan G. Neuroinflammation pathways: a general review. *Int J Neurosci*. 2017;127(7):624–633.
63. Ye L, Huang Y, Zhao L, et al. IL-1β and TNF-α induce neurotoxicity through glutamate production: a potential role for neuronal glutaminase. *J Neurochem*. 2013;125(6):897–908.
64. Viviani B, Boraso M, Marchetti N, Marinovich M. Perspectives on neuroinflammation and excitotoxicity: a neurotoxic conspiracy? *NeuroToxicology*. 2014;43:10–20.
65. Brites D, Fernandes A. Neuroinflammation and depression: microglia activation, extracellular microvesicles and microRNA dysregulation. *Front Cell Neurosci*. 2015;9:476.
66. Savitz J, Drevets WC, Smith CM, et al. Putative neuroprotective and neurotoxic kynurenine pathway metabolites are associated with hippocampal and amygdalar volumes in subjects with major depressive disorder. *Neuropsychopharmacology*. 2015;40(2):463–471.
67. Meier TB, Drevets WC, Wurfel BE, et al. Relationship between neurotoxic kynurenine metabolites and reductions in right medial prefrontal cortical thickness in major depressive disorder. *Brain Behav Immun*. 2016;53:39–48.
68. Ravina K, Briggs DI, Kislal S, et al. Intracerebral delivery of brain-derived neurotrophic factor using HyStem(®)-C hydrogel implants improves functional recovery and reduces neuroinflammation in a rat model of ischemic stroke. *Int J Mol Sci*. 2018;19(12):3782.
69. Cook DJ, Nguyen C, Chun HN, et al. Hydrogel-delivered brain-derived neurotrophic factor promotes tissue repair and recovery after stroke. *J Cereb Blood Flow Metab*. 2017;37(3):1030–1045.
70. Lima Giacobbo B, Doorduin J, Klein HC, Dierckx R, Bromberg E, de Vries EFJ. Brain-derived neurotrophic factor in brain disorders: focus on neuroinflammation. *Mol Neurobiol*. 2019;56(5):3295–3312.

71. Bibel M, Barde Y-A. Neurotrophins: key regulators of cell fate and cell shape in the vertebrate nervous system. *Genes Dev*. 2000;14(23):2919–2937.
72. Meeker RB, Williams KS. The p75 neurotrophin receptor: at the crossroad of neural repair and death. *Neural Regen Res*. 2015;10(5):721–725.
73. Huang EJ, Reichardt LF. Trk receptors: roles in neuronal signal transduction. *Annu Rev Biochem*. 2003;72(1):609–642.
74. Duman RS, Heninger GR, Nestler EJ. A molecular and cellular theory of depression. *Arch Gen Psychiatry*. 1997;54(7):597–606.
75. Wiener CD, de Mello Ferreira S, Pedrotti Moreira F, et al. Serum levels of nerve growth factor (NGF) in patients with major depression disorder and suicide risk. *J Affect Disord*. 2015;184:245–248.
76. Bus BAA, Molendijk ML, Tendolkar I, et al. Chronic depression is associated with a pronounced decrease in serum brain-derived neurotrophic factor over time. *Mol Psychiatry*. 2015;20(5):602–608.
77. Karege F, Bondolfi G, Gervasoni N, Schwald M, Aubry JM, Bertschy G. Low brain-derived neurotrophic factor (BDNF) levels in serum of depressed patients probably results from lowered platelet BDNF release unrelated to platelet reactivity. *Biol Psychiatry*. 2005;57(9):1068–1072.
78. Jiang C, Salton SR. The role of neurotrophins in major depressive disorder. *Transl Neurosci*. 2013;4(1):46–58.
79. Pallavi P, Sagar R, Mehta M, et al. Serum neurotrophic factors in adolescent depression: gender difference and correlation with clinical severity. *J Affect Disord*. 2013;150(2):415–423.
80. Klein AB, Williamson R, Santini MA, et al. Blood BDNF concentrations reflect brain-tissue BDNF levels across species. *Int J Neuropsychopharmacol*. 2011;14(3):347–353.
81. Hock C, Heese K, Muller-Spahn F, et al. Increased cerebrospinal fluid levels of neurotrophin 3 (NT-3) in elderly patients with major depression. *Mol Psychiatry*. 2000;5(5):510–513.
82. Arabska J, Lucka A, Strzelecki D, Wysokinski A. In schizophrenia serum level of neurotrophin-3 (NT-3) is increased only if depressive symptoms are present. *Neurosci Lett*. 2018;684:152–155.
83. Walz JC, Andreazza AC, Frey BN, et al. Serum neurotrophin-3 is increased during manic and depressive episodes in bipolar disorder. *Neurosci Lett*. 2007;415(1):87–89.
84. Kim Y-K, Ham B-J, Han K-M. Interactive effects of genetic polymorphisms and childhood adversity on brain morphologic changes in depression. *Prog Neuro-Psychopharmacol Biol Psychiatry*. 2019;91:4–13.
85. Mitchelmore C, Gede L. Brain derived neurotrophic factor: epigenetic regulation in psychiatric disorders. *Brain Res*. 2014;1586:162–172.
86. Pruunsild P, Kazantseva A, Aid T, Palm K, Timmusk T. Dissecting the human BDNF locus: bidirectional transcription, complex splicing, and multiple promoters. *Genomics*. 2007;90(3):397–406.
87. Egan MF, Kojima M, Callicott JH, et al. The BDNF val66met polymorphism affects activity-dependent secretion of BDNF and human memory and hippocampal function. *Cell*. 2003;112(2):257–269.
88. Frustaci A, Pozzi G, Gianfagna F, Manzoli L, Boccia S. Meta-analysis of the brain-derived neurotrophic factor gene (BDNF) Val66Met polymorphism in anxiety disorders and anxiety-related personality traits. *Neuropsychobiology*. 2008;58(3–4):163–170.
89. Bao MH, Zhu SZ, Gao XZ, Sun HS, Feng ZP. Meta-analysis on the association between brain-derived neurotrophic factor polymorphism rs6265 and ischemic stroke, post-stroke depression. *J Stroke Cerebrovasc Dis*. 2018;27(6):1599–1608.
90. Verhagen M, van der Meij A, van Deurzen PA, et al. Meta-analysis of the BDNF Val66Met polymorphism in major depressive disorder: effects of gender and ethnicity. *Mol Psychiatry*. 2010;15(3):260–271.

91. Pishva E, Rutten BPF, van den Hove D. DNA methylation in major depressive disorder. *Adv Exp Med Biol*. 2017;978:185–196.
92. Gau YT, Liou YJ, Yu YW, et al. Evidence for association between genetic variants of p75 neurotrophin receptor (p75NTR) gene and antidepressant treatment response in Chinese major depressive disorder. *Am J Med Genet B Neuropsychiatr Genet*. 2008;147B(5):594–599.
93. Fujii T, Yamamoto N, Hori H, et al. Support for association between the Ser205Leu polymorphism of p75(NTR) and major depressive disorder. *J Hum Genet*. 2011;56(11):806–809.
94. Colle R, Deflesselle E, Martin S, et al. BDNF/TRKB/P75NTR polymorphisms and their consequences on antidepressant efficacy in depressed patients. *Pharmacogenomics*. 2015;16(9):997–1013.
95. Duman RS, Aghajanian GK, Sanacora G, Krystal JH. Synaptic plasticity and depression: new insights from stress and rapid-acting antidepressants. *Nat Med*. 2016;22(3):238–249.
96. Duman CH. Models of depression. *Vitam Horm*. 2010;82:1–21.
97. Cryan JF, Mombereau C. In search of a depressed mouse: utility of models for studying depression-related behavior in genetically modified mice. *Mol Psychiatry*. 2004;9(4):326–357.
98. Duman RS, Monteggia LM. A neurotrophic model for stress-related mood disorders. *Biol Psychiatry*. 2006;59(12):1116–1127.
99. Allaman I, Papp M, Kraftsik R, Fiumelli H, Magistretti PJ, Martin JL. Expression of brain-derived neurotrophic factor is not modulated by chronic mild stress in the rat hippocampus and amygdala. *Pharmacol Rep*. 2008;60(6):1001–1007.
100. Ueyama T, Kawai Y, Nemoto K, Sekimoto M, Tone S, Senba E. Immobilization stress reduced the expression of neurotrophins and their receptors in the rat brain. *Neurosci Res*. 1997;28(2):103–110.
101. Heine VM, Zareno J, Maslam S, Joëls M, Lucassen PJ. Chronic stress in the adult dentate gyrus reduces cell proliferation near the vasculature and VEGF and Flk-1 protein expression. *Eur J Neurosci*. 2005;21(5):1304–1314.
102. Lindholm JSO, Castrén E. Mice with altered BDNF signaling as models for mood disorders and antidepressant effects. *Front Behav Neurosci*. 2014;8:143.
103. Duman CH, Schlesinger L, Kodama M, Russell DS, Duman RS. A role for MAP kinase signaling in behavioral models of depression and antidepressant treatment. *Biol Psychiatry*. 2007;61(5):661–670.
104. Carola V, Gross C. BDNF moderates early environmental risk factors for anxiety in mouse. *Genes Brain Behav*. 2010;9(4):379–389.
105. Autry AE, Adachi M, Cheng P, Monteggia LM. Gender-specific impact of brain-derived neurotrophic factor signaling on stress-induced depression-like behavior. *Biol Psychiatry*. 2009;66(1):84–90.
106. Li WJ, Yu H, Yang JM, et al. Anxiolytic effect of music exposure on BDNFMet/Met transgenic mice. *Brain Res*. 2010;1347:71–79.
107. Yu H, Wang DD, Wang Y, Liu T, Lee FS, Chen ZY. Variant brain-derived neurotrophic factor Val66Met polymorphism alters vulnerability to stress and response to antidepressants. *J Neurosci*. 2012;32(12):4092–4101.
108. Govindarajan A, Rao BS, Nair D, et al. Transgenic brain-derived neurotrophic factor expression causes both anxiogenic and antidepressant effects. *Proc Natl Acad Sci U S A*. 2006;103(35):13208–13213.
109. Papaleo F, Silverman JL, Aney J, et al. Working memory deficits, increased anxiety-like traits, and seizure susceptibility in BDNF overexpressing mice. *Learn Mem*. 2011;18(8):534–544.
110. Zorner B, Wolfer DP, Brandis D, et al. Forebrain-specific trkB-receptor knockout mice: behaviorally more hyperactive than "depressive". *Biol Psychiatry*. 2003;54(10):972–982.
111. Bergami M, Rimondini R, Santi S, Blum R, Gotz M, Canossa M. Deletion of TrkB in adult progenitors alters newborn neuron integration into hippocampal circuits and increases anxiety-like behavior. *Proc Natl Acad Sci U S A*. 2008;105(40):15570–15575.

112. Koponen E, Vöikar V, Riekki R, et al. Transgenic mice overexpressing the full-length neurotrophin receptor trkB exhibit increased activation of the trkB-PLCγ pathway, reduced anxiety, and facilitated learning. *Mol Cell Neurosci.* 2004;26(1):166–181.
113. Gbyl K, Videbech P. Electroconvulsive therapy increases brain volume in major depression: a systematic review and meta-analysis. *Acta Psychiatr Scand.* 2018;138(3):180–195.
114. Madsen TM, Treschow A, Bengzon J, Bolwig TG, Lindvall O, Tingstrom A. Increased neurogenesis in a model of electroconvulsive therapy. *Biol Psychiatry.* 2000;47(12):1043–1049.
115. Adzic M, Brkic Z, Mitic M, et al. Therapeutic strategies for treatment of inflammation-related depression. *Curr Neuropharmacol.* 2018;16(2):176–209.
116. Ferreira AC, Novais A, Sousa N, Sousa JC, Marques F. Voluntary running rescues the defective hippocampal neurogenesis and behaviour observed in lipocalin 2-null mice. *Sci Rep.* 2019;9(1):1649.
117. Spielman LJ, Little JP, Klegeris A. Physical activity and exercise attenuate neuroinflammation in neurological diseases. *Brain Res Bull.* 2016;125:19–29.
118. Menkes DB, Rasenick MM, Wheeler MA, Bitensky MW. Guanosine triphosphate activation of brain adenylate cyclase: enhancement by long-term antidepressant treatment. *Science.* 1983;219(4580):65–67.
119. Colin SF, Chang HC, Mollner S, et al. Chronic lithium regulates the expression of adenylate cyclase and Gi-protein alpha subunit in rat cerebral cortex. *Proc Natl Acad Sci U S A.* 1991;88(23):10634–10637.
120. Nibuya M, Morinobu S, Duman RS. Regulation of BDNF and trkB mRNA in rat brain by chronic electroconvulsive seizure and antidepressant drug treatments. *J Neurosci.* 1995;15(11):7539–7547.
121. Siuciak JA, Lewis DR, Wiegand SJ, Lindsay RM. Antidepressant-like effect of brain-derived neurotrophic factor (BDNF). *Pharmacol Biochem Behav.* 1997;56(1):131–137.
122. Nibuya M, Nestler EJ, Duman RS. Chronic antidepressant administration increases the expression of cAMP response element binding protein (CREB) in rat hippocampus. *J Neurosci.* 1996;16(7):2365–2372.
123. Brunoni AR, Lopes M, Fregni F. A systematic review and meta-analysis of clinical studies on major depression and BDNF levels: implications for the role of neuroplasticity in depression. *Int J Neuropsychopharmacol.* 2008;11(8):1169–1180.
124. Shirayama Y, Chen AC, Nakagawa S, Russell DS, Duman RS. Brain-derived neurotrophic factor produces antidepressant effects in behavioral models of depression. *J Neurosci.* 2002;22(8):3251–3261.
125. Eisch AJ, Bolanos CA, de Wit J, et al. Brain-derived neurotrophic factor in the ventral midbrain-nucleus accumbens pathway: a role in depression. *Biol Psychiatry.* 2003;54(10):994–1005.
126. Wook Koo J, Labonte B, Engmann O, et al. Essential role of mesolimbic brain-derived neurotrophic factor in chronic social stress-induced depressive behaviors. *Biol Psychiatry.* 2016;80(6):469–478.
127. Seroogy KB, Lundgren KH, Tran TMD, Guthrie KM, Isackson PJ, Gall CM. Dopaminergic neurons in rat ventral midbrain express brain-derived neurotrophic factor and neurotrophin-3 mRNAs. *J Comp Neurol.* 1994;342(3):321–334.
128. Numan S, Seroogy KB. Expression of trkB and trkC mRNAs by adult midbrain dopamine neurons: a double-label in situ hybridization study. *J Comp Neurol.* 1999;403(3):295–308.
129. Altar CA, Boylan CB, Fritsche M, Jackson C, Hyman C, Lindsay RM. The neurotrophins NT-4/5 and BDNF augment serotonin, dopamine, and GABAergic systems during behaviorally effective infusions to the substantia nigra. *Exp Neurol.* 1994;130(1):31–40.
130. Frim DM, Uhler TA, Galpern WR, Beal MF, Breakefield XO, Isacson O. Implanted fibroblasts genetically engineered to produce brain-derived neurotrophic factor

prevent 1-methyl-4-phenylpyridinium toxicity to dopaminergic neurons in the rat. *Proc Natl Acad Sci U S A*. 1994;91(11):5104–5108.
131. Martin-Iverson MT, Altar CA. Spontaneous behaviours of rats are differentially affected by substantia nigra infusions of brain-derived neurotrophic factor and neurotrophin-3. *Eur J Neurosci*. 1996;8(8):1696–1706.
132. Hagg T. Neurotrophins prevent death and differentially affect tyrosine hydroxylase of adult rat nigrostriatal neurons in vivo. *Exp Neurol*. 1998;149(1):183–192.
133. Yurek DM, Seroogy KB. Neurotrophic factor protection of dopaminergic neurons. In: Mochetti I, ed. *Neurobiology of the Neurotrophins*. Johnson City: FP Graham Publishing; 2001:355–397.
134. Adachi M, Barrot M, Autry AE, Theobald D, Monteggia LM. Selective loss of brain-derived neurotrophic factor in the dentate gyrus attenuates antidepressant efficacy. *Biol Psychiatry*. 2008;63(7):642–649.
135. Monteggia LM, Barrot M, Powell CM, et al. Essential role of brain-derived neurotrophic factor in adult hippocampal function. *Proc Natl Acad Sci U S A*. 2004;101(29):10827–10832.
136. Rantamaki T, Hendolin P, Kankaanpaa A, et al. Pharmacologically diverse antidepressants rapidly activate brain-derived neurotrophic factor receptor TrkB and induce phospholipase-Cγ signaling pathways in mouse brain. *Neuropsychopharmacology*. 2007;32(10):2152–2162.
137. Saarelainen T, Hendolin P, Lucas G, et al. Activation of the TrkB neurotrophin receptor is induced by antidepressant drugs and is required for antidepressant-induced behavioral effects. *J Neurosci*. 2003;23(1):349–357.
138. Autry AE, Adachi M, Nosyreva E, et al. NMDA receptor blockade at rest triggers rapid behavioural antidepressant responses. *Nature*. 2011;475(7354):91–95.
139. Dandekar MP, Saxena A, Scaini G, et al. Medial forebrain bundle deep brain stimulation reverses anhedonic-like behavior in a chronic model of depression: importance of BDNF and inflammatory cytokines. *Mol Neurobiol*. 2019;56(6):4364–4380.
140. Tufail Y, Matyushov A, Baldwin N, et al. Transcranial pulsed ultrasound stimulates intact brain circuits. *Neuron*. 2010;66(5):681–694.
141. Ross JA, Gliebus G, Van Bockstaele EJ. Stress induced neural reorganization: a conceptual framework linking depression and Alzheimer's disease. *Prog Neuro-Psychopharmacol Biol Psychiatry*. 2018;85:136–151.
142. Hemmerle AM, Herman JP, Seroogy KB. Stress, depression and Parkinson's disease. *Exp Neurol*. 2012;233(1):79–86.
143. Wang Y, Liu H, Du XD, et al. Association of low serum BDNF with depression in patients with Parkinson's disease. *Parkinsonism Relat Disord*. 2017;41:73–78.
144. Hirsch MA, van Wegen EEH, Newman MA, Heyn PC. Exercise-induced increase in brain-derived neurotrophic factor in human Parkinson's disease: a systematic review and meta-analysis. *Transl Neurodegener*. 2018;7:7.
145. Tuon T, Valvassori SS, Dal Pont GC, et al. Physical training prevents depressive symptoms and a decrease in brain-derived neurotrophic factor in Parkinson's disease. *Brain Res Bull*. 2014;108:106–112.

CHAPTER

7

Neuroprotective roles of neurotrophic growth factors in mood disorders

Cheng Jiang[a], *Stephen R. Salton*[a,b,c]

[a]Fishberg Department of Neuroscience, Icahn School of Medicine at Mount Sinai, New York, NY, United States, [b]Brookdale Department of Geriatrics and Palliative Medicine, Icahn School of Medicine at Mount Sinai, New York, NY, United States, [c]Friedman Brain Institute, Icahn School of Medicine at Mount Sinai, New York, NY, United States

Neurotrophic growth factors and their receptors

The neurotrophin family of secreted neurotrophic growth factors regulates survival, differentiation, growth, and maintenance of neurons in the CNS and PNS.[1–3] Their discovery was based on the identification and characterization of the prototypical family member, nerve growth factor (NGF), a critical regulator of sensory and sympathetic neuronal development and survival, by Rita Levi-Montalcini, Stanley Cohen, and colleagues.[4] The neurotrophin family subsequently underwent expansion following biochemical purification of the protein brain-derived neurotrophic factor (BDNF)[5] and then homology-based cloning of the structurally related neurotrophin 3 (NT-3)[6] and neurotrophin 4/5 (NT-4/5).[7] The function of individual neurotrophins in the CNS or PNS on distinct and overlapping neuronal and non-neuronal target cells is the result of their selective binding to two classes of receptors, the tropomyosin receptor kinase (Trk) family of receptor tyrosine kinases and the p75 neurotrophin receptor (p75NTR).[8, 9] NGF activates TrkA; BDNF, NT-3, and NT-4/5 bind to TrkB; NT-3 activates TrkC[10]; and all four neurotrophins bind to p75NTR.[11] Binding of neurotrophins to their receptors activates different intracellular signaling cascades, including the Ras/ERK (MAPK) pathway, the PI3K/Akt pathway, and the PLCγ pathway (Fig. 7.1),[8] and

https://doi.org/10.1016/B978-0-12-814037-6.00010-0

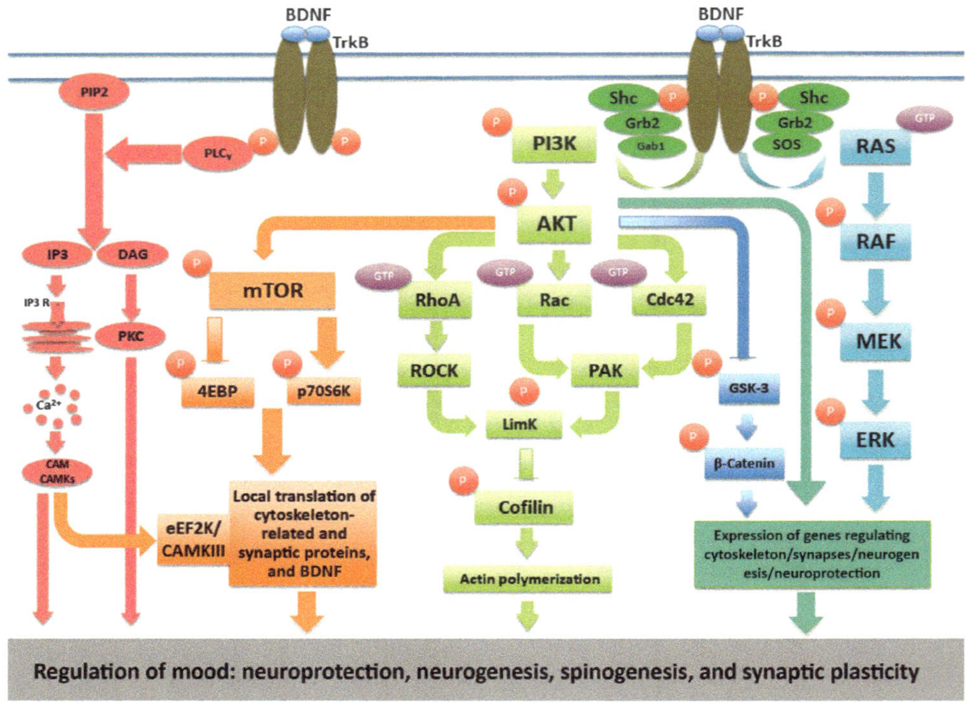

FIG. 7.1 Hippocampal BDNF/TrkB signaling pathways that modulate depression-like behavior and antidepressant efficacy. Binding of BDNF to TrkB activates the PI3K/Akt, Ras/ERK (MAPK), and PLCγ signaling pathways. TrkB phosphorylation (Y490) recruits adaptor proteins including Shc, Grb2, and Gab1, with subsequent activation of the PI3K/Akt pathway and the Rho family of GTPases (RhoA, Rac, and Cdc42), which via downstream signaling including ROCK/PAK, LimK, and Cofilin positively regulate actin polymerization. The PI3K/Akt pathway regulates GSK-3 and its downstream target β-catenin and the mTOR pathway that regulates local protein synthesis and efficacy of rapid-acting antidepressants. PI3K/Akt pathway activation and the Ras/ERK (MAPK) pathway acting through the Rho family of GTPases, mTOR signaling, and synapsin both positively regulate synaptogenesis, spinogenesis, and neurogenesis. TrkB phosphorylation (Y816) recruits PLCγ inducing the hydrolysis of PIP2 to IP3 and DAG, regulating calcium release, and CAMK and PKC activation, with downstream effects on synaptic plasticity. In addition, elongation factor 2 kinase (eEF2K) (alternatively known as calcium/CaM-dependent protein kinase III or CaMKIII), which phosphorylates and inhibits eEF2, dramatically reducing translation, is activated by elevated levels of calcium and binding of calmodulin (CaM), with downstream effects on local protein synthesis.

in addition, immature proneurotrophins, which have not been proteolytically processed, and mature neurotrophins bind to different classes of receptors (e.g., proBDNF to p75NTR and BDNF to TrkB) with distinct biological outcomes, particularly with respect to mood disorders.[12]

NGF, BDNF, and other neurotrophins are released from neuronal and non-neuronal cells via either the regulated or constitutive secretory pathways.[13, 14] These proteins undergo intracellular sorting into dense core vesicles (DCVs) and enzymatic processing, and as a consequence, specific genetic single nucleotide polymorphisms (SNPs) that impact protein sorting signals or processing sites can alter the relative amount of secreted, biologically active neurotrophin. These SNPs and others in promoter regions, intronic regions, and 3′ or 5′ untranslated regions (UTRs) that impact RNA transcription or splicing or translational efficiency are associated with mood disorders, altered cognition, and/or neuropsychiatric disorders, as we discuss in greater detail later and have reviewed previously.[13]

Contributions of BDNF to the pathophysiology of mood disorders

Duman and colleagues[15] initially recognized that BDNF levels in the hippocampus were increased by antidepressant treatment, leading to their formulation of the neurotrophin hypothesis of major depression[16] and the subsequent investigation of the role that neurotrophins play in major depressive disorder (MDD) in human subjects, in depression-like behavior in animal models, and in antidepressant efficacy.

BDNF and other neurotrophin levels in major depressive disorder

Although studies of postmortem human brain tissues have limitations, in particular relatively small sample numbers and potential effects of postmortem interval, their analysis revealed reduced BDNF mRNA and protein levels in the hippocampus, prefrontal cortex (PFC), and amygdala of depressed patients and suicide subjects[17–19] and increased BDNF protein levels in different areas of hippocampus following antidepressant treatment.[20]

Quantification of BDNF protein and RNA levels in plasma therefore might offer a useful, alternative screening measurement if circulating levels correlate with regional BDNF levels in CNS. Indeed, recent studies demonstrate a positive correlation between BDNF levels in blood and hippocampus.[21] Reduced serum and plasma levels of mature BDNF were found in depressed patients,[22–25] in depressed patients who have attempted suicide compared with non-suicidal patients or healthy controls, and in subjects with recurrent or longer lasting episodes of depression.[26, 27]

Interestingly, reduced levels of the BDNF precursor proBDNF were not found in sera or plasma from depressed patients,[22–25] perhaps relevant to their different signaling mechanisms (discussed earlier) and the generally prodepressive function of proBDNF (discussed later).[28] Reduced BDNF content in platelets and decreased BDNF mRNA expression in peripheral blood lymphocytes and mononuclear cells were also noted in depressed patients.[29–31] These reduced plasma BDNF levels were normalized by antidepressant treatment.[32–34] The association of lower serum or plasma BDNF levels with depressive personality traits in healthy human subjects[35] suggests a possible link between BDNF levels and the susceptibility of healthy humans to depression and further that circulating BDNF levels might therefore be a potential biomarker for major depression. Lastly, in bipolar depressed patients resistant to treatment with antidepressants, single-dose ketamine treatment was found to have efficacy in 13 of 25 patients with remission obtained in 8 and 12 patients after 7 and 14 days, respectively.[36] At baseline, there were no differences between responders and non-responders in serum BDNF, NGF, NT3, NT4 and glial-derived neurotrophic factor (GDNF) levels, while serum BDNF alone was significantly reduced after 7 days in non-responders.[36] Interestingly, serum BDNF levels correlate with severity of depression only in female untreated patients, but not males,[37] suggesting that sexual dimorphism should be taken into consideration when employing BDNF as a biomarker.

In a proof of concept study with a relatively small n, significant elevations of BDNF (parietal cortex) and NT3 (parietal, temporal, and occipital cortex; cingulate gyrus; thalamus; putamen; and nucleus caudatus regions) in MDD patients who received antidepressant medication compared with MDD untreated patients and controls were detected.[38] Moreover a significant decrease of NT3 levels was detected in the parietal cortex of untreated patients suffering from MDD compared with healthy individuals.[38]

BDNF genetics in human major depressive disorder

The Val66Met BDNF polymorphism is the most widely studied variation in the human *BDNF* gene, with a single nucleotide polymorphism (SNP) at nucleotide 196 leading to the substitution of methionine (Met) for valine (Val) at codon 66 (refSNP Cluster Report: rs6265). This substitution interferes with BDNF mRNA trafficking to dendrites and with BDNF protein sorting and secretion[39–41] and has been correlated with alterations in BDNF function and hippocampal activity and with depression. In addition, human subjects carrying the Met allele have abnormal hippocampal function and associated memory dysfunction.[39, 42] A number of reports suggest that the BDNF Met allele has a prodepressive role, which has been correlated with smaller volumes of the hippocampus,[43] parahippocampus, amygdala, and frontal gyrus[44] but higher BDNF levels in serum.[45] However, reduced hippocampal volume was observed not only in depressed patients but also in healthy carriers of the Val66Met polymorphism,[46] and no effect of the Val66Met SNP on hippocampal volume and levels of serum and plasma BDNF has also been reported.[47–51] Depressed patients who carry the Met allele (either $BDNF^{Val/Met}$ or $BDNF^{Met/Met}$) are also more likely to suffer from suicidal ideation and to attempt suicide.[52] Met-allele carriers exposed to greater early life stress had smaller hippocampal volumes and more robust syndromal depression.[53] In contrast, early life stress in $BDNF^{Val/Val}$ homozygotes had no effect on depression.[53] Moreover, $BDNF^{Val/Met}$ and $BDNF^{Met/Met}$ subjects have better responses to antidepressant treatment compared with $BDNF^{Val/Val}$ homozygotes.[54, 55] In summary, in human subjects, a deficit in BDNF sorting and secretion alters depression-like behavior and the response to antidepressants, alcohol, and drugs of abuse,[56, 57] likely reflecting in part altered neural plasticity in response to both rewarding and adverse life events.[58] Although there is substantial evidence suggesting that the $BDNF^{Met}$ allele is prodepressive, there is also quite a bit of data that do not support this, and thus the precise contributions of the Val66Met BDNF SNP to mood disorders, cognition, and brain structure remain incompletely understood and a focus of future investigation.

Do peripheral BDNF levels and BDNF genotype provide insight into MDD risk and/or antidepressant efficacy?

As noted earlier, although there are a number of reports that define the utility and validity of using peripheral BDNF levels as a biomarker in mood disorders, many seemingly have contradictory, inconsistent, or poorly reproducible findings that suggest a complex relationship or, alternatively, are the result of different experimental methodologies and/or underpowered studies. For BDNF to be considered a useful biomarker, several key questions need to be addressed: (1) are BDNF levels in serum or plasma a useful state or trait/risk marker of MDD, (2) do peripheral BDNF levels sampled during initial depressive episodes distinguish MDD from bipolar disorder, and (3) are plasma/serum BDNF levels regulated by antidepressants? Recent meta-analytic, longitudinal studies suggest that reduced peripheral plasma/serum BDNF levels are a state marker of MDD[59, 60] and thus are reduced after the onset of depression rather than before. The differential diagnosis of MDD and bipolar disorder, diseases that both often present with initial major depressive episodes, may also be facilitated by measuring plasma BDNF mRNA and protein levels at that time, which are lower in patients who go on to develop bipolar disorder than those with MDD.[61] Antidepressant behavioral responses to electroconvulsive therapy (ECT) were associated with increased peripheral BDNF in some studies[62] but not others,[63, 64] while antidepressant responses of patients with treatment-resistant depression to acute ketamine infusion were associated with increased serum BDNF only at the 1-week time point following infusion, but not after subsequent infusions,[63] and responses to other classes of chemical antidepressants were generally but not always associated with increased plasma/ serum BDNF levels [e.g., Refs. 65, 66]. Thus peripheral BDNF levels in patients with MDD or bipolar disorder may have diagnostic utility, but only in selected patient populations as a state biomarker of depression, while they generally have inconsistent therapeutic utility as an independent indicator of antidepressant efficacy.

Does *BDNF* genotype provide consistent insight into MDD risk and/or antidepressant responses? Meta-analytic pooling of gene association studies of human subjects with the Val66Met BDNF polymorphism confirmed previously reported association of the Met allele with depression in late life only, not in younger subjects (see Ref. 67). However, in additional studies, the Val66Met BDNF polymorphism (1) failed to predict MDD risk[68] and (2) was associated with MDD in males only,[69] while carriers of the Met allele (3) had increased risk for MDD and reduced BDNF protein levels in selected brain regions of depressed patients (unrelated to Val/ Met genotype)[70] and (4) had increased chronicity of MDD, but there was no association with disease recurrence.[71] Meta-analysis further indicated that BDNF Met carriers had an increased antidepressant response compared with Val/Val homozygotes in Asian populations,[60,72] while another study revealed varied responses to different conventional antidepressant drugs by Val/Val homozygotes and Met carriers in depressed Caucasian patients.[73] Moreover, Met carriers showed an impaired response to ketamine.[74] In summary the association of the Val66Met BDNF polymorphism with MDD and antidepressant responses is quite complex.

Analysis of mouse models of the BDNF Val66Met polymorphism

The study of the BDNF Val66Met polymorphism in rodent models, generally via homologous recombination of the human *BDNF* gene that encodes either Val or Met at the Val66Met SNP, into the mouse *Bdnf* locus, followed by inbreeding to generate the appropriate genotypes for analysis (i.e., $BDNF^{Val/Val}$, $BDNF^{Met/Met}$, and $BDNF^{Val/Met}$), has facilitated greater mechanistic understanding of the outcomes of aberrant BDNF sorting and secretion.

$BDNF^{Met/Met}$ mice had increased resilience to chronic social defeat stress, attributable to their reduced BDNF levels in NAc, compared with $BDNF^{Val/Val}$ mice.[75] $BDNF^{Met/Met}$ mice also fail to respond to the administration of conventional and fast-acting antidepressants. $BDNF^{Met/Met}$ mice had lower BDNF protein levels and impaired NMDA receptor-dependent LTP and LTD and decreased survival of newly generated neurons in hippocampus, compared with $BDNF^{Val/Val}$ mice, which were not reversed by fluoxetine.[76,77] BDNF$^{Met/Met}$ mice also had impaired synaptogenesis and longer immobility times in the FST, failing to respond to ketamine administration.[78] BDNF$^{Val/Met}$ mice have increased anxiety- and depressive-like behavior and impaired working memory following restraint stress, additionally showing antidepressant responses to acute administration of desipramine but not fluoxetine, which could be of potential utility when designing effective treatment programs for depressed patients with this genetic variant.[79]

In addition, the BDNF Val66Met polymorphism impaired the beneficial behavioral and neuroplasticity effects induced by voluntary physical exercise in mice, including exercise-induced anxiolytic- and antidepressant-like responses.[80] In other studies, BDNF$^{Val/Val}$ mice were found to have a resilient phenotype in the forced swim test (FST) and developed a robust depressive-like phenotype following chronic late-adolescent corticosterone exposure, to model a history of stress in human populations, while BDNF$^{Met/Met}$ mice subjected to chronic late-adolescent corticosterone treatment had a depression-like phenotype in the FST irrespective of corticosterone treatment.[81]

BDNF and NGF levels in mouse models of depression-like behavior

Regulation of *Bdnf* gene expression in the mouse brain by depression-like behavior is highly region specific. BDNF mRNA and protein levels are down-regulated by acute and chronic stress, including by electrical foot shock, restraint, immobilization, social deprivation, forced swimming, and chronic social defeat, in hippocampus and a number of other brain regions.[82–88] Although BDNF expression is generally reduced in hippocampus in association with the development of depression-like behavior, there are a number of reports that show no change or an increase in hippocampal BDNF mRNA and/or protein levels following acute or chronic stress, perhaps reflecting the different stress paradigms employed, animal age, or possibly protective changes in BDNF gene expression in response to stress.[89–97] In contrast, in the ventral tegmental area (VTA)-nucleus accumbens (NAc) limbic projection, increased BDNF mRNA and protein levels are associated with increased susceptibility to depression-like behavior.[98] In addition to BDNF levels, NGF levels were also reduced in hippocampus of female mice subjected to chronic unpredictable mild stress.[99] Lastly, in PFC, changes in BDNF expression and function, associated with the opposing actions of stress and antidepressant treatment, result in structural alterations, including regulation of neurogenesis, dendrite length, and spine density.[100] A rat model of learned helplessness (LH) had reduced BDNF protein levels in the medial prefrontal cortex (mPFC) and CA3 and dentate gyrus (DG) of the hippocampus, whereas increased

BDNF levels in the NAc were observed.[101] Furthermore, levels of the BDNF precursor proBDNF were higher in the mPFC but lower in the NAc of LH rats,[101] consistent with opposing actions of proBDNF/p75NTR and BDNF/TrkB signaling in depression-like behavior and antidepressant efficacy. In addition, stress exposure has been demonstrated to influence BDNF expression in a sex-specific manner. Maternal deprivation, a stressful event for neonates, reduces BDNF expression in hippocampus and hypothalamus only in male, but not female mice.[102, 103] Restraint stress and chronic foot shock stress, on the other hand, decrease BDNF expression only in hippocampus and PFC of female.[104, 105]

Regulation of depression-like behavior by BDNF/TrkB and proBDNF/p75NTR signaling

As with the regulation of *Bdnf* gene expression by acute and chronic stress, BDNF regulation of depression-like behavior in rodents is also highly region specific in the brain. Localized BDNF ablation in the VTA, projections from which are the likely source of BDNF protein in NAc, reduced depression-like behavior,[98] while infusion of BDNF into VTA produced depression-like behavior.[106] In contrast, and consistent with region- and circuit-specific neurotrophin effects, infusion of BDNF into hippocampal DG reversed depression-like behavior in the LH and FST with a similar efficacy to antidepressant drugs.[107] Moreover, BDNF infusion into the rat midbrain had antidepressant efficacy and reversed impairments in escape latency in a LH paradigm (electrical shock/FST),[108] and intracerebroventricular (icv) BDNF infusion induced sustained antidepressant-like effects.[109] Transgenic overexpression of BDNF in the hippocampus decreased immobility in the FST and protected dendrites in the CA3 region from atrophy,[110] while lentivirus-mediated BDNF overexpression in adult rat hippocampus reduced immobility in the FST and reversed anhedonia, antidepressant-like effects.[111] That the antidepressant or prodepressant efficacy of BDNF/TrkB activation is highly dependent on the brain region and circuits in which it is activated is also buttressed by the finding that intraperitoneal administration of the TrkB agonist 7,8-DHF or the TrkB antagonist ANA-12 each led to an antidepressant response tested 24h after drug administration, in the tail suspension test (TST), FST, or sucrose preference test (SPT), after 10days of chronic social defeat stress.[112] In further support of selective, regional BDNF/TrkB antidepressant responses, a single bilateral infusion of the TrkB agonist 7, 8-DHF, but not a TrkB antagonist ANA-12, into the infralimbic part of the mPFC, DG, and CA3, but not the prelimbic part of the mPFC, exerted antidepressant effects in LH model rats. In contrast a single bilateral infusion of ANA-12, but not 7,8-DHF, into NAc exerted antidepressant-like effects in LH rats.[101] In addition, BDNF regulates depression-like behaviors in a sex-specific manner. For example, female but not male mice with forebrain BDNF knockdown exhibit depressive-like phenotypes at baseline[113] and show increased susceptibility to chronic stress exposure.[114]

Interestingly, icv administration of AAV-proBDNF increased depression-like behavior in naïve rats, icv infused anti-proBDNF antiserum significantly reversed stress-induced depression-like behavior, and the expression of proBDNF/p75NTR/sortilin in neocortex and hippocampus was up-regulated in rats exposed to chronic mild unpredictable stress while that of BDNF/TrkB was down-regulated.[28] Thus the regional balance between proBDNF and mature BDNF and the activation of their respective signaling pathways is likely extremely critical to the regulation of depression-like behavior, suggesting that simple regional manipulation of BDNF gene expression in the treatment of depression may have complex, unintended outcomes.

Peripheral and central inflammation also contribute to depression-like behavior,[115] which has been modeled in mice using lipopolysaccharide (LPS)-induced inflammation. BDNF/TrkB pathway modulation in LPS-treated rodents has been found to elicit antidepressant effects and reduced depression-like behavior.[116]

Regulation of antidepressant efficacy by BDNF expression

Antidepressant efficacy is dependent on BDNF levels and function and was found to be reduced in the FST after deletion of BDNF in the forebrain.[117] Similarly, $BDNF^{+/-}$ germline knockout mice had an impaired response to fluoxetine treatment.[118] In addition to increasing BDNF expression levels, antidepressant drugs promote TrkB receptor activation and the phosphorylation of cyclic AMP response element-binding (CREB) protein via activation of phospholipase C gamma 1 (PLCgamma1) in rodent brain.[119, 120] $TrkB^{+/-}$ heterozygous knockout mice and transgenic mice that overexpress inactive TrkB.T1, both with reduced TrkB activation in the brain, were resistant to the antidepressant effects of imipramine and fluoxetine in the FST.[119, 120] However, BDNF does not necessarily have to be involved in the activation of TrkB receptors by antidepressants,[121] suggesting that antidepressants can regulate alternative signaling pathways that increase TrkB phosphorylation.[122]

In addition, reduced hippocampal BDNF expression blocks the antidepressant effects of ketamine treatment, indicating that BDNF/TrkB signaling in hippocampus is required for the rapid antidepressant efficacy of ketamine.[123] In addition, acute ketamine treatment induces the translation of BDNF[123] and TrkB phosphorylation.[124] $BDNF^{Met/Met}$ mice, which exhibit disrupted BDNF distribution, sorting, trafficking, and secretion,[118] show impaired responses to the rapid-acting antidepressants ketamine, scopolamine, and GLYX-13.[125–127]

Mechanisms by which BDNF and other neurotrophins regulate mood disorders and antidepressant efficacy

Neuroprotective effects and the regulation of neurogenesis by BDNF and other neurotrophins

The generation of functional neurons from neural precursors in the adult brain is restricted to the subgranular zone (SGZ) in the DG of the hippocampus and the subventricular zone (SVZ) of the lateral ventricles in which populations of neural stem cells (NSCs) retain the capacity to proliferate and generate new neurons.[128] Whether or not neurogenesis exists in the adult human hippocampus, however, is highly controversial, based on opposing experimental evidence and conclusions.[129–131] A correlation between neurogenesis and depression was initially shown by the demonstrations in rodents that fluoxetine treatment reversed stress-induced reduction in hippocampal neurogenesis[132] and that impaired hippocampal neurogenesis resulting from irradiation reduced the antidepressant efficacy of fluoxetine and imipramine.[133] Subsequent studies supported these findings, and neurogenesis-independent effects of antidepressant treatment were also identified.[134–136] To determine whether BDNF expression was required for hippocampal neurogenesis, heterozygous BDNF$^{+/-}$ knockout mice were studied: antidepressant-treated BDNF$^{+/-}$ mice had reduced hippocampal progenitor proliferation and survival compared with BDNF$^{+/+}$ mice[137] and reduced hippocampal neurogenesis in response to dietary restriction or environmental enrichment.[137, 138] Using a similar strategy and BDNF$^{+/-}$ mice, BDNF was found not to play a major role in antidepressant-induced increases in hippocampal progenitor proliferation, although it was required for long-term neuronal survival.[139] In addition, NT-3 ablation in hippocampal DG was also found to reduce the number of differentiated neural progenitors.[140] Lastly, using targeted administration of lentivirus expressing short hairpin RNAs for RNA interference, BDNF knockdown in the rat DG but not in CA3 reduced neurogenesis and affected depression-like behaviors.[141]

Conversely the effects of neurotrophin administration or overexpression on hippocampal neurogenesis have also been examined. Adeno-associated virus (AAV)-mediated overexpression of BDNF enhanced hippocampal neurogenesis.[142] Moreover, intrahippocampal BDNF, icv BDNF, or icv NGF infusion also enhanced neurogenesis or survival of newborn neurons in hippocampus,[143] olfactory bulb,[144] or adult DG,[145] respectively. Both short-term environmental enrichment[146] and exercise[147] were also found to increase hippocampal NGF levels,[146, 147] to improve memory, and increase neuronal survival and synaptogenesis in DG.[146] Lastly, intravenous BDNF administration in a rat stroke model improved recovery and increased neurogenesis in the DG.[148]

The Trk receptor and its downstream signaling pathways PI3K/Akt and Ras/ERK (MAPK) converge on the transcription factor CREB,[149] which regulates expression of genes involved in neurogenesis including cyclin A (CCNA) and cyclin D2 (CCND2).[150] With relevance to the role that Akt signaling plays in depression, impaired neurogenesis has also been observed in GSK-3 α/β 21A/21A/9A/9A knockin mice[151] and in GSK-3β overexpressing mice.[152] In many ways similar to the opposing functions of proBDNF and BDNF, proNGF inhibited proliferation and oligodendrogenesis of postnatal hippocampal neural progenitors (neural stem cells or NSCs) in vitro through p75NTR, and also decreased levels of ERK1/2 phosphorylation in a p75NTR-dependent manner.[153]

Regulation of neuronal plasticity by BDNF

Disruption of neuronal plasticity has been linked to mood disorders, including major depressive disorder (MDD) and bipolar disorder, and to impaired learning and memory.[154] Data from clinical studies support a reduction in hippocampal volumes in depressed patients.[155–160] Experimental evidence from animal studies has indicated that stress induces retraction of hippocampal pyramidal cell dendrites[161, 162] and loss of pyramidal neuron dendritic spines in medial prefrontal cortex.[163, 164] In contrast, antidepressant treatments including electroconvulsive therapy (ECT) and pharmacotherapy increase synaptic plasticity by increasing hippocampal dendritic sprouting[165] and dendritic spine density[166] and promoting synaptogenesis.[167]

To investigate the role of BDNF in the regulation of neuronal synaptic plasticity, heterozygous germline $BDNF^{+/-}$ knockout and $hBDNF^{Val/Met}$ knockin mice have been studied. Dendritic retraction in the hippocampus and reduced hippocampal volume were found in $BDNF^{+/-}$ knockout mice,[118, 162] while $hBDNF^{Val/Met}$ knockin mice had impaired spine formation, decreased spine density and dendritic length, and impaired synaptic function,[78, 79, 118] in the hippocampus, amygdala, and/or PFC. In addition to aberrant sorting and secretion of BDNF, BDNF mRNA trafficking in dendrites is also impacted in $hBDNF^{Val/Met}$ carriers, potentially driving changes in neuronal plasticity,[41] which in turn impair local protein translation that is required for synaptogenesis and spine maintenance.[168] Moreover, using the inverse approach, widespread two- to threefold BDNF overexpression in transgenic mice, driven by the actin promoter, enhanced the dendritic complexity of dentate granule cells,[169] increased dendritic density in basolateral amygdala, and prevented stress-induced dendritic atrophy in CA3.[110]

Pathways downstream of TrkB, including Ras/ERK (MAPK) and PI3K/Akt (see Fig. 7.1), modulate depressive behavior and antidepressant responses at least in part via modulation of synaptogenesis and spinogenesis.

Overexpression of PI3K or Akt led to increased soma size, dendritic complexity, and the production of filopodia-like protrusions, whereas mTOR inhibition had the opposite effects.[170] In addition, the Rho family of GTPases, including RhoA, Rac, and Cdc42, which are downstream of the Ras/ERK (MAPK) and PI3K/Akt pathways, regulates actin polymerization and thus dendritic spine plasticity.[171–173] Moreover, synaptogenesis is further regulated by Ras/ERK (MAPK) pathway-dependent phosphorylation of synapsin.[174] Lastly the Ras/ERK (MAPK) and PI3K/Akt pathways are required for ketamine-induced activation of mTOR signaling, which is functionally correlated with synapse formation.[175] Notably, BDNF/TrkB signaling stabilizes CA3 clustered synapses, which requires the proBDNF to BDNF converting enzyme matrix metalloproteinase-9 and enhances neuronal activity in the mature brain, while proBDNF/p75NTR down-regulates out-of-sync synapses,[176] contributing to the precisely connected networks that regulate mood, memory, and behavior. Taken together, these studies support the importance of downstream BDNF/TrkB and proBDNF/p75NTR signaling pathways in the brain region-specific regulation of neural circuit plasticity by antidepressants and in response to prodepressant learned helplessness and stress paradigms.

Regulation of excitatory-inhibitory balance by BDNF

Considerable evidence implicates deficits in gamma-amino butyric acid (GABA)ergic neurotransmission in MDD.[177] Reduced levels of GABA(A) Rs were identified in parahippocampal gyrus and lateral superior temporal gyrus using positron emission tomography (PET).[178] Reduction in the levels of GABA-related transcripts including somatostatin (SST) and neuropeptide Y (NPY) was observed in subgenual anterior cingulate cortex, dorsolateral prefrontal cortex, and amygdala of MDD subjects.[19, 179–181] Moreover, similar patterns of altered gene expression in PFC were demonstrated using chronic stress models in rodents.[182] Interestingly, many of these changes correlated with the reduction of BDNF and/or TrkB levels in postmortem tissue analysis and were found in $BDNF^{+/-}$ heterozygous knockout and BDNF exon-IV knockout mouse models,[19, 181] suggesting a potential link between the neurotrophic[16] and GABA[177] hypotheses of depression.

SST is coexpressed in GABAergic interneurons that target dendrites.[183] Disinhibition of SST-expressing interneurons has been shown to produce antidepressant-like effects.[184] A recent study by Oh et al. demonstrated that the levels of BDNF transcripts that include a long 3′ untranslated region (long 3′ UTR), which regulates local BDNF translation in the dendrites of pyramidal neurons, are reduced in the PFC of MDD subjects.[185] They further showed that the suppression of the long 3′ UTR BDNF transcript leads to depression-like behaviors, reduction of SST, and shrinkage of dendrites, suggesting a causal link between reduced dendritic BDNF expression and the behavioral, molecular, and structural phenotypes of depression. Indeed, BDNF facilitates the functional maturation of GABAergic interneurons, which in turn promotes pyramidal neuron dendritic maturation.[186] Because GABAergic interneurons do not express BDNF, at least in the visual cortex,[187] they generally rely on the neurotrophic support provided by other cell populations, including pyramidal neurons. In the event of exposure to stress, dendritic BDNF expression in pyramidal neurons and thus the BDNF supply available to interneurons targeting these dendrites would be reduced, disrupting function of these interneurons and altering the balance between excitation and inhibition that is received by the pyramidal neurons, leading to mood dysregulation.[185]

Rapid-acting antidepressants regulate local translation and activity-dependent secretion of BDNF

The antidepressant efficacy of the ionotropic glutamatergic *N*-methyl-D-aspartate (NMDA) receptor blocker ketamine requires BDNF/TrkB signaling and protein synthesis but not transcription, which suggests that this BDNF is locally translated in dendrites and undergoes regulated secretion.[123, 125, 188] Two other rapid-acting antidepressants, GLYX-13, a novel glutamatergic compound and NMDA modulator,[127] and scopolamine, a non-selective muscarinic acetylcholine receptor antagonist,[126] also are mechanistically dependent on the rapid desuppression of translation of select mRNAs, including those that encode BDNF. Local synthesis and secretion of BDNF from dendrites is in part the result of translation of a specific BDNF mRNA isoform that is selectively transported to distal dendrites by virtue of targeting sequences within its long 3′ untranslated region (UTR).[189] Selective ablation of the long 3′ UTR BDNF mRNA profoundly reduces activity-dependent BDNF secretion,[189, 190] resulting in impaired GABA-dependent neurogenesis,[191] impaired long-term potentiation (LTP) at dendritic synapses of hippocampal CA1 neurons,[189] and obesity resulting from increased food intake.[192, 193] Consistent with a critical requirement for local translation, the antidepressant efficacies of ketamine, GLYX-13, and scopolamine require activation of eukaryotic elongation factor 2 kinase (eEF2K)[123, 194] and are reduced by inhibitors of mammalian target of rapamycin (mTOR).[175, 195, 196] It is important to note that BDNF also rapidly activates eEF2 (reduced phosphorylation) (see Fig. 7.1) and local protein translation, with mTOR dependence, in axonal growth cones of dorsal root ganglion neurons,[197] consistent with the critical role that we hypothesize an autoregulatory feedback loop plays in antidepressant efficacy (as described and shown in Fig. 7.2).

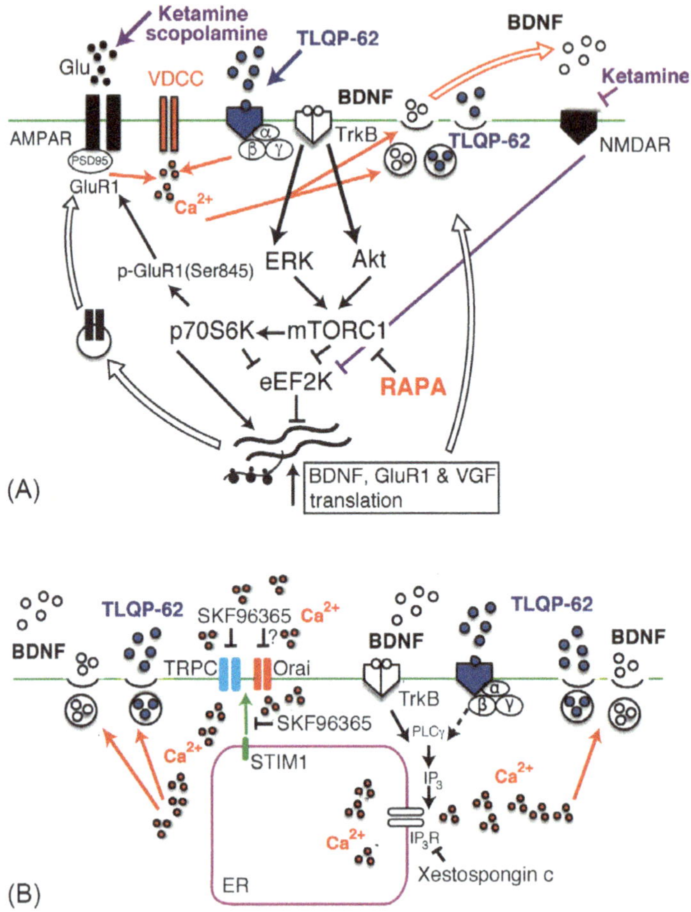

FIG. 7.2 A VGF/BDNF/TrkB autoregulatory feedback loop regulates rapid-acting antidepressant efficacy. In panel (A), dendritic BDNF/TrkB signaling pathways that contribute to rapid antidepressant responses and that stimulate changes in synaptic plasticity and local protein synthesis are shown. Rapid-acting antidepressants ketamine, scopolamine, GLYX-13, and the neuropeptide TLQP-62 trigger BDNF release and increase TrkB signaling, relieving translational suppression via mTOR and/or eEF2 pathways. Local synthesis and insertion of GluR1 and local synthesis and secretion of VGF (TLQP-62) and BDNF are rapidly stimulated, increasing neuronal synaptic plasticity. In panel (B) the role that Ca^{2+} mobilization plays in the BDNF and TLQP-62 autoregulatory loop is explored. Via one hypothesized pathway, TLQP-62 binds a putative GPCR and induces IP_3-mediated Ca^{2+} release from the endoplasmic reticulum (ER) (this pathway is blocked by xestospongin C). Depletion of the Ca^{2+} pool in the ER is sensed by stromal interaction molecule 1 (STIM1), which then activates Orai and/or TRPC Ca^{2+} channels on the cell surface, allowing Ca^{2+} influx via store-operated calcium entry (SOCE) and/or TRPC-mediated extracellular calcium entry (Ca^{2+} channels blocked by SKF96365). Increased levels of intracellular Ca^{2+} stimulate BDNF secretion and subsequent BDNF/TrkB signaling, activating the mTOR pathway, resulting in rapid local synthesis and secretion of VGF (TLQP-62) and BDNF, and synthesis, phosphorylation, and insertion of GluR1, reinforcing activation and synaptic plasticity. *RAPA*, rapamycin; *VDCC*, voltage-dependent calcium channel.

Ketamine potentially functions by at least two independent, complementary mechanisms[198, 199] that require BDNF/TrkB signaling. These include (1) blockade of NMDA-R that triggers burst-like glutamate release, AMPA-R activation, BDNF secretion, TrkB activation, and increased local protein synthesis that requires mTOR pathway activation[175, 200] and/or (2) suppression of spontaneous NMDA-R-mediated neurotransmission that potentiates hippocampal CA1 synaptic responses, inhibits activity of eEF2K, and desuppresses translation of BDNF.[123, 194, 201, 202]

Targets of BDNF/TrkB signaling with potential roles in mood disorders

BDNF/TrkB signaling modulates antidepressant responses and depression-like behavior via downstream activation of gene expression and/or activated TrkB-initiated posttranslational modifications of target proteins (e.g., phosphorylation). Several genes that are transcriptionally regulated are induced with immediate early gene kinetics. Relevant to mood disorders, these include activity-regulated cytoskeleton-associated protein (Arc), which is also regulated by synaptic activity and is localized to neuronal dendrites,[203, 204] and VGF (non-acronymic), a secreted protein and neuropeptide precursor that is robustly regulated by BDNF and NT-3 in CNS neurons.[205]

Arc mRNA expression was found to be increased in hippocampus and parietal cortex by antidepressant and electroconvulsive treatments, and in hippocampus and frontal cortex, Arc mRNA levels were transiently down-regulated by chronic mild stress,[206] as they were also reduced in the medial prefrontal cortex of mice susceptible to chronic social defeat[207] and in three rodent models with depression-like behavior.[206, 208, 209] These findings support a correlation between increased Arc expression and antidepressant effects and decreased Arc levels and depression-like behavior, although several studies have reported the opposite.[97, 210–212]

With similarities to Arc, hippocampal VGF levels were found to be reduced by learned helplessness and the forced swim test (FST)[213] and robustly increased by prolonged voluntary exercise, a potent antidepressant in rodents.[214] Recent studies further demonstrate that VGF is down-regulated in dorsal hippocampus (male/female) and prefrontal cortex (male/female) while it is up-regulated in nucleus accumbens (male), in depressed human subjects, and in mice subjected to chronic social defeat stress and/or chronic restraint stress.[215, 216] In the hippocampus and prefrontal cortex, the antidepressant response to ketamine is associated with rapid VGF translation and is impaired by reduced VGF expression[215–217] and requires coincident, rapid translation and release of BDNF[123, 125] and TrkB phosphorylation.[124] Of note, chronic or acute administration of the

VGF-derived C-terminal peptide TLQP-62 (by convention named by N-terminal 4 amino acids and length) to mouse dorsal hippocampus or prefrontal cortex has antidepressant efficacy that requires BDNF/TrkB signaling[215, 216, 218–220] and calcium mobilization,[216] and as a consequence a positive autoregulatory feedback loop of rapidly translated BDNF and VGF that is critically involved in the antidepressant mechanism of action of ketamine and TLQP-62 has been proposed (Fig. 7.2).[215, 216, 219]

Overview

BDNF modulates mood disorders and antidepressant actions via regulation of neuroprotection, neurogenesis, neuronal and synaptic plasticity, excitatory-inhibitory balance, and local protein synthesis, controlled at least in part by a number of downstream targets of BDNF/TrkB signaling in the brain. Association of BDNF expression levels in the brain with depression-like behavior or antidepressant responses is highly region specific, while in the periphery a number of longitudinal studies indicate that BDNF plasma or serum levels are generally reduced in depressed MDD patients, most likely representing a biomarker of their depressed behavioral state rather than of their disease risk. In addition, behavioral responses to antidepressant treatment are generally associated with increased peripheral BDNF levels. Thus BDNF/TrkB pathways are critically involved in the pathophysiology of MDD and the mechanism of action of antidepressants, while the measurement of circulating BDNF levels has not yet provided consistent predictive insight into disease causality or antidepressant efficacy.

References

1. Lewin GR, Barde YA. Physiology of the neurotrophins. *Annu Rev Neurosci*. 1996;19:289–317.
2. Huang EJ, Reichardt LF. Neurotrophins: roles in neuronal development and function. *Annu Rev Neurosci*. 2001;24:677–736.
3. Bibel M, Barde YA. Neurotrophins: key regulators of cell fate and cell shape in the vertebrate nervous system. *Genes Dev*. 2000;14(23):2919–2937.
4. Levi-Montalcini R. The nerve growth factor: thirty-five years later. *Biosci Rep*. 1987;7(9):681–699.
5. Barde YA, Edgar D, Thoenen H. Purification of a new neurotrophic factor from mammalian brain. *EMBO J*. 1982;1(5):549–553.
6. Maisonpierre PC, Belluscio L, Squinto S, et al. Neurotrophin-3: a neurotrophic factor related to NGF and BDNF. *Science*. 1990;247(4949 Pt 1):1446–1451.
7. Ip NY, Ibanez CF, Nye SH, et al. Mammalian neurotrophin-4: structure, chromosomal localization, tissue distribution, and receptor specificity. *Proc Natl Acad Sci U S A*. 1992;89(7):3060–3064.
8. Reichardt LF. Neurotrophin-regulated signalling pathways. *Philos Trans R Soc Lond Ser B Biol Sci*. 2006;361(1473):1545–1564.

9. Hennigan A, O'Callaghan RM, Kelly AM. Neurotrophins and their receptors: roles in plasticity, neurodegeneration and neuroprotection. *Biochem Soc Trans.* 2007;35(Pt 2):424–427.
10. Barbacid M. Neurotrophic factors and their receptors. *Curr Opin Cell Biol.* 1995;7(2):148–155.
11. Chao MV, Hempstead BL. p75 and Trk: a two-receptor system. *Trends Neurosci.* 1995;18(7):321–326.
12. Castren E, Kojima M. Brain-derived neurotrophic factor in mood disorders and antidepressant treatments. *Neurobiol Dis.* 2017;97(Pt B):119–126.
13. Lin WJ, Salton SR. The regulated secretory pathway and human disease: insights from gene variants and single nucleotide polymorphisms. *Front Endocrinol (Lausanne).* 2013;4:96.
14. Leal G, Bramham CR, Duarte CB. BDNF and hippocampal synaptic plasticity. *Vitam Horm.* 2017;104:153–195.
15. Nibuya M, Morinobu S, Duman RS. Regulation of BDNF and trkB mRNA in rat brain by chronic electroconvulsive seizure and antidepressant drug treatments. *J Neurosci.* 1995;15(11):7539–7547.
16. Duman RS, Heninger GR, Nestler EJ. A molecular and cellular theory of depression. *Arch Gen Psychiatry.* 1997;54(7):597–606.
17. Dwivedi Y, Rizavi HS, Conley RR, Roberts RC, Tamminga CA, Pandey GN. Altered gene expression of brain-derived neurotrophic factor and receptor tyrosine kinase B in postmortem brain of suicide subjects. *Arch Gen Psychiatry.* 2003;60(8):804–815.
18. Karege F, Vaudan G, Schwald M, Perroud N, La Harpe R. Neurotrophin levels in postmortem brains of suicide victims and the effects of antemortem diagnosis and psychotropic drugs. *Brain Res Mol Brain Res.* 2005;136(1–2):29–37.
19. Guilloux JP, Douillard-Guilloux G, Kota R, et al. Molecular evidence for BDNF- and GABA-related dysfunctions in the amygdala of female subjects with major depression. *Mol Psychiatry.* 2012;17(11):1130–1142.
20. Chen B, Dowlatshahi D, MacQueen GM, Wang JF, Young LT. Increased hippocampal BDNF immunoreactivity in subjects treated with antidepressant medication. *Biol Psychiatry.* 2001;50(4):260–265.
21. Klein AB, Williamson R, Santini MA, et al. Blood BDNF concentrations reflect brain-tissue BDNF levels across species. *Int J Neuropsychopharmacol.* 2011;14(3):347–353.
22. Karege F, Perret G, Bondolfi G, Schwald M, Bertschy G, Aubry JM. Decreased serum brain-derived neurotrophic factor levels in major depressed patients. *Psychiatry Res.* 2002;109(2):143–148.
23. Yoshida T, Ishikawa M, Niitsu T, et al. Decreased serum levels of mature brain-derived neurotrophic factor (BDNF), but not its precursor proBDNF, in patients with major depressive disorder. *PLoS ONE.* 2012;7(8):e42676.
24. Lee BH, Kim H, Park SH, Kim YK. Decreased plasma BDNF level in depressive patients. *J Affect Disord.* 2007;101(1–3):239–244.
25. Karege F, Bondolfi G, Gervasoni N, Schwald M, Aubry JM, Bertschy G. Low brain-derived neurotrophic factor (BDNF) levels in serum of depressed patients probably results from lowered platelet BDNF release unrelated to platelet reactivity. *Biol Psychiatry.* 2005;57(9):1068–1072.
26. Kim YK, Lee HP, Won SD, et al. Low plasma BDNF is associated with suicidal behavior in major depression. *Prog Neuro-Psychopharmacol Biol Psychiatry.* 2007;31(1):78–85.
27. Birkenhager TK, Geldermans S, Van den Broek WW, van Beveren N, Fekkes D. Serum brain-derived neurotrophic factor level in relation to illness severity and episode duration in patients with major depression. *J Psychiatr Res.* 2012;46(3):285–289.
28. Bai YY, Ruan CS, Yang CR, et al. ProBDNF signaling regulates depression-like behaviors in rodents under chronic stress. *Neuropsychopharmacology.* 2016;41(12):2882–2892.

29. Lee BH, Kim YK. Reduced platelet BDNF level in patients with major depression. *Prog Neuro-Psychopharmacol Biol Psychiatry*. 2009;33(5):849–853.
30. Lee BH, Kim YK. BDNF mRNA expression of peripheral blood mononuclear cells was decreased in depressive patients who had or had not recently attempted suicide. *J Affect Disord*. 2010;125(1–3):369–373.
31. Pandey GN, Dwivedi Y, Rizavi HS, Ren X, Zhang H, Pavuluri MN. Brain-derived neurotrophic factor gene and protein expression in pediatric and adult depressed subjects. *Prog Neuro-Psychopharmacol Biol Psychiatry*. 2010;34(4):645–651.
32. Shimizu E, Hashimoto K, Okamura N, et al. Alterations of serum levels of brain-derived neurotrophic factor (BDNF) in depressed patients with or without antidepressants. *Biol Psychiatry*. 2003;54(1):70–75.
33. Deuschle M, Gilles M, Scharnholz B, Lederbogen F, Lang UE, Hellweg R. Changes of serum concentrations of brain-derived neurotrophic factor (BDNF) during treatment with venlafaxine and mirtazapine: role of medication and response to treatment. *Pharmacopsychiatry*. 2012;46(2):54–58.
34. Lee HY, Kim YK. Plasma brain-derived neurotrophic factor as a peripheral marker for the action mechanism of antidepressants. *Neuropsychobiology*. 2008;57(4):194–199.
35. Lang UE, Hellweg R, Gallinat J. BDNF serum concentrations in healthy volunteers are associated with depression-related personality traits. *Neuropsychopharmacology*. 2004;29(4):795–798.
36. Rybakowski JK, Permoda-Osip A, Skibinska M, Adamski R, Bartkowska-Sniatkowska A. Single ketamine infusion in bipolar depression resistant to antidepressants: are neurotrophins involved? *Hum Psychopharmacol*. 2013;28(1):87–90.
37. Kreinin A, Lisson S, Nesher E, et al. Blood BDNF level is gender specific in severe depression. *PLoS ONE*. 2015;10(5):e0127643.
38. Sheldrick A, Camara S, Ilieva M, Riederer P, Michel TM. Brain-derived neurotrophic factor (BDNF) and neurotrophin 3 (NT3) levels in post-mortem brain tissue from patients with depression compared to healthy individuals—a proof of concept study. *Eur Psychiatry*. 2017;46:65–71.
39. Egan MF, Kojima M, Callicott JH, et al. The BDNF val66met polymorphism affects activity-dependent secretion of BDNF and human memory and hippocampal function. *Cell*. 2003;112(2):257–269.
40. Chen ZY, Bath K, McEwen B, Hempstead B, Lee F. Impact of genetic variant BDNF (Val66Met) on brain structure and function. *Novartis Found Symp*. 2008;289:180–188 [discussion 188–195].
41. Chiaruttini C, Vicario A, Li Z, et al. Dendritic trafficking of BDNF mRNA is mediated by translin and blocked by the G196A (Val66Met) mutation. *Proc Natl Acad Sci U S A*. 2009;106(38):16481–16486.
42. Hariri AR, Goldberg TE, Mattay VS, et al. Brain-derived neurotrophic factor val66met polymorphism affects human memory-related hippocampal activity and predicts memory performance. *J Neurosci*. 2003;23(17):6690–6694.
43. Frodl T, Schule C, Schmitt G, et al. Association of the brain-derived neurotrophic factor Val66Met polymorphism with reduced hippocampal volumes in major depression. *Arch Gen Psychiatry*. 2007;64(4):410–416.
44. Montag C, Weber B, Fliessbach K, Elger C, Reuter M. The BDNF Val66Met polymorphism impacts parahippocampal and amygdala volume in healthy humans: incremental support for a genetic risk factor for depression. *Psychol Med*. 2009;39(11):1831–1839.
45. Lang UE, Hellweg R, Sander T, Gallinat J. The Met allele of the BDNF Val66Met polymorphism is associated with increased BDNF serum concentrations. *Mol Psychiatry*. 2009;14(2):120–122.
46. Hajek T, Kopecek M, Hoschl C. Reduced hippocampal volumes in healthy carriers of brain-derived neurotrophic factor Val66Met polymorphism: meta-analysis. *World J Biol Psychiatry*. 2012;13(3):178–187.

47. Jessen F, Schuhmacher A, von Widdern O, et al. No association of the Val66Met polymorphism of the brain-derived neurotrophic factor with hippocampal volume in major depression. *Psychiatr Genet*. 2009;19(2):99–101.
48. Terracciano A, Martin B, Ansari D, et al. Plasma BDNF concentration, Val66Met genetic variant and depression-related personality traits. *Genes Brain Behav*. 2010;9(5):512–518.
49. Yoshimura R, Kishi T, Suzuki A, et al. The brain-derived neurotrophic factor (BDNF) polymorphism Val66Met is associated with neither serum BDNF level nor response to selective serotonin reuptake inhibitors in depressed Japanese patients. *Prog Neuro-Psychopharmacol Biol Psychiatry*. 2011;35(4):1022–1025.
50. Chen SL, Lee SY, Chang YH, et al. The BDNF Val66Met polymorphism and plasma brain-derived neurotrophic factor levels in Han Chinese patients with bipolar disorder and schizophrenia. *Prog Neuro-Psychopharmacol Biol Psychiatry*. 2014;51:99–104.
51. Harrisberger F, Smieskova R, Schmidt A, et al. BDNF Val66Met polymorphism and hippocampal volume in neuropsychiatric disorders: a systematic review and meta-analysis. *Neurosci Biobehav Rev*. 2015;55:107–118.
52. Sarchiapone M, Carli V, Roy A, et al. Association of polymorphism (Val66Met) of brain-derived neurotrophic factor with suicide attempts in depressed patients. *Neuropsychobiology*. 2008;57(3):139–145.
53. Gatt JM, Nemeroff CB, Dobson-Stone C, et al. Interactions between BDNF Val66Met polymorphism and early life stress predict brain and arousal pathways to syndromal depression and anxiety. *Mol Psychiatry*. 2009;14(7):681–695.
54. Choi MJ, Kang RH, Lim SW, Oh KS, Lee MS. Brain-derived neurotrophic factor gene polymorphism (Val66Met) and citalopram response in major depressive disorder. *Brain Res*. 2006;1118(1):176–182.
55. Tsai SJ, Hong CJ, Liou YJ. Effects of BDNF polymorphisms on antidepressant action. *Psychiatry Investig*. 2010;7(4):236–242.
56. Wojnar M, Brower KJ, Strobbe S, et al. Association between Val66Met brain-derived neurotrophic factor (BDNF) gene polymorphism and post-treatment relapse in alcohol dependence. *Alcohol Clin Exp Res*. 2009;33(4):693–702.
57. Jia W, Shi JG, Wu B, Ao L, Zhang R, Zhu YS. Polymorphisms of brain-derived neurotrophic factor associated with heroin dependence. *Neurosci Lett*. 2011;495(3):221–224.
58. Chen J, Li X, McGue M. Interacting effect of BDNF Val66Met polymorphism and stressful life events on adolescent depression. *Genes Brain Behav*. 2012;11(8):958–965.
59. Ihara K, Yoshida H, Jones PB, et al. Serum BDNF levels before and after the development of mood disorders: a case-control study in a population cohort. *Transl Psychiatry*. 2016;6:e782.
60. Kishi T, Yoshimura R, Ikuta T, Iwata N. Brain-derived neurotrophic factor and major depressive disorder: evidence from meta-analyses. *Front Psychol*. 2018;8:308.
61. Li Z, Zhang C, Fan J, et al. Brain-derived neurotrophic factor levels and bipolar disorder in patients in their first depressive episode: 3-year prospective longitudinal study. *Br J Psychiatry J Ment Sci*. 2014;205(1):29–35.
62. Rocha RB, Dondossola ER, Grande AJ, et al. Increased BDNF levels after electroconvulsive therapy in patients with major depressive disorder: a meta-analysis study. *J Psychiatr Res*. 2016;83:47–53.
63. Allen AP, Naughton M, Dowling J, et al. Serum BDNF as a peripheral biomarker of treatment-resistant depression and the rapid antidepressant response: a comparison of ketamine and ECT. *J Affect Disord*. 2015;186:306–311.
64. Sorri A, Jarventausta K, Kampman O, et al. Effect of electroconvulsive therapy on brain-derived neurotrophic factor levels in patients with major depressive disorder. *Brain Behav*. 2018;8(11):e01101.
65. Sen S, Duman R, Sanacora G. Serum brain-derived neurotrophic factor, depression, and antidepressant medications: meta-analyses and implications. *Biol Psychiatry*. 2008;64(6):527–532.

66. Zhou C, Zhong J, Zou B, et al. Meta-analyses of comparative efficacy of antidepressant medications on peripheral BDNF concentration in patients with depression. *PLoS ONE*. 2017;12(2):e0172270.
67. Tsang RS, Mather KA, Sachdev PS, Reppermund S. Systematic review and meta-analysis of genetic studies of late-life depression. *Neurosci Biobehav Rev*. 2017;75:129–139.
68. Gyekis JP, Yu W, Dong S, et al. No association of genetic variants in BDNF with major depression: a meta- and gene-based analysis. *Am J Med Genet B Neuropsychiatr Genet*. 2013;162B(1):61–70.
69. Verhagen M, van der Meij A, van Deurzen PA, et al. Meta-analysis of the BDNF Val66Met polymorphism in major depressive disorder: effects of gender and ethnicity. *Mol Psychiatry*. 2010;15(3):260–271.
70. Youssef MM, Underwood MD, Huang YY, et al. Association of BDNF Val66Met polymorphism and brain BDNF levels with major depression and suicide. *Int J Neuropsychopharmacol*. 2018;21(6):528–538.
71. Lee Y, Lim SW, Kim SY, et al. Association between the BDNF Val66Met polymorphism and chronicity of depression. *Psychiatry Investig*. 2013;10(1):56–61.
72. Zou YF, Ye DQ, Feng XL, Su H, Pan FM, Liao FF. Meta-analysis of BDNF Val66Met polymorphism association with treatment response in patients with major depressive disorder. *Eur Neuropsychopharmacol*. 2010;20(8):535–544.
73. Colle R, Gressier F, Verstuyft C, et al. Brain-derived neurotrophic factor Val66Met polymorphism and 6-month antidepressant remission in depressed Caucasian patients. *J Affect Disord*. 2015;175:233–240.
74. Laje G, Lally N, Mathews D, et al. Brain-derived neurotrophic factor Val66Met polymorphism and antidepressant efficacy of ketamine in depressed patients. *Biol Psychiatry*. 2012;72(11):e27–e28.
75. Krishnan V, Han MH, Graham DL, et al. Molecular adaptations underlying susceptibility and resistance to social defeat in brain reward regions. *Cell*. 2007;131(2):391–404.
76. Bath KG, Jing DQ, Dincheva I, et al. BDNF Val66Met impairs fluoxetine-induced enhancement of adult hippocampus plasticity. *Neuropsychopharmacology*. 2012;37(5):1297–1304.
77. Ninan I, Bath KG, Dagar K, et al. The BDNF Val66Met polymorphism impairs NMDA receptor-dependent synaptic plasticity in the hippocampus. *J Neurosci*. 2010;30(26):8866–8870.
78. Liu RJ, Lee FS, Li XY, Bambico F, Duman RS, Aghajanian GK. Brain-derived neurotrophic factor Val66Met allele impairs basal and ketamine-stimulated synaptogenesis in prefrontal cortex. *Biol Psychiatry*. 2012;71(11):996–1005.
79. Yu H, Wang DD, Wang Y, Liu T, Lee FS, Chen ZY. Variant brain-derived neurotrophic factor Val66Met polymorphism alters vulnerability to stress and response to antidepressants. *J Neurosci*. 2012;32(12):4092–4101.
80. Ieraci A, Madaio AI, Mallei A, Lee FS, Popoli M. Brain-derived neurotrophic factor Val66Met human polymorphism impairs the beneficial exercise-induced neurobiological changes in mice. *Neuropsychopharmacology*. 2016;41(13):3070–3079.
81. Notaras M, Du X, Gogos J, van den Buuse M, Hill RA. The BDNF Val66Met polymorphism regulates glucocorticoid-induced corticohippocampal remodeling and behavioral despair. *Transl Psychiatry*. 2017;7(9):e1233.
82. Smith MA, Makino S, Kvetnansky R, Post RM. Stress and glucocorticoids affect the expression of brain-derived neurotrophic factor and neurotrophin-3 mRNAs in the hippocampus. *J Neurosci*. 1995;15(3 Pt 1):1768–1777.
83. Ueyama T, Kawai Y, Nemoto K, Sekimoto M, Tone S, Senba E. Immobilization stress reduced the expression of neurotrophins and their receptors in the rat brain. *Neurosci Res*. 1997;28(2):103–110.
84. Rasmusson AM, Shi L, Duman R. Downregulation of BDNF mRNA in the hippocampal dentate gyrus after re-exposure to cues previously associated with footshock. *Neuropsychopharmacology*. 2002;27(2):133–142.

85. Luo C, Xu H, Li XM. Post-stress changes in BDNF and Bcl-2 immunoreactivities in hippocampal neurons: effect of chronic administration of olanzapine. *Brain Res.* 2004;1025(1–2):194–202.
86. Pizarro JM, Lumley LA, Medina W, et al. Acute social defeat reduces neurotrophin expression in brain cortical and subcortical areas in mice. *Brain Res.* 2004;1025(1–2):10–20.
87. Berry A, Bellisario V, Capoccia S, et al. Social deprivation stress is a triggering factor for the emergence of anxiety- and depression-like behaviours and leads to reduced brain BDNF levels in C57BL/6J mice. *Psychoneuroendocrinology.* 2012;37(6):762–772.
88. Gronli J, Bramham C, Murison R, et al. Chronic mild stress inhibits BDNF protein expression and CREB activation in the dentate gyrus but not in the hippocampus proper. *Pharmacol Biochem Behav.* 2006;85(4):842–849.
89. Allaman I, Papp M, Kraftsik R, Fiumelli H, Magistretti PJ, Martin JL. Expression of brain-derived neurotrophic factor is not modulated by chronic mild stress in the rat hippocampus and amygdala. *Pharmacol Rep.* 2008;60(6):1001–1007.
90. Lucca G, Comim CM, Valvassori SS, et al. Chronic mild stress paradigm reduces sweet food intake in rats without affecting brain derived neurotrophic factor protein levels. *Curr Neurovasc Res.* 2008;5(4):207–213.
91. Marmigere F, Givalois L, Rage F, Arancibia S, Tapia-Arancibia L. Rapid induction of BDNF expression in the hippocampus during immobilization stress challenge in adult rats. *Hippocampus.* 2003;13(5):646–655.
92. Larsen MH, Mikkelsen JD, Hay-Schmidt A, Sandi C. Regulation of brain-derived neurotrophic factor (BDNF) in the chronic unpredictable stress rat model and the effects of chronic antidepressant treatment. *J Psychiatr Res.* 2010;44(13):808–816.
93. Bulygina VV, Shishkina GT, Berezova IV, Dygalo NN. BDNF protein expression in the hippocampus following exposure of rats to forced swimming stress. *Dokl Biol Sci.* 2011;437:82–84.
94. Shi SS, Shao SH, Yuan BP, Pan F, Li ZL. Acute stress and chronic stress change brain-derived neurotrophic factor (BDNF) and tyrosine kinase-coupled receptor (TrkB) expression in both young and aged rat hippocampus. *Yonsei Med J.* 2010;51(5): 661–671.
95. Uysal N, Sisman AR, Dayi A, et al. Acute footshock-stress increases spatial learning-memory and correlates to increased hippocampal BDNF and VEGF and cell numbers in adolescent male and female rats. *Neurosci Lett.* 2012;514(2):141–146.
96. Neto FL, Borges G, Torres-Sanchez S, Mico JA, Berrocoso E. Neurotrophins role in depression neurobiology: a review of basic and clinical evidence. *Curr Neuropharmacol.* 2011;9(4):530–552.
97. Coppens CM, Siripornmongcolchai T, Wibrand K, et al. Social defeat during adolescence and adulthood differentially induce BDNF-regulated immediate early genes. *Front Behav Neurosci.* 2011;5:72.
98. Berton O, McClung CA, Dileone RJ, et al. Essential role of BDNF in the mesolimbic dopamine pathway in social defeat stress. *Science.* 2006;311(5762):864–868.
99. Filho CB, Jesse CR, Donato F, et al. Chronic unpredictable mild stress decreases BDNF and NGF levels and Na(+),K(+)-ATPase activity in the hippocampus and prefrontal cortex of mice: antidepressant effect of chrysin. *Neuroscience.* 2015;289:367–380.
100. Duman RS, Li N. A neurotrophic hypothesis of depression: role of synaptogenesis in the actions of NMDA receptor antagonists. *Philos Trans R Soc Lond Ser B Biol Sci.* 2012;367(1601):2475–2484.
101. Shirayama Y, Yang C, Zhang JC, Ren Q, Yao W, Hashimoto K. Alterations in brain-derived neurotrophic factor (BDNF) and its precursor proBDNF in the brain regions of a learned helplessness rat model and the antidepressant effects of a TrkB agonist and antagonist. *Eur Neuropsychopharmacol.* 2015;25(12):2449–2458.
102. Kikusui T, Ichikawa S, Mori Y. Maternal deprivation by early weaning increases corticosterone and decreases hippocampal BDNF and neurogenesis in mice. *Psychoneuroendocrinology.* 2009;34(5):762–772.

103. Viveros MP, Diaz F, Mateos B, Rodriguez N, Chowen JA. Maternal deprivation induces a rapid decline in circulating leptin levels and sexually dimorphic modifications in hypothalamic trophic factors and cell turnover. *Horm Behav*. 2010;57(4–5):405–414.
104. Yamaura K, Bi Y, Ishiwatari M, Oishi N, Fukata H, Ueno K. Sex differences in stress reactivity of hippocampal BDNF in mice are associated with the female preponderance of decreased locomotor activity in response to restraint stress. *Zool Sci*. 2013;30(12):1019–1024.
105. Lin Y, Ter Horst GJ, Wichmann R, et al. Sex differences in the effects of acute and chronic stress and recovery after long-term stress on stress-related brain regions of rats. *Cereb Cortex*. 2009;19(9):1978–1989.
106. Eisch AJ, Bolanos CA, de Wit J, et al. Brain-derived neurotrophic factor in the ventral midbrain-nucleus accumbens pathway: a role in depression. *Biol Psychiatry*. 2003;54(10):994–1005.
107. Shirayama Y, Chen AC, Nakagawa S, Russell DS, Duman RS. Brain-derived neurotrophic factor produces antidepressant effects in behavioral models of depression. *J Neurosci*. 2002;22(8):3251–3261.
108. Siuciak JA, Lewis DR, Wiegand SJ, Lindsay RM. Antidepressant-like effect of brain-derived neurotrophic factor (BDNF). *Pharmacol Biochem Behav*. 1997; 56(1):131–137.
109. Hoshaw BA, Malberg JE, Lucki I. Central administration of IGF-I and BDNF leads to long-lasting antidepressant-like effects. *Brain Res*. 2005;1037(1–2):204–208.
110. Govindarajan A, Rao BS, Nair D, et al. Transgenic brain-derived neurotrophic factor expression causes both anxiogenic and antidepressant effects. *Proc Natl Acad Sci U S A*. 2006;103(35):13208–13213.
111. Taliaz D, Loya A, Gersner R, Haramati S, Chen A, Zangen A. Resilience to chronic stress is mediated by hippocampal brain-derived neurotrophic factor. *J Neurosci*. 2011;31(12):4475–4483.
112. Zhang JC, Yao W, Dong C, et al. Comparison of ketamine, 7,8-dihydroxyflavone, and ANA-12 antidepressant effects in the social defeat stress model of depression. *Psychopharmacology*. 2015;232(23):4325–4335.
113. Monteggia LM, Luikart B, Barrot M, et al. Brain-derived neurotrophic factor conditional knockouts show gender differences in depression-related behaviors. *Biol Psychiatry*. 2007;61(2):187–197.
114. Autry AE, Adachi M, Cheng P, Monteggia LM. Gender-specific impact of brain-derived neurotrophic factor signaling on stress-induced depression-like behavior. *Biol Psychiatry*. 2009;66(1):84–90.
115. Miller AH, Maletic V, Raison CL. Inflammation and its discontents: the role of cytokines in the pathophysiology of major depression. *Biol Psychiatry*. 2009;65(9):732–741.
116. Zhang JC, Yao W, Hashimoto K. Brain-derived neurotrophic factor (BDNF)-TrkB signaling in inflammation-related depression and potential therapeutic targets. *Curr Neuropharmacol*. 2016;14(7):721–731.
117. Monteggia LM, Barrot M, Powell CM, et al. Essential role of brain-derived neurotrophic factor in adult hippocampal function. *Proc Natl Acad Sci U S A*. 2004;101(29):10827–10832.
118. Chen ZY, Jing D, Bath KG, et al. Genetic variant BDNF (Val66Met) polymorphism alters anxiety-related behavior. *Science*. 2006;314(5796):140–143.
119. Saarelainen T, Hendolin P, Lucas G, et al. Activation of the TrkB neurotrophin receptor is induced by antidepressant drugs and is required for antidepressant-induced behavioral effects. *J Neurosci*. 2003;23(1):349–357.
120. Rantamaki T, Hendolin P, Kankaanpaa A, et al. Pharmacologically diverse antidepressants rapidly activate brain-derived neurotrophic factor receptor TrkB and induce phospholipase-Cgamma signaling pathways in mouse brain. *Neuropsychopharmacology*. 2007;32(10):2152–2162.

121. Rantamaki T, Vesa L, Antila H, et al. Antidepressant drugs transactivate TrkB neurotrophin receptors in the adult rodent brain independently of BDNF and monoamine transporter blockade. *PLoS ONE*. 2011;6(6):e20567.
122. Jeanneteau F, Chao MV. Promoting neurotrophic effects by GPCR ligands. *Novartis Found Symp*. 2006;276:181–189 [discussion 189–192, 233–187, 275–181].
123. Autry AE, Adachi M, Nosyreva E, et al. NMDA receptor blockade at rest triggers rapid behavioural antidepressant responses. *Nature*. 2011;475(7354):91–95.
124. Reus GZ, Abaleira HM, Titus SE, et al. Effects of ketamine administration on the phosphorylation levels of CREB and TrKB and on oxidative damage after infusion of MEK inhibitor. *Pharmacol Rep*. 2016;68(1):177–184.
125. Lepack AE, Fuchikami M, Dwyer JM, Banasr M, Duman RS. BDNF release is required for the behavioral actions of ketamine. *Int J Neuropsychopharmacol*. 2014;18(1). https://doi.org/10.1093/ijnp/pyu033.
126. Ghosal S, Bang E, Yue W, et al. Activity-dependent brain-derived neurotrophic factor release is required for the rapid antidepressant actions of scopolamine. *Biol Psychiatry*. 2018;83(1):29–37.
127. Kato T, Fogaca MV, Deyama S, Li XY, Fukumoto K, Duman RS. BDNF release and signaling are required for the antidepressant actions of GLYX-13. *Mol Psychiatry*. 2017;23(10):2007–2017.
128. Lledo PM, Alonso M, Grubb MS. Adult neurogenesis and functional plasticity in neuronal circuits. *Nat Rev Neurosci*. 2006;7(3):179–193.
129. Boldrini M, Underwood MD, Mann JJ, Arango V. Serotonin-1A autoreceptor binding in the dorsal raphe nucleus of depressed suicides. *J Psychiatr Res*. 2008;42(6):433–442.
130. Kempermann G, Gage FH, Aigner L, et al. Human adult neurogenesis: evidence and remaining questions. *Cell Stem Cell*. 2018;23(1):25–30.
131. Sorrells SF, Paredes MF, Cebrian-Silla A, et al. Human hippocampal neurogenesis drops sharply in children to undetectable levels in adults. *Nature*. 2018;555(7696):377–381.
132. Malberg JE, Duman RS. Cell proliferation in adult hippocampus is decreased by inescapable stress: reversal by fluoxetine treatment. *Neuropsychopharmacology*. 2003;28(9):1562–1571.
133. Santarelli L, Saxe M, Gross C, et al. Requirement of hippocampal neurogenesis for the behavioral effects of antidepressants. *Science*. 2003;301(5634):805–809.
134. Surget A, Saxe M, Leman S, et al. Drug-dependent requirement of hippocampal neurogenesis in a model of depression and of antidepressant reversal. *Biol Psychiatry*. 2008;64(4):293–301.
135. Overstreet DH, Fredericks K, Knapp D, Breese G, McMichael J. Nerve growth factor (NGF) has novel antidepressant-like properties in rats. *Pharmacol Biochem Behav*. 2010;94(4):553–560.
136. David DJ, Samuels BA, Rainer Q, et al. Neurogenesis-dependent and -independent effects of fluoxetine in an animal model of anxiety/depression. *Neuron*. 2009;62(4):479–493.
137. Lee J, Duan W, Mattson MP. Evidence that brain-derived neurotrophic factor is required for basal neurogenesis and mediates, in part, the enhancement of neurogenesis by dietary restriction in the hippocampus of adult mice. *J Neurochem*. 2002;82(6):1367–1375.
138. Rossi C, Angelucci A, Costantin L, et al. Brain-derived neurotrophic factor (BDNF) is required for the enhancement of hippocampal neurogenesis following environmental enrichment. *Eur J Neurosci*. 2006;24(7):1850–1856.
139. Sairanen M, Lucas G, Ernfors P, Castren M, Castren E. Brain-derived neurotrophic factor and antidepressant drugs have different but coordinated effects on neuronal turnover, proliferation, and survival in the adult dentate gyrus. *J Neurosci*. 2005;25(5):1089–1094.
140. Shimazu K, Zhao M, Sakata K, et al. NT-3 facilitates hippocampal plasticity and learning and memory by regulating neurogenesis. *Learn Mem*. 2006;13(3):307–315.

141. Taliaz D, Stall N, Dar DE, Zangen A. Knockdown of brain-derived neurotrophic factor in specific brain sites precipitates behaviors associated with depression and reduces neurogenesis. *Mol Psychiatry*. 2010;15(1):80–92.
142. Henry RA, Hughes SM, Connor B. AAV-mediated delivery of BDNF augments neurogenesis in the normal and quinolinic acid-lesioned adult rat brain. *Eur J Neurosci*. 2007;25(12):3513–3525.
143. Scharfman H, Goodman J, Macleod A, Phani S, Antonelli C, Croll S. Increased neurogenesis and the ectopic granule cells after intrahippocampal BDNF infusion in adult rats. *Exp Neurol*. 2005;192(2):348–356.
144. Zigova T, Pencea V, Wiegand SJ, Luskin MB. Intraventricular administration of BDNF increases the number of newly generated neurons in the adult olfactory bulb. *Mol Cell Neurosci*. 1998;11(4):234–245.
145. Frielingsdorf H, Simpson DR, Thal LJ, Pizzo DP. Nerve growth factor promotes survival of new neurons in the adult hippocampus. *Neurobiol Dis*. 2007;26(1):47–55.
146. Birch AM, McGarry NB, Kelly AM. Short-term environmental enrichment, in the absence of exercise, improves memory, and increases NGF concentration, early neuronal survival, and synaptogenesis in the dentate gyrus in a time-dependent manner. *Hippocampus*. 2013;23(6):437–450.
147. Chae CH, Jung SL, An SH, et al. Swimming exercise stimulates neuro-genesis in the subventricular zone via increase in synapsin I and nerve growth factor levels. *Biol Sport*. 2014;31(4):309–314.
148. Schabitz WR, Steigleder T, Cooper-Kuhn CM, et al. Intravenous brain-derived neurotrophic factor enhances poststroke sensorimotor recovery and stimulates neurogenesis. *Stroke*. 2007;38(7):2165–2172.
149. Fournier NM, Lee B, Banasr M, Elsayed M, Duman RS. Vascular endothelial growth factor regulates adult hippocampal cell proliferation through MEK/ERK- and PI3K/Akt-dependent signaling. *Neuropharmacology*. 2012;63(4):642–652.
150. Peltier J, O'Neill A, Schaffer DV. PI3K/Akt and CREB regulate adult neural hippocampal progenitor proliferation and differentiation. *Dev Neurobiol*. 2007;67(10):1348–1361.
151. Eom TY, Jope RS. Blocked inhibitory serine-phosphorylation of glycogen synthase kinase-3alpha/beta impairs in vivo neural precursor cell proliferation. *Biol Psychiatry*. 2009;66(5):494–502.
152. Sirerol-Piquer M, Gomez-Ramos P, Hernandez F, et al. GSK3beta overexpression induces neuronal death and a depletion of the neurogenic niches in the dentate gyrus. *Hippocampus*. 2011;21(8):910–922.
153. Guo J, Wang J, Liang C, et al. proNGF inhibits proliferation and oligodendrogenesis of postnatal hippocampal neural stem/progenitor cells through p75NTR in vitro. *Stem Cell Res*. 2013;11(2):874–887.
154. Duman RS. Pathophysiology of depression: the concept of synaptic plasticity. *Eur Psychiatry*. 2002;17(Suppl 3):306–310.
155. Sheline YI, Wang PW, Gado MH, Csernansky JG, Vannier MW. Hippocampal atrophy in recurrent major depression. *Proc Natl Acad Sci U S A*. 1996;93(9):3908–3913.
156. Bremner JD, Narayan M, Anderson ER, Staib LH, Miller HL, Charney DS. Hippocampal volume reduction in major depression. *Am J Psychiatry*. 2000;157(1):115–118.
157. Frodl T, Meisenzahl EM, Zetzsche T, et al. Hippocampal changes in patients with a first episode of major depression. *Am J Psychiatry*. 2002;159(7):1112–1118.
158. MacQueen GM, Campbell S, McEwen BS, et al. Course of illness, hippocampal function, and hippocampal volume in major depression. *Proc Natl Acad Sci U S A*. 2003;100(3):1387–1392.
159. Campbell S, Macqueen G. The role of the hippocampus in the pathophysiology of major depression. *J Psychiatry Neurosci*. 2004;29(6):417–426.

160. Videbech P, Ravnkilde B. Hippocampal volume and depression: a meta-analysis of MRI studies. *Am J Psychiatry*. 2004;161(11):1957–1966.
161. Fuchs E, Flugge G, Czeh B. Remodeling of neuronal networks by stress. *Front Biosci*. 2006;11:2746–2758.
162. Magarinos AM, Li CJ, Gal Toth J, et al. Effect of brain-derived neurotrophic factor haploinsufficiency on stress-induced remodeling of hippocampal neurons. *Hippocampus*. 2011;21(3):253–264.
163. Radley JJ, Rocher AB, Miller M, et al. Repeated stress induces dendritic spine loss in the rat medial prefrontal cortex. *Cereb Cortex*. 2006;16(3):313–320.
164. Izquierdo A, Wellman CL, Holmes A. Brief uncontrollable stress causes dendritic retraction in infralimbic cortex and resistance to fear extinction in mice. *J Neurosci*. 2006;26(21):5733–5738.
165. Lamont SR, Paulls A, Stewart CA. Repeated electroconvulsive stimulation, but not antidepressant drugs, induces mossy fibre sprouting in the rat hippocampus. *Brain Res*. 2001;893(1–2):53–58.
166. Norrholm SD, Ouimet CC. Altered dendritic spine density in animal models of depression and in response to antidepressant treatment. *Synapse*. 2001;42(3):151–163.
167. Hajszan T, MacLusky NJ, Leranth C. Short-term treatment with the antidepressant fluoxetine triggers pyramidal dendritic spine synapse formation in rat hippocampus. *Eur J Neurosci*. 2005;21(5):1299–1303.
168. Duman RS, Aghajanian GK. Synaptic dysfunction in depression: potential therapeutic targets. *Science*. 2012;338(6103):68–72.
169. Tolwani RJ, Buckmaster PS, Varma S, et al. BDNF overexpression increases dendrite complexity in hippocampal dentate gyrus. *Neuroscience*. 2002;114(3):795–805.
170. Kumar V, Zhang MX, Swank MW, Kunz J, Wu GY. Regulation of dendritic morphogenesis by Ras-PI3K-Akt-mTOR and Ras-MAPK signaling pathways. *J Neurosci*. 2005;25(49):11288–11299.
171. Rodgers EE, Theibert AB. Functions of PI 3-kinase in development of the nervous system. *Int J Dev Neurosci*. 2002;20(3–5):187–197.
172. Golden SA, Russo SJ. Mechanisms of psychostimulant-induced structural plasticity. *Cold Spring Harb Perspect Med*. 2012;2(10). https://doi.org/10.1101/cshperspect.a011957
173. Dietz DM, Dietz KC, Nestler EJ, Russo SJ. Molecular mechanisms of psychostimulant-induced structural plasticity. *Pharmacopsychiatry*. 2009;42(Suppl 1):S69–S78.
174. Giachello CN, Fiumara F, Giacomini C, et al. MAPK/Erk-dependent phosphorylation of synapsin mediates formation of functional synapses and short-term homosynaptic plasticity. *J Cell Sci*. 2010;123(Pt 6):881–893.
175. Li N, Lee B, Liu RJ, et al. mTOR-dependent synapse formation underlies the rapid antidepressant effects of NMDA antagonists. *Science*. 2010;329(5994):959–964.
176. Niculescu D, Michaelsen-Preusse K, Guner U, van Dorland R, Wierenga CJ, Lohmann C. A BDNF-mediated push-pull plasticity mechanism for synaptic clustering. *Cell Rep*. 2018;24(8):2063–2074.
177. Luscher B, Shen Q, Sahir N. The GABAergic deficit hypothesis of major depressive disorder. *Mol Psychiatry*. 2011;16(4):383–406.
178. Klumpers UM, Veltman DJ, Drent ML, et al. Reduced parahippocampal and lateral temporal GABAA-[11C]flumazenil binding in major depression: preliminary results. *Eur J Nucl Med Mol Imaging*. 2010;37(3):565–574.
179. Sibille E, Morris HM, Kota RS, Lewis DA. GABA-related transcripts in the dorsolateral prefrontal cortex in mood disorders. *Int J Neuropsychopharmacol*. 2011; 14(6):721–734.
180. Tripp A, Kota RS, Lewis DA, Sibille E. Reduced somatostatin in subgenual anterior cingulate cortex in major depression. *Neurobiol Dis*. 2011;42(1):116–124.

181. Tripp A, Oh H, Guilloux JP, Martinowich K, Lewis DA, Sibille E. Brain-derived neurotrophic factor signaling and subgenual anterior cingulate cortex dysfunction in major depressive disorder. *Am J Psychiatry*. 2012;169(11):1194–1202.
182. Banasr M, Lepack A, Fee C, et al. Characterization of GABAergic marker expression in the chronic unpredictable stress model of depression. *Chronic Stress (Thousand Oaks)*. 2017;1. https://doi.org/10.1177/2470547017720459.
183. Melchitzky DS, Lewis DA. Dendritic-targeting GABA neurons in monkey prefrontal cortex: comparison of somatostatin- and calretinin-immunoreactive axon terminals. *Synapse*. 2008;62(6):456–465.
184. Fuchs T, Jefferson SJ, Hooper A, Yee PH, Maguire J, Luscher B. Disinhibition of somatostatin-positive GABAergic interneurons results in an anxiolytic and antidepressant-like brain state. *Mol Psychiatry*. 2017;22(6):920–930.
185. Oh H, Piantadosi SC, Rocco BR, Lewis DA, Watkins SC, Sibille E. The role of dendritic brain-derived neurotrophic factor transcripts on altered inhibitory circuitry in depression. *Biol Psychiatry*. 2018;85(6):517–526.
186. Luscher B, Fuchs T. GABAergic control of depression-related brain states. *Adv Pharmacol*. 2015;73:97–144.
187. Gorba T, Wahle P. Expression of TrkB and TrkC but not BDNF mRNA in neurochemically identified interneurons in rat visual cortex in vivo and in organotypic cultures. *Eur J Neurosci*. 1999;11(4):1179–1190.
188. Lepack AE, Bang E, Lee B, Dwyer JM, Duman RS. Fast-acting antidepressants rapidly stimulate ERK signaling and BDNF release in primary neuronal cultures. *Neuropharmacology*. 2016;111:242–252.
189. An JJ, Gharami K, Liao GY, et al. Distinct role of long 3′ UTR BDNF mRNA in spine morphology and synaptic plasticity in hippocampal neurons. *Cell*. 2008; 134(1):175–187.
190. Lau AG, Irier HA, Gu J, et al. Distinct 3′UTRs differentially regulate activity-dependent translation of brain-derived neurotrophic factor (BDNF). *Proc Natl Acad Sci U S A*. 2010;107(36):15945–15950.
191. Waterhouse EG, An JJ, Orefice LL, et al. BDNF promotes differentiation and maturation of adult-born neurons through GABAergic transmission. *J Neurosci*. 2012;32(41):14318–14330.
192. Liao GY, An JJ, Gharami K, et al. Dendritically targeted Bdnf mRNA is essential for energy balance and response to leptin. *Nat Med*. 2012;18(4):564–571.
193. Liao GY, Bouyer K, Kamitakahara A, et al. Brain-derived neurotrophic factor is required for axonal growth of selective groups of neurons in the arcuate nucleus. *Mol Metab*. 2015;4(6):471–482.
194. Nosyreva E, Szabla K, Autry AE, Ryazanov AG, Monteggia LM, Kavalali ET. Acute suppression of spontaneous neurotransmission drives synaptic potentiation. *J Neurosci*. 2013;33(16):6990–7002.
195. Liu RJ, Duman C, Kato T, et al. GLYX-13 produces rapid antidepressant responses with key synaptic and behavioral effects distinct from ketamine. *Neuropsychopharmacology*. 2017;42(6):1231–1242.
196. Voleti B, Navarria A, Liu RJ, et al. Scopolamine rapidly increases mammalian target of rapamycin complex 1 signaling, synaptogenesis, and antidepressant behavioral responses. *Biol Psychiatry*. 2013;74(10):742–749.
197. Hoshi O, Sugizaki A, Cho Y, Takei N. BDNF reduces eEF2 phosphorylation and enhances novel protein synthesis in the growth cones of dorsal root ganglia neurons. *Neurochem Res*. 2018;43(6):1242–1249.
198. Zanos P, Gould TD. Mechanisms of ketamine action as an antidepressant. *Mol Psychiatry*. 2018;23(4):801–811.
199. Zanos P, Thompson SM, Duman RS, Zarate Jr. CA, Gould TD. Convergent mechanisms underlying rapid antidepressant action. *CNS Drugs*. 2018;32(3):197–227.

200. Abdallah CG, Sanacora G, Duman RS, Krystal JH. Ketamine and rapid-acting antidepressants: a window into a new neurobiology for mood disorder therapeutics. *Annu Rev Med*. 2015;66:509–523.
201. Bjorkholm C, Monteggia LM. BDNF—a key transducer of antidepressant effects. *Neuropharmacology*. 2016;102:72–79.
202. Miller OH, Yang L, Wang CC, et al. GluN2B-containing NMDA receptors regulate depression-like behavior and are critical for the rapid antidepressant actions of ketamine. *elife*. 2014;3:e03581.
203. Lyford GL, Yamagata K, Kaufmann WE, et al. Arc, a growth factor and activity-regulated gene, encodes a novel cytoskeleton-associated protein that is enriched in neuronal dendrites. *Neuron*. 1995;14(2):433–445.
204. Link W, Konietzko U, Kauselmann G, et al. Somatodendritic expression of an immediate early gene is regulated by synaptic activity. *Proc Natl Acad Sci U S A*. 1995;92(12):5734–5738.
205. Salton SR, Ferri GL, Hahm S, et al. VGF: a novel role for this neuronal and neuroendocrine polypeptide in the regulation of energy balance. *Front Neuroendocrinol*. 2000;21(3):199–219.
206. Elizalde N, Pastor PM, Garcia-Garcia AL, et al. Regulation of markers of synaptic function in mouse models of depression: chronic mild stress and decreased expression of VGLUT1. *J Neurochem*. 2010;114(5):1302–1314.
207. Covington 3rd HE, Lobo MK, Maze I, et al. Antidepressant effect of optogenetic stimulation of the medial prefrontal cortex. *J Neurosci*. 2010;30(48):16082–16090.
208. Molteni R, Calabrese F, Maj PF, et al. Altered expression and modulation of activity-regulated cytoskeletal associated protein (Arc) in serotonin transporter knockout rats. *Eur Neuropsychopharmacol*. 2009;19(12):898–904.
209. Eriksson TM, Delagrange P, Spedding M, et al. Emotional memory impairments in a genetic rat model of depression: involvement of 5-HT/MEK/Arc signaling in restoration. *Mol Psychiatry*. 2012;17(2):173–184.
210. Molteni R, Calabrese F, Mancini M, Racagni G, Riva MA. Basal and stress-induced modulation of activity-regulated cytoskeletal associated protein (Arc) in the rat brain following duloxetine treatment. *Psychopharmacology*. 2008;201(2):285–292.
211. Ons S, Marti O, Armario A. Stress-induced activation of the immediate early gene Arc (activity-regulated cytoskeleton-associated protein) is restricted to telencephalic areas in the rat brain: relationship to c-fos mRNA. *J Neurochem*. 2004;89(5):1111–1118.
212. Mikkelsen JD, Larsen MH. Effects of stress and adrenalectomy on activity-regulated cytoskeleton protein (Arc) gene expression. *Neurosci Lett*. 2006;403(3):239–243.
213. Thakker-Varia S, Krol JJ, Nettleton J, et al. The neuropeptide VGF produces antidepressant-like behavioral effects and enhances proliferation in the hippocampus. *J Neurosci*. 2007;27(45):12156–12167.
214. Hunsberger JG, Newton SS, Bennett AH, et al. Antidepressant actions of the exercise-regulated gene VGF. *Nat Med*. 2007;13(12):1476–1482.
215. Jiang C, Lin WJ, Sadahiro M, et al. VGF function in depression and antidepressant efficacy. *Mol Psychiatry*. 2018;23(7):1632–1642.
216. Jiang C, Lin WJ, Labonte B, et al. VGF and its C-terminal peptide TLQP-62 in ventromedial prefrontal cortex regulate depression-related behaviors and the response to ketamine. *Neuropsychopharmacology*. 2018;44(5):971–981.
217. Shen M, Lv D, Liu X, et al. Essential roles of neuropeptide VGF regulated TrkB/mTOR/BICC1 signaling and phosphorylation of AMPA receptor subunit GluA1 in the rapid antidepressant-like actions of ketamine in mice. *Brain Res Bull*. 2018;143:58–65.
218. Lin P, Wang C, Xu B, et al. The VGF-derived peptide TLQP62 produces antidepressant-like effects in mice via the BDNF/TrkB/CREB signaling pathway. *Pharmacol Biochem Behav*. 2014;120:140–148.

219. Jiang C, Lin WJ, Salton SR. Role of a VGF/BDNF/TrkB autoregulatory feedback loop in rapid-acting antidepressant efficacy. *J Mol Neurosci*. 2018;68(3):504–509.
220. Lv D, Chen Y, Shen M, et al. Mechanisms underlying the rapid-acting antidepressant-like effects of neuropeptide VGF (non-acronymic) C-terminal peptide TLQP-62. *Neuropharmacology*. 2018;143:317–326.

CHAPTER

8

Heme metabolism, mitochondria, and complex I in neuropsychiatric disorders

Lee S. Ifhar, Dorit Ben-Shachar

Laboratory of Psychobiology, Department of Psychiatry, Rambam Health Care Campus, B. Rappaport Faculty of Medicine, Technion IIT, Haifa, Israel

Introduction

Traditionally, heme was considered as an oxygen transfer due to its great abundance in blood hemoglobin. Nevertheless, later studies revealed its involvement in multiple complicated cellular processes such as the regulation of transcription and translation,[1] signal transduction,[2] protein assembly[3] and synthesis,[4] and posttranscriptional processing.[5–7] When roaming, unloose heme is cytotoxic, especially in the brain. Recently, alterations of heme metabolism have been associated with a growing number of neurodegenerative disorders.[8] This article will consider three neuropsychiatric disorders: Alzheimer's disease (AD), Parkinson's disease (PD), and schizophrenia (SZ). AD and PD are the two most prevalent neurodegenerative diseases.[9] SZ, although considered by some as neurodegenerative, does not fulfill the criteria for apparent cell death and specific histological changes that normally characterize neurodegenerative diseases.[10, 11] SZ, AD, and PD are common brain diseases with distinct neuropathology and symptoms. However, they share several clinical symptoms such as impairment in cognitive functions including reduction in memory and perception, psychosis, paranoia, and delusions. Electrophysiological findings reveal deficits in the fast-spiking, parvalbumin-positive (PV+) GABA interneurons in AD and SZ, a feature that is critically important for complex information processing in the hippocampus and neocortex.[12] Imbalances in the activity of these hippocampus and neocortex pathways are hypothesized to underlie movement disorders, including PD.[13, 14]

https://doi.org/10.1016/B978-0-12-814037-6.00008-2

In addition, studies suggest that gene expression signature is shared by patients with AD and SZ at the superior temporal gyrus, and autophagy dysfunction is also a commonality in both.[15–17] Mitochondrial dysfunction is another common feature among the three disorders.[18–21] This finding is not surprising as mitochondria are central to ATP generation and neurons are particularly sensitive to decrease in ATP due to their high energy demand.[22] Abnormal cellular energy state can lead to alterations in neuronal function, distorted synaptic plasticity, and connectivity of neuronal networks, thereby causing cognitive and behavioral deviations,[23] as observed in AD, SZ, and to a lesser extent PD. Heme of which the first and three last steps of synthesis occur in the mitochondria has been linked to these disorders. Recent studies shed light on alteration in heme level in AD,[24] SZ,[25] and potentially PD.[26] This chapter will compare and contrast mitochondrial alterations in these three disorders and provide evidence for their possible involvement in heme impaired metabolism.

Heme

Heme (iron protoporphyrin IX) is a nitrogen-containing cyclic tetrapyrrole. Its prosthetic group consists of an iron atom in the center of a protoporphyrin composed of four pyrrole rings linked together by a methane bridge, four methyl groups, two vinyl groups, and two propionic acid side chains (Fig. 8.1). A lone pair of electrons around this central atom enables the chelation of metal ions to heme.[27] Magnesium binding to the porphyrin will result in chlorophyll, whereas the binding of iron will result in heme. In the last few decades, heme was shown to take part in intricate processes essential for cell survival.[3, 28] Heme being small hydrophobic molecule (~620 Da) easily enters the nucleus, which is often aided by the help of enzyme carriers,[29] and binds transcription factors and repressors, enhancing or inhibiting their actions.[30, 31] In addition, heme can regulate mRNA posttranscriptional processing and translation.[5–7, 32] The latter occurs via heme-regulated inhibitor (HRI), a protein kinase that is one of four protein kinases known in mammals to phosphorylate the highly conserved eukaryotic translation initiation factor 2-alpha (eIF2α).[33–35] HRI is a homodimer that binds two heme molecules in each of its N-terminus domains to dimerize and undergo autophosphorylation at threonine 485 (Thr485).[36] These two heme molecules are an integral part of HRI and cannot be removed. When heme levels are high, two additional heme molecules occupy a second heme-binding domain, the C-terminus kinase insertion domain, and upon binding reversibly inhibit HRI's kinase activity.[37, 38] In this state, HRI is unable to autophosphorylate and becomes inactive, resulting in failure to phosphorylate eIF2α at Ser-51.[36, 38] The nonphosphorylated eIFf2α is then activated

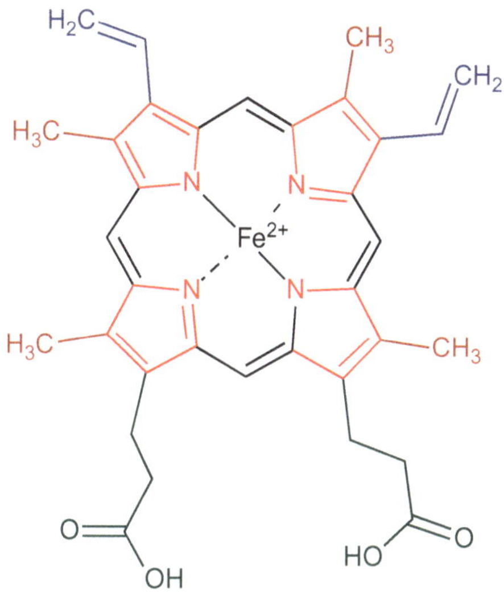

FIG. 8.1 Heme (iron protoporphyrin IX) structure. Four pyrrole rings, *orange*; two propinoic acid groups, *green*; two vinyl groups, *blue*; four methyl groups, *red*.

and initiates global protein synthesis. When heme levels are low, only the N-terminus domain is occupied, resulting in free HRI kinase domain. Autophosphorylation at the Thr485 residue occur, and the kinase domain phosphorylates eIF2α (Fig. 8.2). Phosphorylated eIF2α (peIFf2α) is inactive and hinders global translation by inhibiting Met-tRNA delivery to the translation initiation complex, preventing the selection of the start codon.[39] eIF2α not only regulates its own levels and the levels of heme-regulated proteins but also controls the translation of numerous additional proteins including stress-response proteins and basic leucine zipper (bZIP) DNA-binding proteins.[40] Thus as one regulator of eIF2α, heme is involved in the modulation of numerous cellular pathways including those responding to stress.

Heme biosynthesis pathway

The biosynthesis of heme is carried out in all cells and is highly conserved. Composed of eight distinct steps, the first and last three steps take place in the mitochondria (Fig. 8.3A). The reaction begins with the condensation of glycine and succinyl CoA, resulting in the heme precursor aminolevulinic acid (Δ-ALA). Catalyzing this reaction is the enzyme Δ-ALA synthase (Δ-ALAS), which is the rate-limiting enzyme of the heme biosynthesis pathway.[41, 42] Δ-ALAS belongs to a family of a pyridoxal phosphate (PLP)-dependent enzymes, requiring this cofactor to function as a homodimer. In mammals, regulation of heme biosynthesis is achieved

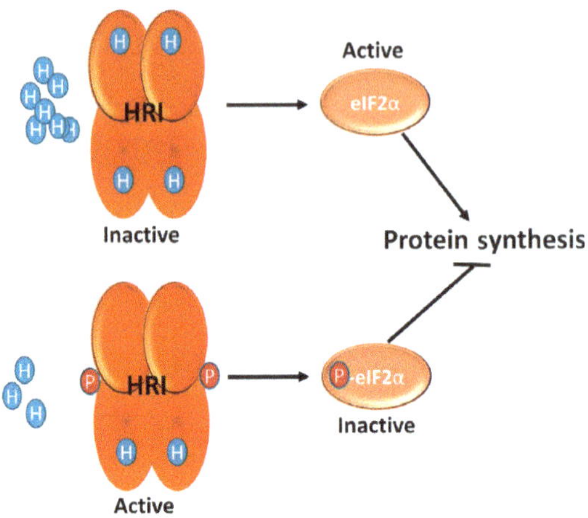

FIG. 8.2 HRI regulation by heme concentrations and its downstream effect on eIF2α-controlled translation. At low heme levels, the N-terminus domain of HRI is occupied, resulting in its free kinase domain. Autophosphorylation at the Thr485 residue occurs, and the kinase domain phosphorylates eIF2α at Ser-51. Phosphorylated eIF2α is inactive and hinders global translation. At high heme levels, two additional heme molecules occupy a second heme binding domain, the C-terminus kinase insertion domain, inhibiting its kinase activity, autophosphorylation, and thereby phosphorylation of eIF2α enabling the initiation of protein translation.

by a feedback-loop manner in two ways: The first is by transcriptional regulation of Δ-ALAS; the second is by inhibiting the transport of the enzyme's precursor into the mitochondria. Δ-ALAS precursor (71 kDa) is synthesized in the cytosol and is cleaved to form a 65-kDa enzyme that is translocated into the mitochondria upon demand. Once heme levels are high, heme can bind several heme regulatory motifs (HRMs) within Δ-ALAS leading sequence, disabling its efficient transport into the mitochondria.[43] Δ-ALAS product, Δ-ALA, is produced in the mitochondria and is exported to the cytosol for further processing. The exact mechanism of its export is unknown; however, the involvement of ATP-binding cassette protein-B10 (ABCB10) is suggested.[44] Once in the cytoplasm, two Δ-ALA molecules are condensed to form porphobilinogen (PBG) by the enzyme 5-aminolaevulinic acid dehydratase (Δ-ALAD).[45] Eukaryotic Δ-ALAD requires eight zinc atoms as cofactors, four of which have a catalytic role, while the other four are of structural nature.[46] Subsequently, porphobilinogen deaminase condenses four PBG molecules to create a linear tetrapyrrole called preuroporphyrinogen.[47] In the next step, uroporphyrinogen III (UPG) synthase catalyzes disulfide bridges and rearranges the molecule, giving rise to a cyclic molecule named uroporphyrinogen III.[41] Next the enzyme uroporphyrinogen III (UPG) decarboxylase, the only known decarboxylase without a prosthetic groups, catalyzes decarboxylation of four sequential uroporphyrinogen III acetates.[48]

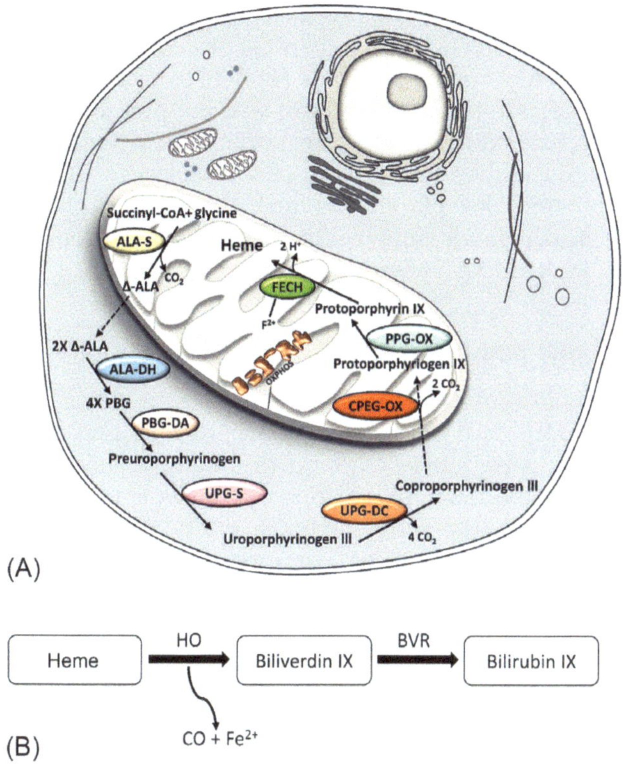

FIG. 8.3 Heme metabolism. (A) The eight steps in heme biosynthesis; (1) Glycine and succinyl CoA condensed by Δ-ALAS to Δ-ALA. (2) Once in the cytosole, two Δ-ALA molecules are condensed to form PBG molecules by Δ-ALADH. (3) Four PBG are then condensed by PBG-DA to preuroporphyrinogen. (4) Preuroporphyrinogen is rearranged by UPG-S to a cyclic uroporphyrinogen III. (5) Coproporphyrinogen III is formed by the decarboxylase UPG-DC. (6) Following import into the mitochondria, coproporphyrinogen III is oxidized and decarboxylated by CPFG-OX to protoporphyrinogen IX. (7) PPG-OX converts the colorless protoporphyrinogen IX to the dark colored protoporphyrin IX. (8) Heme is produced by FECH catalyzing the insertion of a ferrous ion into protoporphyrin IX. (B) Heme catabolism to CO and bilirubin IX. Δ-ALA, aminolevulinic acid; Δ-ALADH, 5-aminolaevulinic acid dehydratase; Δ-ALAS, Δ-ALA synthase; CPFG-OX, coproporphyrinogen III oxidase; FECH, ferrochelatase; PBG, porphobilinogen; PBG-DA, porphobilinogen deaminase; PPG-OX, protoporphyrinogen IX oxidase; UPG-DC, uroporphyrinogen III decarboxylase; UPG-S, uroporphyrinogen III synthase.

The product, coproporphyrinogen III, is imported back into the mitochondria for further processing, possibly by the ATP-binding cassette transporter, ATP-binding cassette subfamily B member 6 (ABCB6).[49, 50] The enzyme coproporphyrinogen III (CPFG) oxidase (also known as coproporphyrinogen III dehydrogenase) has an unusually long leading sequence directing it to the intermembranous mitochondrial space.[51, 52] CPFG oxidase catalyzes the oxidative decarboxylation of coproporphyrinogen III to yield protoporphyrinogen IX, producing two H_2O_2 and CO_2 molecules. Next to last is the conversion of the colorless protoporphyrinogen IX to the dark colored protoporphyrin IX. This reaction is catalyzed by the flavin adenine dinucleotide (FAD) containing enzyme protoporphyrinogen IX (PPG) oxidase, located within the outer surface of the inner mitochondrial membrane, and involves the removal of six electrons.[53] The final step in heme biosynthesis takes place in the matrix of the mitochondria, where the enzyme ferrochelatase catalyzes the insertion of a ferrous ion into protoporphyrin IX, yielding a heme molecule.[41]

Heme catabolic pathway

Heme binds iron in its reduced form, Fe^{2+} (ferrous ion), to allow the reversible binding to molecular oxygen.[54] To maintain the Fe^{2+} at its oxidative state, heme must be bound to a hemoprotein or chaperone, guarding it from oxidation by various molecules abundant in cells. An unbound-heme molecule will quickly react with oxygen to form an oxidized Fe^{3+} (ferric ion), a molecule commonly known as hemin.[55] Under physiological conditions, free heme is in the form of hematin (Fe^{3+} protoporphyrin IX-OH), forming complexes that are virtually insoluble. Free heme is highly toxic as it triggers lysis of cells,[56] with neurons showing greater vulnerability.[57] Free heme generates redox reactive substances such as H_2O_2, superoxide radicals, and hydroxyl radicals, which can cause lipid peroxidation.[58, 59] Therefore, heme cannot exist in the body in its free form and must be eliminated as quickly as possible.[60] Heme is degraded in the cytosol to biliverdin IX by the microsomal enzyme heme oxygenase (HO). Electrons donated by reduced nicotinamide adenine dinucleotide phosphate (NADPH) cytochrome P450 reductase are required for HO catalytic activity.[61] HO catalyzes the oxidative cleavage of a methenyl bridge between pyrroles I and II forming biliverdin IX, releasing in the process carbon monoxide (CO) and an atom of iron.[62] Finally, biliverdin reductase (BVR) reduces the central methenyl bridge of biliverdin IX to produce bilirubin IX (Fig. 8.3B). Two HO isoforms exist. The first, HO1, is induced by a variety of stimuli including free oxygen radicals, heavy metals, bacterial lipopolysaccharides, hydrogen peroxide, ultraviolet light, and levels of heme.[60, 63] The second, HO2, is noninducible and is constitutively expressed in most tissues and cells.[64] Heme degradation is not only important for the removal of cytotoxic heme but also essential for the preservation of iron in the body and the generation of CO, which is an important second messenger and neurotransmitter, and in mammals is mainly produced by heme.[65–68] CO is an endogenous modulator of nitric oxide–cyclic GMP signaling system. HO1 is also a known regulator of the JNK signaling pathway[69] and is induced by ROS-JNK.[70]

Heme in neurons

Heme is key to proper functioning of the CNS, as changes in heme levels lead to changes in brain metabolism, oxygen sensing, and neuronal survival.[71–73] Altered heme metabolism can affect the nervous system in several ways, one of which is via NMDA receptors, which enhance inhibitory γ-aminobutyric acid (GABA) transmission. Animal studies have shown that an increase in heme precursor Δ-ALA levels was linked to diminished NMDA receptors activity in the cerebellum and reduced receptor affinity in the cortex.[74,75] Decrease in heme was also shown to suppress NMDA receptor expression in primary cortical neurons and instigate neurite damage.[76,77] Heme involvement in several cellular pathways may also contribute to neuronal sensitivity to alterations in heme levels. For example, heme deficiency in a neuroendocrine cell line induced apoptosis via the JNK signaling pathway and inactivated the prosurvival Ras-ERK1/2 signaling pathway.[78,79] Two pathways that are key for the survival and differentiation of neurons are extracellular signal–regulated kinase (ERK)/phosphoinositide 3-kinase (PI3K)/protein kinase B (AKT) and soluble guanylyl cyclase (sGC)/cGMP-dependent protein kinase (PKG) signaling pathways, enhancing the expression of growth factors brain-derived neurotrophic factor (BDNF) and glial cell-derived neurotrophic factor (GDNF), respectively.[80] The catabolic action of HO1 and specifically its by-products CO and bilirubin were found to play a major role in activating these pathways enhancing the expression of BDNF in dopaminergic neurons and of GDNF in glia.[81–83] Downstream effects of these pathways are the inhibition of apoptosis and promotion of cell survival and growth, suggesting that the degradation of heme may have a neuroprotective role.[84,85] Both CO and bilirubin show a dual protective or destructive effect, depending on their concentration, cellular environment, and the signaling pathways involved, inducing antioxidant/prooxidant and anti-inflammatory/proinflammatory processes.[86] Overexpression of HO1 was found to render neuroprotection in the brains of rats[87] and offer defense against neuronal and glial oxidative injury caused by neurotoxins, glutamate, and hydrogen peroxide.[88–91] The mechanism responsible for HO1 neuroprotection or neurotoxicity is not entirely clear, but it is suggested that a moderate generation of CO, which can act as a neurotransmitter, is critical for proper neurological functions and affects several intracellular pathways as a regulatory molecule.[68,92,93]

Additional pathways that are affected by changes in heme level are the Ras-mitogen-activated protein kinase (MAPK) and the cyclic AMP (cAMP)-response element-binding protein (CREB), both contributing to neuronal differentiation. Heme deficiency decreases the activation of signaling intermediates such as cAMP and its downstream target CREB,[94] resulting in decreased neuronal differentiation.[95] These findings are in line with previously observed roles of heme in preventing neuronal degeneration[96, 97] and indicate that adequate heme levels and controlled rates of its degradation are critical for neuronal cell signaling and their appropriate functioning.

Alterations in heme metabolism in neuropsychiatric disorders

Inappropriate heme biosynthesis and accumulation of its metabolites have long been linked to neurological compromise. High levels of heme precursor Δ-ALA, for example, were historically associated with neuropsychiatric manifestations of porphyria. Porphyria is a group of disorders caused by abnormalities in the biosynthesis of heme. While abdominal pain is the most common symptom of porphyria, 10% of patients develop neuroporphyria. Often misdiagnosed as psychiatric patients,[98] these patient may experience psychosis, anxiety, insomnia, depression, hallucination, and paranoia.[99, 100] The underlying etiology of porphyria involves Δ-ALA aggregation in the mitochondria, which in turn blocks heme synthesis and results in an overall decrease in heme levels. Elevation in Δ-ALA is caused by either a decrease in Δ-ALAD,[101] increase in Δ-ALAS,[102] or lead (Pb) exposure.[103] Neuroporphyria is not the only psychiatric disorder associated with high Δ-ALA levels. Elevated prenatal levels of Δ-ALA in rats were found to increase the risk of adulthood schizophrenia-like symptoms by almost twofold.[104] When located outside of the mitochondria, Δ-ALA can affect neurons by enhancing their GABAergic neurotransmission. This association was first found as muscimol and barbiturates, the first a potent GABA agonist and the latter GABA facilitators, were found to induce neuroporphyria in patients.[105] This was later explained by the striking similarity in the molecular structure of Δ-ALA and GABA, selectively competing for the binding to GABA receptors, mimicking its action.[106] The neuropathic effect of Δ-ALA is not limited to GABA, but is attributed in part to its toxic effect on neurons and glia at concentrations as low as 10 μM.[107] Δ-ALA was also found to block peripheral myelin formation, with increased vulnerability of glial mitochondria.[27, 108] Inhibition of Na^+/K^+ ATPase activity by Δ-ALA in neurons was also found to play a role in porphyria.[109] While Δ-ALA is the rate-limiting enzyme in heme biosynthesis pathway, changes in the heme catabolic pathway have also been implicated with neuropsychiatric diseases. A growing number of studies report alteration of HO1 expression and activity in neuropsychiatric disorders, as HO1 is highly inducible by numerous stimuli in both neuronal and glial cells.[110] Overexpression of HO1 was also reported in the cerebral cortex and hippocampus of individuals with Alzheimer's disease (AD),[111, 112] substantia nigra of individuals diagnosed with

Parkinson's disease (PD),[113] and prefrontal cortex of patients with SZ[21] suggesting that an increase in the enzyme's activity is associated with some aspects of neuropsychiatric disease. It is important to note that whether higher HO expression necessarily means increased HO1 activity[28] and whether elevation of HO is coupled to heme increase of synthesis or to its degrading products are unclear.[114] Nevertheless, the data described hitherto support an association between heme metabolism and neuropsychiatric disorders.

Heme pyrrole ring as a marker for mental illness

Interest in pyrroles that are also substrates in heme biosynthesis[115] as markers for mental illness emerged in the 1960s as a pyrrole-derived molecule commonly referred to as "Mauve Factor" (later identified as hydroxyhemopyrrolin-2-one, or HPL) was found elevated in the urine of SZ patients[116–118] and patients with other mental disorders.[119] The mechanism of action by which HPL acts upon the nervous system and its exact role are still unclear; however, studies find the inhibition of local electrically evoked action potentials in isolated nerves of rats[120] and mice[121] treated with HPL. HPL's molecular structure is strikingly similar to the enzyme Δ-ALAD, suggesting it may be involved in the second step of heme biosynthesis reaction, possibly as a derivative, agonist, or antagonist.[122] In addition, urine and blood samples from patients suggest elevated HPL is associated with a total decrease in heme[123], and elevation in the heme biosynthesis rate-limiting enzyme Δ-ALAS.[124, 125] The product of Δ-ALAS's enzymatic action, Δ-ALA, was also positively correlated with HPL levels when measured in the urine of psychiatric patients (N=128).[123] Measurements of heme in these studies were either hepatic tissue-specific or total heme count, but depressed heme levels in neurons of psychiatric patients are predicted as well.[126] During the 1970s and 1980s, administration of vitamin B_6 and zinc to psychiatric patients with high-HPL became a common practice, as B_6 and zinc were found to decrease HPL levels and SZ-associated symptoms.[127, 128] Zinc and B_6 are essential cofactors in the heme biosynthesis pathway—zinc is present in a cofactor for the enzyme Δ-ALAD,[129] while vitamin B_6 is the substrate for the synthesis of PLP, a vital cofactor of Δ-ALAS.[130] To maintain adequate levels of Δ-ALAS, B_6 must be constantly converted to PLP; hence, shortage of B_6 may lead to reduction in the levels of the enzyme and thereby in overall heme levels. Though HPL as a marker for mental illness and the zinc/B_6 treatment held a great promise once, they failed to stand the test of time. Later studies of HPL in the mental disorders could not replicate the initial studies. HPL levels in the urine of SZ patients, for example, did not differ from those of controls.[131–133] HPL was then neglected as an etiological factor in mental illness. Nevertheless, it is noteworthy to remark this early interest of researchers and clinicians for many years in heme metabolism and SZ.

Alzheimer's disease and the mitochondria

AD, the most common neurodegenerative disease in Western countries, is characterized by cognitive impairments and progressive neuronal loss.[134] In 1906 German psychiatrist Alouis Alzheimer described a 51-year-old patient with "severe disease process of the cerebral cortex," the first case of what would later be known as "Alzheimer's disease."[135] Although a century had passed, the underlying mechanism of the disease is still unknown. In recent years, numerous reports support the claim that mitochondria are involved in AD pathogenesis. Similarly, to other pathologies involving mitochondrial dysfunction such as SZ, imaging studies report hypometabolism in AD. Cerebral 2-[^{18}F]fluoro-2-deoxy-D-glucose (FDG) positron emission tomography (PET) imaging reveals a specific pattern of reductions in glucose metabolic rate in the parietotemporal posterior cingulate cortexes of patients with AD.[136–138] The involvement of frontal association cortices grows as the disease progresses.[139] In addition, hypometabolism was witnessed in healthy offspring of females with AD, suggesting a "maternal effect," a hallmark of mitochondrial hereditary pattern.[140] Neurons especially vulnerable to hypometabolism and mitochondrial insult are those with the highest metabolic rates. Such neurons with large projections and long axons are the most damaged in AD.[141] It is generally accepted that amyloid β (Aβ) is among the factors that initiate neurodegeneration in AD.[142] The seemingly harmless amyloid protein precursor (APP)[143] is cleaved by β and λ secretases producing toxic Aβ peptides. The association between Aβ load and hypometabolism is disputable; some suggest positive correlation,[144–146] and others find no association.[147–149] However, studies do show a connection between Aβ accumulation and mitochondrial dysfunction, though it is unclear which preceded which. When maintained in the presence of Aβ, cells in culture presented reduced electron transport chain enzyme activities,[150] and isolated mitochondria had decreased respiratory chain function.[151] Furthermore, decreased mitochondrial membrane potential has been witnessed in animal models and in human cortical neurons.[152–154] Both APP and Aβ were shown to interact with mitochondrial proteins and affect mitochondrial fusion/fission dynamics, increasing dynamin-1-like protein (Drp1)[155] and fission 1 protein (FIS1) activity.[156] Aβ-Drp1 interaction also mediated mitochondrial fragmentation and exacerbate mtDNA mutations.[157] The third most important role of the mitochondria after ATP production and intracellular Ca^{2+} buffering is the handling of reactive oxygen species (ROS). As the electron transport chain is coupled with constant production of ROS in the form of superoxide anion/hydrogen peroxide, the mitochondria are protected by specialized antioxidant systems.[158] Evidence for ROS damage consequences such as lipid peroxidation, protein oxidation, and DNA/RNA oxidation observed in the hippocampus of AD patients[159–161] further support the role of mitochondria in AD. In recent years the mitochondria have been designated as an AD therapeutic target, and providing alternative energy sources and enhancing mitochondrial proteostasis[162, 163] have been studied.

Complex I and IV in Alzheimer's disease

Of the five complexes responsible for mitochondrial oxidative phosphorylation, the forth complex cytochrome C oxidase (Co-IV) is the most studied in AD. Co-IV contains a unique heme-a molecule, which together with (Copper) Cu_B forms the catalytic site that binds and reduces molecular oxygen.[164] Decrease in Co-IV was reported in brains[165–167] and platelets[168, 169] from AD patients. Assembly of Co-IV subunits is also reduced in AD.[170, 171] A recent metaanalysis showed strong evidence for deficit in both Co-IV and Co-I in AD.[172] These results may be explained in part by previous works outlining mutual relationship between the two complexes and their involvement with complex assembly and stability.[173, 174] Co-I does not contain heme molecule; it uses flavin mononucleotide (FMN) and iron-sulfur clusters, as a reduction center instead.[175] Nevertheless, there are evidence for Co-I dysfunction in AD. Co-I dysfunction was observed in posterior cingulate astrocytes of AD patients,[176] and its nuclear DNA-encoded subunits NADH dehydrogenase [ubiquinone] flavoprotein 2 (NDUFV2) and NADH dehydrogenase (ubiquinone) Fe-S protein 1 (NDUFS1) were reduced in the temporal, occipital, and parietal cortices of AD subjects.[177] The mitochondrial DNA-encoded subunit NADH-ubiquinone oxidoreductase chain 6 (ND6) was reduced by 50% in AD brains compared with healthy subjects.[178] Shortage of subunit will potentially inhibit complex assembly. Studies in animal models of AD support alteration in Co-I. One study, for example, compared Co-I levels in three strains of AD mice single-transgenic pR5, double-transgenic APP/presenilin (PS2), and a crossbreeding (^{triple}AD). At 8 months, Co-I activity was only decreased in pR5 mice. At 12 months, all three transgenic mouse models exhibited a significant decrease of Co-I activity when compared with control. Interestingly, at the age of 12 months, content of Co-I was increased in ^{triple}AD mice, suggesting a compensatory upregulation in response to functional deficits. The increase persisted in ^{triple}AD mice cortical mitochondria, a model that exhibits the strongest AD pathology, both plaques and tangles.[179] The nature of Co-I involvement in AD is still unclear. One study suggests that mild inhibition of mitochondrial Co-I with the small-molecule palmarumycin CP2 reduces levels of both Aβ and pTau and averts the development of cognitive and behavior phenotype in three mouse models of familial AD. It suggests that this mild inhibition of Co-I involves the restoration of AMP-dependent tau protein kinase (AMPK) activity.[180] AMPK is a major player in metabolic function,[181] and its abnormal phosphorylation and aggregation was found to be associated with AD.[182–184] Indeed, emerging evidence demonstrates that Co-I and/or AMPK activation are a potential target for improving perturbed brain energy metabolism in AD.[185]

Heme metabolism in Alzheimer's disease

Altered heme homeostasis and heme deposits in Aβ plaques are a frequently observed in AD brains.[186] Nevertheless, the significance of heme in AD pathogenesis has been unappreciated in the past due to the lack of detailed understanding of the interaction of heme and Aβ peptides.[187] During the last decade, however, researchers were able to shed some light on the nature and kinetics of heme involvement in the pathogenesis of AD. The expression of HO1, heme catabolic enzyme often used as a marker of oxidative stress,[188] was increased in frontal and temporal cortices and hippocampus[189] of AD patients. HO1 expression was also increased in the hippocampi of the AD APP/PSI mice model.[190] Interestingly, in leukocytes and lymphocytes, HO1 activity was lower in AD-derived tissue versus controls.[191, 192] A possible explanation for this discrepancy is that enhanced action of HO1 will be observed at areas of oxidative stress damage, a well-known phenomenon in AD,[193, 194] and thus will not occur in the periphery as much as in the central nervous system (CNS).[195] The etiology and reason for increased HO1 are still unclear. HO1 is induced by oxidative stress[110]; thus, increase in HO1 may well be an outcome phenomenon. Alternatively, this may be a compensatory event as HO1 is a well-known neuroprotective agent.[196] Indeed, HO1 activity was found to suppress tau expression and CO, a product of heme metabolism by HO1, prevents the activation of AMPK, a downstream target of Aβ.[197] In addition, it was reported that heme binds Aβ probably via three or more histidine resudies.[198, 199] Heme-Aβ complex may serve as defense mechanism to reduce Aβ aggregation[200]; however, it also causes heme deficiency,[24, 201] enhanced peroxidase activity,[202] and the formation of partially reduced oxygen species. The latter are biomolecules that bind heme causing partial oxidation or reduction. Examples of such molecules include serotonin (5-HT),[203] apomyoglobin,[204] aponeuroglobin,[205] nitric oxide (NO), and Cu.[204] Heme deficiency also contributes to oxidative stress by triggering the release of oxidants such as H_2O_2 from mitochondria due to the loss of Co-IV.[202] In all, despite the accumulated evidence for the involvement of heme and HO1 in AD, the mechanism by which it contributes to either pathology or its prevention is still unclear.

Parkinson's disease and the mitochondria

Parkinson's disease is the second most common neurodegenerative disease after AD.[206] The physiopathology of PD is associated with the selective loss of dopaminergic neurons in the substantia nigra pars compacta and the corpus striatum of the nigrostriatal DA pathway in the brain.[207] Symptoms include bradykinesia, resting tremor, muscle rigidity, and postural instability as well as sleep disturbances, depression, and cognitive deficits.[208] The exact etiology of PD remains unknown; however, the presence of Lewy bodies[209] and excess of ROS[210] were implicated. Mitochondrial dysfunction is increasingly appreciated as key determinants of dopaminergic neuronal susceptibility in PD disease.[211, 212] The initial connection between PD and the mitochondria was reported in the 1980s, when drug abusers developed parkinsonian symptoms. It was soon discovered that 1-methyl-4-phenyl-1,2,3,6-tetrahydropyridine (MPTP), a neurotoxin derivative, can induce parkinsonian symptoms via the inhibition of Co-I.[213–216] In the following years, it was discovered that a number of proteins that cause familial PD, such as PTEN-induced kinase 1 (PINK1), protein/nucleic acid deglycase (DJ-1), synuclein, and parkin, are either mitochondrial or associated with mitochondrial function.[217] Furthermore, mutations in parkin or PINK1 are the most common cause of recessive familial parkinsonism.[218] Both proteins act to selectively isolate and eliminate unhealthy mitochondria by mitophagy, implying that failure of mitophagy may play a role in the pathogenesis of PD.[219] Trafficking is an important feature of mitochondrial dynamics. During trafficking, mitochondria are more inclined to fuse as opposed to when they are stationary.[220] PINK1 was found to form a mitochondrial multiprotein complex with atypical GTPase Miro and Milton, both involved in mitochondrial trafficking.[221] Animal model studies support this claim as knockdown of *PINK1* in *drosophila* axons resulted in enhanced mitochondrial anterograde transport[222] and overexpression of *PINK1* and *Parkin* in neurons from HeLa, SH-SY5Y, and mouse cells resulted in the inhibition of mitochondrial trafficking.[223, 224]

Complex I and Parkinson's disease

A number of studies points toward Co-I as an important player in PD pathogenesis.

Significant reduction in Co-I level was found in the brains of PD patients, specifically in the substantia nigra[225–227] and frontal cortex.[228]

Low Co-I activity was also found in platelet mitochondria of PD patients.[229] Moreover, using cytoplasmic hybrid models in which mitochondria from PD patients were transferred into mitochondria-less cells, Co-I activity was found to be lower than in normal cells.[230] Substances that inhibit Co-I provide additional evidence for its involvement in the pathophysiology of PD. MPTP and rotenone, two substances known to induce parkinsonism in human, inhibits mitochondrial Co-I, of the dopaminergic neurons in particular.[231, 232] Though not fully clear, the inhibition of Co-I initiates a process that results in superoxide formation.[233] In addition to the rotenone/MPTP models of PD, the prototypical model of 6-hydroxydopamine (6-OHDA) also supports the Co-I theory of PD[234] as 20% decrease in substantial nigra Co-I activity was observed due to 6-OHDA exposure.[235] Lastly the PINK1- Parkin double-deficient Medaka fish model of PD should be considered.[236] Though very different from their human counterparts, this model is phenotypically similar to human PD as they present late-onset locomotor dysfunction and deterioration of dopaminergic neurons. In this model a decrease in mitochondrial membrane potential and Co-I activity were observed.[236]

As a vital player in the pathogenesis of PD, Co-I is used as a therapeutic target. Replacing Co-I with a Ndi, a yeast NADH dehydrogenase that is insensitive to ROS was found to restore mitochondrial respiratory chain function when tested in the brains of MPTP and rotenone treated animal models.[237] Whether Co-I inhibition is a cause or just downstream consequence of the disease is unknown. As subunits of Co-I displayed oxidative damage, it was suggested that ROS precedes complex-I disassembly and decreased function.[238]

Heme metabolism in Parkinson's disease

Heme levels were never assessed in PD; however, levels of other heme-related metabolites offer some insight. Iron, for example, is a heme component. Each molecule of heme contains one iron atom. Ferrochelatase, the mitochondrial enzyme that is responsible for the last step in heme biosynthesis, adds iron to protoporphyrin IX to create one whole heme molecule. Abnormal iron accumulation in the basal ganglia of PD brain was noted as early as 1924.[239] Since then, multiple imaging[240–242] and histological[243, 244] studies found elevated iron levels in PD. While the majority of studies found the substantia nigra as the main site for iron accumulation in PD, other brain areas such as the globus pallidus had decreased iron levels.[226] It is unclear whether iron accumulation is a secondary effect or an etiological factor in PD. Supporting the former, studies show iron accumulated mostly in the advanced stages of the disease.[245, 246] Supporting the latter, intranigral iron injection induced lesion of dopamine neurons resulting in parkinsonism in rats.[247] In addition, dopaminergic neurons taken from the substantia nigra of PD patients were found to contain neuromelanin-iron complex, which were absent from control brains.[248] Dopamine, a neurotransmitter known to decrease in PD, can penetrate the coordination sphere of even exceptionally stable iron chelates[249] causing iron accumulation. All the previously mentioned can potentially cause a decrease in bioavailability of the atom to other molecules, such as heme. Upregulation of HO1 was observed in astrocytes of the substantia nigra and neuronal Lewy bodies,[113] as well as in blood serum of patients with PD as compared with age-matched controls[250] and patients with AD.[251] HO1 may also influence iron homeostasis. Increased deposits of glial iron in hippocampus and other subcortical region were found in glial fibrillary acidic protein/heme oxygenase 1 (GFAP.HMOX) mice who are overexpressing HO1.[252] HO1 also downregulates micro RNAs 153 and 223, which in turn upregulates alpha-synuclein,[253] a major ingredient of Lewi bodies.[254] The major role of iron and mitochondria in PD calls for a thorough study of heme metabolism in this disorder.

Schizophrenia and The mitochondria

SZ is a devastating mental disorder with a lifetime prevalence of approximately 1%, which commonly follows a chronic course with an onset at late adolescence. The symptoms encompass abnormalities in multiple modalities of higher brain function, including perception (hallucinations), ideation, reality testing, thought content (delusions), thought processes (loose associations), affect (flatness or inappropriate affect), behavior (catatonia or disorganization), attention, concentration, motivation (avolition or impaired intention and planning), and judgment. These psychological and behavioral characteristics are associated with a variety of impairments in occupational and social functioning. No single symptom is pathognomonic of schizophrenia.[255, 256]

Current evidence point at inadequate development of the brain instigated by interwoven genetic and environmental factors,[257] as leading to abnormalities in synaptic plasticity and neural connectivity in SZ. Multiple neuronal systems show abnormal transmission specifically the dopaminergic, glutamatergic, and GABAergic systems.[258] We were among the first to suggest mitochondrial dysfunction as a pathological factor in SZ. Core functions of mitochondria involve energy supply and Ca^{2+} buffering, both of which are crucial for neuronal function, explicitly to meet their high energy expenditure and to maintain Ca^{2+} homeostasis. Multifaceted structural, biochemical, molecular, and genetic abnormalities in the mitochondria of brain and peripheral cells are persistently shown in SZ patients.[259–262] Notably, we were able to reverse mitochondrial deficits in SZ and improve neuronal differentiation by transplanting isolated active normal mitochondria (IAN-MIT). Transplanting IAN-MIT into SZ-derived induced pluripotent stem cells (SZ-iPSCs) improved their mitochondrial function and differentiation into glutamatergic neurons.[263, 264] In an animal model of SZ, intraprefrontal cortex transplantation of IAN-MIT, in adolescent rats, restored mitochondrial impairments in neurons and attentional deficit, in adulthood.[264] In control rats, however, in vivo mitochondrial transplantation worsened mitochondrial function and impaired attentional behavior. These findings are in favor of a direct link between mitochondria and SZ related cellular and behavioral impairments yet emphasize the risks in using mitochondrial transplantation in the clinics.

Complex I in schizophrenia

Scrutiny of data on various mitochondrial impairments in SZ has uncovered deficits in Co-I as a major cause for mitochondrial dysfunction in the disorder.[262] Investigation of the mitochondrial electron transport chain in SZ revealed alterations in the activities of several complexes but primarily that of Co-I in postmortem brain specimens and peripheral blood cells and a pathological interaction between Co-I and dopamine, a major factor in the disease.[265] Genetic variations observed in Co-I suggest this complex as a risk factor for SZ.[266, 267] Accumulating biochemical and molecular studies further substantiate the malfunction of Co-I in the disease demonstrating disease state-dependent alteration in its activity and impaired expression of several subunits of the complex and their assembly into the holocomplex.[268–271] Co-I abnormalities were associated with impairments in cellular respiration, mitochondrial membrane potential ($\Delta\psi m$), and network dynamics in SZ-derived cells, such as Epstein-Barr virus (EBV) transformed lymphocytes (LCL), hair-follicle keratinocytes, SZ-iPSC and their differentiated dopaminergic and glutamatergic neurons.[272, 273]

Heme metabolism in schizophrenia

After the failure in the initial interest in HPL as a biomarker for SZ (described earlier), other heme metabolites were investigated in SZ, though very scarcely. Iron deficiency, for example, was found in correlation with psychotic symptoms,[274] and offspring of iron deficient mother were found to have increased rates of SZ.[275, 276] Whether iron is indicatory of heme levels is still an open question. Therefore, we sought to assess heme levels and metabolism in SZ. Intercellular heme levels in SZ patient-derived EBV LCL were 20% lower than control. Reduced heme levels were associated with an increase in HO1, while no change were observed in ΔALAS1 or mitochondrial heme levels, suggesting deceased cytosolic heme degradation with no effect on its synthesis in SZ cells. The findings of elevated HO1 levels are supported by proteomic study of SZ brain tissue.[21] In addition, the GFAP.HMOX mice model in which astrocytic HO1 is overexpressed exhibits SZ-like behaviors, mitochondrial damage, subcortical oxidative stress, augmented dopamine and serotonin levels, and reduced dopamine receptor 1 (D_1) binding in nucleus accumbens, suggesting that HO1 may be a pathological feature of SZ.[25] The decrease in heme was associated with an increase in its downstream target phosphorylated eIf2α (the inactive form), suggesting that this mild decrease in heme can affect cell protein translation. The same changes in heme metabolism were also observed in vivo in the polyinosinic:polycytidylic acid (Poly I:C)-induced maternal immune activation model of SZ, albeit in a brain region-dependent manner.[277] These results correspond with prior reports of reduced protein synthesis in SZ patient-derived olfactory cells[278] and in vivo in PERK-deficient mice. eIF2α phosphorylation was diminished in these mice and associated with SZ-like behaviors. Interestingly, treatment with glycine transporter inhibitor normalized eIF2α phosphorylation and behavioral flexibility in PERK-deficient mice.[279]

As mentioned earlier, Co-I has been repeatedly reported in SZ. To study whether Co-I reduced levels are associated with impaired heme metabolism, we mimic SZ-like decrease in Co-I activity by applying the Co-I inhibitor rotenone to healthy subject-derived LCL. We observed an overall decrease in heme levels associated with an increase in HO1 and in phosphorylated eIf2α.[277] These results suggest that Co-I is involved in heme metabolism, which is in line with two previous studies indicating correlation between decrease in Co-I activity and heme levels.[280, 281]

Conclusions

Co-I, the first and largest ETC mitochondrial complex, is the key entry point for electrons to the respiratory chain. As such it is considered to be rate limiting in overall cellular respiration.[282] Deficit in Co-I is a common feature shared between AD, PD, and SZ. Co-I is among the main producers of superoxide and H_2O_2 in the mitochondria,[283, 284] and a decrease in its activity is known to cause oxidative stress.[285, 286] However, the reverse is also true, and oxidative stress can cause Co-I damage. Despite its rapid dismutation by peroxidase, superoxide can still attack the iron-sulfur clusters of Co-I before it is neutralized causing both Co-I deficiency and the release of iron that reacts with other molecules to create toxic radicals.[287] Therefore, it is unclear whether Co-I and mitochondrial dysfunction precedes ROS damage or vice versa. However, it is suggested that the two processes are intertwined and lead a vicious cycle that leads to neuronal dysfunction.[196] Regardless of etiology, mitochondrial dysfunction is a well-documented feature of AD,[288] PD,[289] and SZ.[21] Studies suggest that heme metabolism, which heavily relies on intact mitochondria, is altered in AD,[187, 290] PD,[251] and SZ.[277] Defective heme metabolism can have hazardous consequences for cells. Firstly, ALAS1, which is a heme synthesis rate-limiting enzyme, is transported in and out of the mitochondria, and its accumulation outside of the mitochondria was reported to effect neurons negatively.[108] Secondly, iron is a heme breakdown product that may lead to free radical production. Indeed, elevated iron levels were reported in AD[252, 291] and PD,[240] while a deficiency of iron was associated with SZ.[274, 275] Thirdly, HO1, a heme catabolism rate-limiting enzyme, has neuroprotective role,[83, 93] and its expression is induced by ROS.[70] This may be beneficial to the cell in the short term, but long-term HO1 activation lowers the level of its substrate, heme, causing unfavorable heme deficiency that can lead to further chemical imbalance including increased levels of CO and bilirubin, all having various neuroregulatory roles.[28] Indeed, studies report elevated HO1 levels in AD,[111, 292] PD,[251] and SZ.[25, 277] The prevalence and functional importance of these findings should be further investigated in light of the potential of heme and mitochondrial Co-I to become additional therapeutic targets that will improve the efficiency of the current treatments. Many studies have suggested improving mitochondrial function and reducing ROS production and apoptosis in congenital mitochondrial diseases and chronic disorders such as type 1 and 2 diabetes, cardiovascular disorders, neuropsychiatric disorders, and cancer. These treatment strategies range from diet and nutrients through exercise training to pharmacological treatments including antioxidants, peroxisome proliferator-activated receptor gamma coactivator 1-alpha (PGC-1α)-activating drugs, and mitochondrial transplantation.[293–296] Free heme, which may occur in various pathophysiological conditions, at high levels can exert toxic effects via prooxidant, proinflammatory, and cytotoxic effects. However, one of its metabolites, CO has been shown to regulate inflammation, cell survival, and growth. Indeed a large number of studies have shown that the HO1/CO system exhibits potent antioxidative, antiinflammatory, antiapoptotic, and cytoprotective activities under different pathological conditions primarily ischemia-reperfusion injury. The main therapeutic attempts are via the upregulation of HO1 by pharmacological drugs such as thioredoxin and herbal compounds such as senkyunolide I and resveratrol and by genetic overexpression of HO1.[297] Bearing in mind the link between heme metabolism and mitochondria, it is not surprising that most pharmacological compounds that stimulate HO1/CO system activate nuclear factor erythroid-derived 2-related factor 2 (Nrf2) and PPARα pathways, both important regulators of mitochondrial biogenesis.[298, 299] Finally, based on the evidence described hitherto, we hypothesize that deficit in heme metabolism and mitochondria are interwoven and play an important role in the pathophysiology of neuropsychiatric diseases.

Acknowledgment

Funding was provided by Israel Science Foundation-ISF grant 1517/1.

References

1. Hou S, Reynolds MF, Horrigan FT, Heinemann SH, Hoshi T. Reversible binding of heme to proteins in cellular signal transduction. *Acc Chem Res*. 2006;39(12):918–924. https://doi.org/10.1021/ar040020w.
2. Bibb LA, King ND, Kunkle CA, Schmitt MP. Analysis of a heme-dependent signal transduction system in Corynebactenum diphtheriae: deletion of the chrAS genes results in heme sensitivity and diminished heme-dependent activation of the hmuO promoter. *Infect Immun*. 2005; https://doi.org/10.1128/IAI.73.11.7406-7412.2005.
3. Yang F, Wang E-D. Heme regulates protein homeostasis at transcription, protein translation, and degradation levels. *Front Biol (Beijing)*. 2011;5(6):516–523. https://doi.org/10.1007/s11515-010-7700-5.
4. Raffel C, Stein S, Kaempfer R. Role for heme in mammalian protein synthesis: activation of an initiation factor. *Proc Natl Acad Sci U S A*. 1974;71(10):4020–4024. http://www.ncbi.nlm.nih.gov/pubmed/4530282.
5. Yamamoto M, Kure S, Engel JD, Hiraga K. Structure, turnover, and heme-mediated suppression of the level of mRNA encoding rat liver delta-aminolevulinate synthase. *J Biol Chem*. 1988;263(31):15973–15979. http://www.ncbi.nlm.nih.gov/pubmed/3182776.
6. Hamilton JW, Bement WJ, Sinclair PR, Sinclair JF, Alcedo JA, Wetterhahn KE. Heme regulates hepatic 5-aminolevulinate synthase mRNA expression by decreasing mRNA half-life and not by altering its rate of transcription. *Arch Biochem Biophys*. 1991;289(2):387–392. http://www.ncbi.nlm.nih.gov/pubmed/1898078.
7. Zheng J, Shan Y, Lambrecht RW, Donohue SE, Bonkovsky HL. Differential regulation of human ALAS1 mRNA and protein levels by heme and cobalt protoporphyrin. *Mol Cell Biochem*. 2008;319(1–2):153–161. https://doi.org/10.1007/s11010-008-9888-0.
8. Chiabrando D, Fiorito V, Petrillo S, Tolosano E. Unraveling the role of Heme in neurodegeneration. *Front Neurosci*. 2018;12:712. https://doi.org/10.3389/fnins.2018.00712.
9. Checkoway H, Lundin JI, Kelada SN. Neurodegenerative diseases. *IARC Sci Publ*. 2011;(163):407–419. http://www.ncbi.nlm.nih.gov/pubmed/22997874.
10. Lieberman JA. Is schizophrenia a neurodegenerative disorder? A clinical and neurobiological perspective. *Biol Psychiatry*. 1999;46(6):729–739. https://doi.org/10.1016/S0006-3223(99)00147-X.
11. Christopoulos I, Massouri G, Fotopoulos V. A neurodegenerative perspective on schizophrenia. *Ann General Psychiatry*. 2006;2:1–2. https://doi.org/10.1186/1744-859X-5-S1-S261.
12. Kann O. The interneuron energy hypothesis: implications for brain disease. *Neurobiol Dis*. 2016;90:75–85. https://doi.org/10.1016/j.nbd.2015.08.005.
13. Nelson AB, Gittis AH, Palop JJ, Thwin MT, Kreitzer AC. Distinct roles of GABAergic interneurons in the regulation of striatal output pathways. *J Neurosci*. 2010; https://doi.org/10.1523/jneurosci.4870-09.2010.
14. Galvan A, Wichmann T. GABAergic circuits in the basal ganglia and movement disorders. *Prog Brain Res*. 2007;160:287–312. https://doi.org/10.1016/S0079-6123(06)60017-4.
15. Horesh Y, Katsel P, Haroutunian V, Domany E. Gene expression signature is shared by patients with Alzheimer's disease and schizophrenia at the superior temporal gyrus. *Eur J Neurol*. 2011; https://doi.org/10.1111/j.1468-1331.2010.03166.x.
16. Moreira PI, Siedlak SL, Wang X, et al. Erratum: increased autophagic degradation of mitochondria in Alzheimer disease (autophagy). *Autophagy*. 2007; https://doi.org/10.4161/auto.4872.

17. Yuan Y, Wang H, Wei Z, Li W. Impaired autophagy in hilar mossy cells of the dentate gyrus and its implication in schizophrenia. *J Genet Genomics*. 2015; https://doi.org/10.1016/j.jgg.2014.12.001.
18. Bose A, Beal MF. Mitochondrial dysfunction in Parkinson's disease. *J Neurochem*. 2016; https://doi.org/10.1111/jnc.13731.
19. Lin MT, Beal MF. Mitochondrial dysfunction and oxidative stress in neurodegenerative diseases. *Nature*. 2006; https://doi.org/10.1038/nature05292.
20. Reddy PH, Beal MF. Are mitochondria critical in the pathogenesis of Alzheimer's disease? *Brain Res Rev*. 2005; https://doi.org/10.1016/j.brainresrev.2005.03.004.
21. Prabakaran S, Swatton JE, Ryan MM, et al. Mitochondrial dysfunction in schizophrenia: evidence for compromised brain metabolism and oxidative stress. *Mol Psychiatry*. 2004;9(7):684–697. https://doi.org/10.1038/sj.mp.4001511.
22. Johri A, Beal MF. Mitochondrial dysfunction in neurodegenerative diseases. *J Pharmacol Exp Ther*. 2012;342(3):619–630. https://doi.org/10.1124/jpet.112.192138.
23. Ben-Shachar D. Mitochondrial dysfunction in schizophrenia: a possible linkage to dopamine. *J Neurochem*. 2002;83:1241–1251. https://doi.org/10.1046/j.1471-4159.2002.01263.x.
24. Atamna H, Frey WH. A role for heme in Alzheimer's disease: heme binds amyloid beta and has altered metabolism. *Proc Natl Acad Sci U S A*. 2004;101(30):11153–11158. https://doi.org/10.1073/pnas.0404349101.
25. Song W, Zukor H, Lin S-H, et al. Schizophrenia-like features in transgenic mice overexpressing human HO-1 in the astrocytic compartment. *J Neurosci*. 2012;32(32):10841–10853. https://doi.org/10.1523/JNEUROSCI.6469-11.2012.
26. Gozzelino R. Heme inflammation and Parkinson's disease. *Eur Hematol Cent*. 2015; https://learningcenter.ehaweb.org/eha/2015/20th/102997/raffaella.gozzelino.heme.inflammation.and.parkinson.s.disease.html.
27. Martin Warren AS. *Tetrapyrroles: Birth, Life and Death*. vol. illustrate, Springer Science & Business Media; 2009.
28. Smith AG, Raven EL, Chernova T. The regulatory role of heme in neurons. *Metallomics*. 2011;3(10):955–962. https://doi.org/10.1039/c1mt00085c.
29. Tudor C, Lerner-Marmarosh N, Engelborghs Y, Gibbs PEM, Maines MD. Biliverdin reductase is a transporter of haem into the nucleus and is essential for regulation of HO-1 gene expression by haematin. *Biochem J*. 2008;413(3):405–416. https://doi.org/10.1042/BJ20080018.
30. Ogawa K, Sun J, Taketani S, et al. Heme mediates derepression of Maf recognition element through direct binding to transcription repressor Bach1. *EMBO J*. 2001;20(11):2835–2843. https://doi.org/10.1093/emboj/20.11.2835.
31. Sun J, Hoshino H, Takaku K, et al. Hemoprotein Bach1 regulates enhancer availability of heme oxygenase-1 gene. *EMBO J*. 2002;21(19):5216–5224. https://doi.org/10.1093/emboj/cdf516.
32. Quick-Cleveland J, Jacob JP, Weitz SH, Shoffner G, Senturia R, Guo F. The DGCR8 RNA-binding heme domain recognizes primary microRNAs by clamping the hairpin. *Cell Rep*. 2014;7(6):1994–2005. https://doi.org/10.1016/j.celrep.2014.05.013.
33. Kozak M. Initiation of translation in prokaryotes and eukaryotes. *Gene*. 1999;234(2):187–208. http://www.ncbi.nlm.nih.gov/cgi-bin/Entrez/referer?http://www.elsevier.com:80/cgi-bin/cas/tree/store/gene/cas%7B_%7Dsub/browse/browse.cgi%7B%25%7D3F-year=1999%7B&%7Dvolume=234%7B&%7Dissue=2%7B&%7Daid=11932.
34. Koumenis C, Naczki C, Koritzinsky M, et al. Regulation of protein synthesis by hypoxia via activation of the endoplasmic reticulum kinase PERK and phosphorylation of the translation initiation factor eIF2alpha. *Mol Cell Biol*. 2002;22(21):7405–7416. https://doi.org/10.1128/MCB.22.21.7405.
35. Bauer BN, Rafie-Kolpin M, Lu L, Han A, Chen JJ. Multiple autophosphorylation is essential for the formation of the active and stable homodimer of heme-regulated eIF2α kinase. *Biochemistry*. 2001;40(38):11543–11551. https://doi.org/10.1021/bi010983s.

36. Chefalo PJ, Oh J, Rafie-Kolpin M, Kan B, Chen JJ. Heme-regulated eIF-2α kinase purifies as a hemoprotein. *Eur J Biochem*. 1998;258(2):820–830. https://doi.org/10.1046/j.1432-1327.1998.2580820.x.
37. Miksanova M, Igarashi J, Minami M, et al. Characterization of heme-regulated eIF2α kinase: roles of the N-terminal domain in the oligomeric state, heme binding, catalysis, and inhibition. *Biochemistry*. 2006;45(32):9894–9905. https://doi.org/10.1021/bi060556k.
38. Chen JJ, London IM. Regulation of protein synthesis by heme-regulated eIF-2 alpha kinase. *Trends Biochem Sci*. 1995;20(3):105–108. https://doi.org/10.1016/S0968-0004(00)88975-6 [pii: S0968000400889756].
39. Sonenberg N, John WB, Hershey MM, eds. *Translational Control of Gene Expression*. 2nd ed.Cold Spring Harbor Laboratory Press; 2000.
40. Wek RC, Jiang H-Y, Anthony TG. Coping with stress: eIF2 kinases and translational control. *Biochem Soc Trans*. 2006;34(Pt 1):7–11. https://doi.org/10.1042/BST20060007.
41. Jordan PM, ed. *Biosynthesis of Tetrapyrolles*. New Comprehensive Biochemistryvol. 19. 1991.
42. Ferreira GC. 5-Aminolevulinate synthase and mammalian heme biosynthesis. In: *Iron Metabolism*. Weinheim: Wiley-VCH Verlag GmbH; 1999:15–34. https://doi.org/10.1002/9783527613700.ch02.
43. Lathrop JT, Timko MP. Regulation by heme of mitochondrial protein transport through a conserved amino acid motif. *Science*. 1993;259(5094):522–525. https://doi.org/10.1126/science.8424176.
44. Bayeva M, Khechaduri A, Wu R, et al. ATP-binding cassette B10 regulates early steps of heme synthesis. *Circ Res*. 2013;113(3):279–287. https://doi.org/10.1161/CIRCRESAHA.113.301552.
45. Jaffe EK. Porphobilinogen synthase, the first source of Heme's asymmetry. *J Bioenerg Biomembr*. 1995;27(2):169–179. https://doi.org/10.1007/BF02110032.
46. Dent AJ, Beyersmann D, Block C, Hasnain SS. Two different zinc sites in bovine 5-aminolevulinate dehydratase distinguished by extended X-ray absorption fine structure. *Biochemistry*. 1990;29(34):7822–7828. https://doi.org/10.1021/bi00486a007.
47. Bogorad L. Intermediates in biosynthesis of porphyrins from porphobilinogen. *Science*. 1955;121:878–879.
48. Shoolingin-Jordan PM. The biosynthesis of coproporphyrinogen III. In: Kadish KM, Smith KMGR, eds. *The Porphyrin Handbook*. vol. 12. New York: Elsevier; 2003:33–74.
49. Chavan H, Khan MMT, Tegos G, Krishnamurthy P. Efficient purification and reconstitution of ATP binding cassette transporter B6 (ABCB6) for functional and structural studies. *J Biol Chem*. 2013;288(31):22658–22669. https://doi.org/10.1074/jbc.M113.485284.
50. Krishnamurthy PC, Du G, Fukuda Y, et al. Identification of a mammalian mitochondrial porphyrin transporter. *Nature*. 2006;443(7111):586–589. https://doi.org/10.1038/nature05125.
51. Elder GH, Evans JO, Thomas N, et al. The primary enzyme defect in hereditary coproporphyria. *Lancet*. 1976;308(7997):1217–1219. https://doi.org/10.1016/S0140-6736(76)91143-0.
52. Grandchamp B, Phung N, Nordmann Y. The mitochondrial localization of coproporphyrinogen III oxidase. *Biochem J*. 1978;176(1):97–102. http://www.pubmedcentral.nih.gov/articlerender.fcgi?artid=1186208%7B&%7Dtool=pmcentrez%7B&%7Drendertype=abstract.
53. Che FS, Watanabe N, Iwano M, et al. Molecular characterization and subcellular localization of protoporphyrinogen oxidase in spinach chloroplasts. *Plant Physiol*. 2000;124(1):59–70. https://doi.org/10.1104/pp.124.1.59.
54. Berg JM, Tymoczko JL, Stryer L. Oxygen transporters and oxidative phosphorylation. In: *Biochemistry*. 5th ed.New York: W. H. Freeman & Co; 2002.
55. Aich A, Freundlich M, Vekilov PG. The free heme concentration in healthy human erythrocytes. *Blood Cells Mol Dis*. 2015;55(4):402–409. https://doi.org/10.1016/j.bcmd.2015.09.003.

56. Fitch CD, Chevli R, Kanjananggulpan P, Dutta P, Chevli K, Chou AC. Intracellular ferriprotoporphyrin IX is a lytic agent. *Blood*. 1983;62(6):1165–1168. http://www.ncbi.nlm.nih.gov/pubmed/6640106.
57. Goldstein L, Teng ZP, Zeserson E, Patel M, Regan RF. Hemin induces an iron-dependent, oxidative injury to human neuron-like cells. *J Neurosci Res*. 2003;73(1):113–121. https://doi.org/10.1002/jnr.10633.
58. Becker K, Tilley L, Vennerstrom JL, Roberts D, Rogerson S, Ginsburg H. Oxidative stress in malaria parasite-infected erythrocytes: host-parasite interactions. *Int J Parasitol*. 2004;34(2):163–189. https://doi.org/10.1016/j.ijpara.2003.09.011.
59. Klouche K, Morena M, Canaud B, Descomps B, Beraud JJ, Cristol JP. Mechanism of in vitro heme-induced LDL oxidation: effects of antioxidants. *Eur J Clin Investig*. 2004;34(9):619–625. https://doi.org/10.1111/j.1365-2362.2004.01395.x.
60. Maines MD. The heme oxygenase system: a regulator of second messenger gases. *Annu Rev Pharmacol Toxicol*. 1997;37(1):517–554. https://doi.org/10.1146/annurev.pharmtox.37.1.517.
61. Rodgers PA, Stevenson DK. Developmental biology of heme oxygenase. *Clin Perinatol*. 1990;17(2):275–291.
62. Bissell DM. Heme catabolism and bilirubin formation. In: Ostrow JD, ed. *Bile Pigments and Jaundice. Molecular, Metabolic and Medical Aspects*. New York: Marcel Dekker Inc.; 1986:133–156.
63. Nath KA. The functional significance of induction of heme oxygenase by oxidant stress. *J Lab Clin Med*. 1994;123(4):461–463. http://www.ncbi.nlm.nih.gov/pubmed/8144993.
64. Nakagami T, Toyomura K, Kinoshita T, Morisawa S. A beneficial role of bile pigments as an endogenous tissue protector: anti-complement effects of biliverdin and conjugated bilirubin. *Biochim Biophys Acta*. 1993;1158(2):189–193. http://www.ncbi.nlm.nih.gov/pubmed/8399320.
65. Valaes TN, Harvey-Wilkes K. Pharmacologic approaches to the prevention and treatment of neonatal hyperbilirubinemia. *Clin Perinatol*. 1990;17(2):245–273.
66. Leffler CW, Nasjletti A, Yu C, Johnson RA, Fedinec AL, Walker N. Carbon monoxide and cerebral microvascular tone in newborn pigs. *Am J Phys*. 1999;276(5 Pt 2):H1641–H1646.
67. Rodgers PA, Vreman HJ, Dennery PA, Stevenson DK. Sources of carbon monoxide (CO) in biological systems and applications of CO detection technologies. *Semin Perinatol*. 1994;18(1):2–10. http://www.ncbi.nlm.nih.gov/pubmed/8209283.
68. Chien W-L, Lee T-R, Hung S-Y, Kang K-H, Lee M-J, Fu W-M. Impairment of oxidative stress-induced heme oxygenase-1 expression by the defect of Parkinson-related gene of PINK1. *J Neurochem*. 2011;117(4):643–653. https://doi.org/10.1111/j.1471-4159.2011.07229.x.
69. Song YJ, Zong ZM, Liu HZ, et al. Heme oxygenase-1 regulates the JNK signaling pathway through the MLK3-MKK7-JNK3 signaling module in brain ischemia injury. *Brain Res*. 2012; https://doi.org/10.1016/j.brainres.2011.10.021.
70. Lin CY, Hsiao WC, Huang CJ, Kao CF, Hsu GSW. Heme oxygenase-1 induction by the ROS-JNK pathway plays a role in aluminum-induced anemia. *J Inorg Biochem*. 2013; https://doi.org/10.1016/j.jinorgbio.2013.07.026.
71. Shibahara S, Han F, Li B, Takeda K. Hypoxia and heme oxygenases: oxygen sensing and regulation of expression. *Antioxid Redox Signal*. 2007;9(12):2209–2225. https://doi.org/10.1089/ars.2007.1784.
72. Wagner KR, Sharp FR, Ardizzone TD, Lu A, Clark JF. Heme and iron metabolism: role in cerebral hemorrhage. *J Cereb Blood Flow Metab*. 2003;23(6):629–652. https://doi.org/10.1097/01.WCB.0000073905.87928.6D.
73. DeGracia DJ, Kumar R, Owen CR, Krause GS, White BC. Molecular pathways of protein synthesis inhibition during brain reperfusion: implications for neuronal survival or death. *J Cereb Blood Flow Metab*. 2002;22(2):127–141. https://doi.org/10.1097/00004647-200202000-00001.

74. Emanuelli T, Pagel FW, Porciúncula LO, Souza DO. Effects of 5-aminolevulinic acid on the glutamatergic neurotransmission. *Neurochem Int*. 2003;42(2):115–121. http://www.ncbi.nlm.nih.gov/pubmed/12421591.
75. Lavandera JV, Fossati M, Azcurra J, Batlle A, Buzaleh AM. Glutamatergic system: another target for the action of porphyrinogenic agents. *Cell Mol Biol (Noisy-le-Grand)*. 2009;55(1):23–28. http://www.ncbi.nlm.nih.gov/pubmed/19267998.
76. Chernova T, Nicotera P, Smith AG. Heme deficiency is associated with senescence and causes suppression of N-methyl-D-aspartate receptor subunits expression in primary cortical neurons. *Mol Pharmacol*. 2006;69(3):697–705. https://doi.org/10.1124/mol.105.016675.uct.
77. Chernova T, Steinert JR, Guerin CJ, Nicotera P, Forsythe ID, Smith AG. Neurite degeneration induced by heme deficiency mediated via inhibition of NMDA receptor-dependent extracellular signal-regulated kinase 1/2 activation. *J Neurosci*. 2007;27(32):8475–8485. https://doi.org/10.1523/JNEUROSCI.0792-07.2007.
78. Ryan KR, Sirenko O, Parham F, et al. Neurite outgrowth in human induced pluripotent stem cell-derived neurons as a high-throughput screen for developmental neurotoxicity or neurotoxicity. *Neurotoxicology*. 2016;53:271–281. https://doi.org/10.1016/j.neuro.2016.02.003.
79. Sengupta A, Hon T, Zhang L. Heme deficiency suppresses the expression of key neuronal genes and causes neuronal cell death. *Brain Res Mol Brain Res*. 2005;137(1–2):23–30. https://doi.org/10.1016/j.molbrainres.2005.02.007.
80. Hung SY, Liou HC, Fu WM. The mechanism of heme oxygenase-1 action involved in the enhancement of neurotrophic factor expression. *Neuropharmacology*. 2010;58(2):321–329. https://doi.org/10.1016/j.neuropharm.2009.11.003.
81. Ollinger R, Kogler P, Troppmair J, et al. Bilirubin inhibits tumor cell growth via activation of ERK. *Cell Cycle*. 2007;6(24):3078–3085. https://doi.org/10.4161/cc.6.24.5022.
82. Mendoza MC, Er EE, Blenis J. The Ras-ERK and PI3K-mTOR pathways: cross-talk and compensation. *Trends Biochem Sci*. 2011;36(6):320–328. https://doi.org/10.1016/j.tibs.2011.03.006.
83. Hung S, Liou H, Kang K, Wu R-M, Wen C, Fu W-M. Overexpression of heme oxygenase-1 protects dopaminergic neurons against 1-methyl-4-phenylpyridinium-induced neurotoxicity. *Mol Pharmacol*. 2008;74(6):1564–1575. https://doi.org/10.1124/mol.108.048611 [pii:mol.108.048611].
84. McKinstry R, Qiao L, Yacoub A, et al. Inhibitors of MEK1/2 interact with UCN-01 to induce apoptosis and reduce colony formation in mammary and prostate carcinoma cells. *Cancer Biol Ther*. 2002;1(3):243–253.
85. Roberts PJ, Der CJ. Targeting the Raf-MEK-ERK mitogen-activated protein kinase cascade for the treatment of cancer. *Oncogene*. 2007;26(22):3291–3310. https://doi.org/10.1038/sj.onc.1210422.
86. Lee H, Choi YK. Regenerative effects of heme oxygenase metabolites on neuroinflammatory diseases. *Int J Mol Sci*. 2018;20(1):https://doi.org/10.3390/ijms20010078.
87. Panahian N, Yoshiura M, Maines MD. Overexpression of heme oxygenase-1 is neuroprotective in a model of permanent middle cerebral artery occlusion in transgenic mice. *J Neurochem*. 1999;72(3):1187–1203. https://doi.org/10.1111/j.1471-4159.1999.721187.x.
88. Schipper HM, Song W. A heme oxygenase-1 transducer model of degenerative and developmental brain disorders. *Int J Mol Sci*. 2015;16(3):5400–5419. https://doi.org/10.3390/ijms16035400.
89. Colín-González AL, Orozco-Ibarra M, Chánez-Cárdenas ME, et al. Heme oxygenase-1 (HO-1) upregulation delays morphological and oxidative damage induced in an excitotoxic/pro-oxidant model in the rat striatum. *Neuroscience*. 2013;231:91–101. https://doi.org/10.1016/j.neuroscience.2012.11.031.

90. Teng YC, Tai YI, Lee YH, Lin AMY. Role of HO-1 in the arsenite-induced neurotoxicity in primary cultured cortical neurons. *Mol Neurobiol*. 2013;48:281–287. https://doi.org/10.1007/s12035-013-8492-9.
91. Ramalho-Santos J, Rodrigues AS. From oocytes and pluripotent stem cells to fully differentiated fates: (also) a mitochondrial odyssey. In: *Mitochondrial DNA, Mitochondria, Disease and Stem Cells*. Totowa, NJ: Humana Press; 2013:69–86. https://doi.org/10.1007/978-1-62703-101-1_4.
92. Ryter SW, Choi AMK. Cytoprotective and anti-inflammatory actions of carbon monoxide in organ injury and sepsis models. *Novartis Found Symp*. 2007;280:165–181. http://www.ncbi.nlm.nih.gov/pubmed/17380794.
93. Mahan VL. Neuroprotective, neurotherapeutic, and neurometabolic effects of carbon monoxide. *Med Gas Res*. 2012;2(1):32. https://doi.org/10.1186/2045-9912-2-32.
94. Varone CL, Giono LE, Ochoa A, Zakin MM, Cánepa ET. Transcriptional regulation of 5-aminolevulinate synthase by phenobarbital and cAMP-dependent protein kinase. *Arch Biochem Biophys*. 1999;372(2):261–270. https://doi.org/10.1006/abbi.1999.1470.
95. Zhu Y, Hon T, Ye W, Zhang L. Heme deficiency interferes with the Ras-mitogen-activated protein kinase signaling pathway and expression of a subset of neuronal genes. *Cell Growth Differ*. 2002;13(9):431–439.
96. Whetsell WO. Porphyrin-heme biosynthesis in organotypic cultures of mouse dorsal root ganglia. Effects of heme and lead on porphyrin synthesis and peripheral myelin. *J Clin Invest*. 1984;74(2):600–607. https://doi.org/10.1172/JCI111457.
97. Nibuya M, Nestler EJ, Duman RS. Chronic antidepressant administration increases the expression of cAMP response element binding protein (CREB) in rat hippocampus. *J Neurosci*. 1996;16(7):2365–2372. http://www.ncbi.nlm.nih.gov/entrez/query.fcgi?cmd=Retrieve&db=PubMed&dopt=Citation&list_uids=8601816.
98. Ellencweig N, Schoenfeld N, Zemishlany Z. Acute intermittent porphyria: psychosis as the only clinical manifestation. *Isr J Psychiatry Relat Sci*. 2006;43(1):52–56. http://www.ncbi.nlm.nih.gov/pubmed/16910386.
99. Anderson KE, Sassa S, Bishop DF, Desnick RJ. Disorders of heme biosynthesis: X-linked sideroblastic anemia and the porphyrias. In: Scriver CR, Beaudt AL, Sly WS, Valle D, Barton C, Kinzler KW, Vogelstein B, eds. *The Metabolic and Molecular Bases of Inherited Disease*. vol. 2. New York: The McGraw-Hill Companies, Inc; 2001:2991–3062.
100. Kappas A, Sassa S, Galbraith RA, Nordmann Y. The porphyrias. In: Scriver CR, Beaudt AL, Sly WS, Valle D, eds. *The Metabolic and Molecular Bases of Inherited Disease*. vol. 2. The McGraw-Hill Companies, Inc; 1995:2103–2169.
101. Gross U, Sassa S, Jacob K, et al. 5-Aminolevulinic acid dehydratase deficiency porphyria: a twenty-year clinical and biochemical follow-up. *Clin Chem*. 1998;44(9):1892–1896.
102. Strand LJ, Meyer UA, Felsher BF, Redeker AG, Marver HS. Decreased red cell uroporphyrinogen I synthetase activity in intermittent acute porphyria. *J Clin Invest*. 1972;51(10):2530–2536. https://doi.org/10.1172/JCI107068.
103. Stein Y, Finkelstein Y, Levy-Nativ O, Bonne O, Aschner M, Richter ED. Exposure and susceptibility: schizophrenia in a young man following prolonged high exposures to organic solvents. *Neurotoxicology*. 2010;31(5):603–607. https://doi.org/10.1016/j.neuro.2010.06.010.
104. Opler MGA, Buka SL, Groeger J, et al. Prenatal exposure to lead, delta-aminolevulinic acid, and schizophrenia: further evidence. *Environ Health Perspect*. 2008;116(11):1586–1590. https://doi.org/10.1289/ehp.10464.
105. Müller WE, Snyder SH. Delta-aminolevulinic acid: influences on synaptic GABA receptor binding may explain CNS symptoms of porphyria. *Ann Neurol*. 1977;2(4):340–342. https://doi.org/10.1002/ana.410020415.
106. Brennan MJ, Cantrill RC. Delta-aminolaevulinic acid is a potent agonist for GABA autoreceptors. *Nature*. 1979;280(5722):514–515. http://www.ncbi.nlm.nih.gov/pubmed/223065.

107. Taljaard JJ, Lamm MC, Truter L, McCarthy BW, Percy VA, Neethling AC. Mechanism of delta-aminolevulinic acid neurotoxicity. *S Afr Med J*. 1981;60(5):180–183. http://www.ncbi.nlm.nih.gov/pubmed/6114570.
108. Felitsyn N, McLeod C, Shroads AL, Stacpoole PW, Notterpek L. The heme precursor delta-aminolevulinate blocks peripheral myelin formation. *J Neurochem*. 2008;106(5):2068–2079. https://doi.org/10.1111/j.1471-4159.2008.05552.x.
109. Russell VA, Lamm MC, Taljaard JJ. Inhibition of Na^+, K^+-ATPase activity by delta-aminolevulinic acid. *Neurochem Res*. 1983;8(11):1407–1415. http://www.ncbi.nlm.nih.gov/pubmed/6140649.
110. Maines MD. The heme oxygenase system and its functions in the brain. *Cell Mol Biol*. 2000;46(3):573–585.
111. Smith MA, Kutty RK, Richey PL, et al. Heme oxygenase-1 is associated with the neurofibrillary pathology of Alzheimer's disease. *Am J Pathol*. 1994;145(1):42–47.
112. Barone E, Di Domenico F, Sultana R, et al. Heme oxygenase-1 posttranslational modifications in the brain of subjects with Alzheimer disease and mild cognitive impairment. *Free Radic Biol Med*. 2012;52(11–12):2292–2301. https://doi.org/10.1016/j.freeradbiomed.2012.03.020.
113. Schipper HM, Liberman A, Stopa EG. Neural heme oxygenase-1 expression in idiopathic Parkinson's disease. *Exp Neurol*. 1998;150(1):60–68. https://doi.org/10.1006/exnr.1997.6752.
114. Kannan M, Steinert JR, Forsythe ID, Smith AG, Chernova T. Mevastatin accelerates loss of synaptic proteins and neurite degeneration in aging cortical neurons in a heme-independent manner. *Neurobiol Aging*. 2010;31(9):1543–1553. https://doi.org/10.1016/j.neurobiolaging.2008.09.004.
115. Mikirova N. Cross-sectional analysis of pyrroles in psychiatric disorders: association with nutritional and immunological markers. *J Orthomol Med*. 2015;30(1):25–32.
116. Irvine DG. Apparently non-indolic Ehrlich-positive substances related to mental illnesses. *J Neuropsychiatr*. 1961;2:292–305. http://www.ncbi.nlm.nih.gov/pubmed/13717893.
117. Hoffer A, Osmond H. Malvaria: a new psychiatric disease. *Acta Psychiatr Scand*. 1963;39:335–366. http://www.ncbi.nlm.nih.gov/pubmed/14078702.
118. Hoffer A. The presence of malvaria in some mentally retarded children. *Am J Ment Defic*. 1963;67:730–732. http://www.ncbi.nlm.nih.gov/pubmed/13963913.
119. O'Reilly PO, Ernest M, Hughes G. The incidence of malvaria. *Br J Psychiatry*. 1965;111:741–744. http://www.ncbi.nlm.nih.gov/pubmed/14337423.
120. Gorchein A, Mitchell LD, Rogers AT. Pharmacological properties of kryptopyrrole and its oxidation products on isolated sciatic nerve of rat and on guinea-pig ileum. *Br J Pharmacol*. 1979;65(1):23–27. https://doi.org/10.1111/j.1476-5381.1979.tb17329.x.
121. Wetterberg L. Pharmacological and toxic effects of kryptopyrrole in mice. *Ups J Med Sci*. 1973;78(1):78–80. http://www.ncbi.nlm.nih.gov/pubmed/4702335.
122. Irvine DG. Autotransfer chromatography in the characterization of pyrroles. Chemistry of multiple-spot phenomena. *J Chromatogr*. 1976;123(1):69–78. http://www.ncbi.nlm.nih.gov/pubmed/950360.
123. Irvine D. *Clinical, EEG and biochemical correlates of hydroxyhemopyrrolenone excretion*. In: *Proceedings of the 22nd Annual Psychiatric Research Meeting* Saskatoon, SK: Saskatoon Psychiatric Research Division; 1977:59–60.
124. Moore MR, Graham DJ. Monopyrroles in porphyria, psychosis and lead exposure. *Int J Biochem*. 1980;12(5–6):827–832. http://www.ncbi.nlm.nih.gov/pubmed/6893824.
125. Graham DJ, Thompson GG, Moore MR, Goldberg A. The effects of selected monopyrroles on various aspects of heme biosynthesis and degradation in the rat. *Arch Biochem Biophys*. 1979;197(1):132–138. http://www.ncbi.nlm.nih.gov/pubmed/583487.
126. McGinnis WR, Audhya T, Walsh WJ, et al. Discerning the mauve factor, part 1. *Altern Ther Health Med*. 2008;14(2):40–50. http://www.ncbi.nlm.nih.gov/pubmed/18383989.

127. Pfeiffer CC, Sohler A, Jenney CH, Iliev V. Treatment of pyroluric schizophrenia (Malvaria) with large doses of pyridoxine and a dietary supplement of zinc. *J Orthomol Psychiatry*. 1974;3(4):292–300.
128. Pfeiffer CC, Iliev V. Pyroluria, urinary mauve factor causes double deficiency of B6 and zinc in schizophrenics. *Fed Am Soc Exp Biol*. 1973;32:276.
129. Ames BN, Atamna H, Killilea DW. Mineral and vitamin deficiencies can accelerate the mitochondrial decay of aging. *Mol Asp Med*. 2005;26(4–5):363–378. https://doi.org/10.1016/j.mam.2005.07.007.
130. Scholnick PL, Hammaker LE, Marver HS. Soluble-aminolevulinic acid synthetase of rat liver. II. Studies related to the mechanism of enzyme action and hemin inhibition. *J Biol Chem*. 1972;247(13):4132–4137. http://www.ncbi.nlm.nih.gov/pubmed/5035685.
131. Gendler PL, Duhan HA, Rapoport H. Hemopyrrole and kryptopyrrole are absent from the urine of schizophrenics and normal persons. *Clin Chem*. 1978;24(2): 230–233.
132. Cruz R, Vogel WH. Pyroluria: a poor marker in chronic schizophrenia. *Am J Psychiatry*. 1978;135(10):1239–1240. https://doi.org/10.1176/ajp.135.10.1239.
133. Jacobson SJ, Rapoport H, Ellman GL. The nonoccurrence of hemo- and kryptopyrrole in urine of schizophrenics. *Biol Psychiatry*. 1975;10(1):91–93. http://www.ncbi.nlm.nih.gov/pubmed/1120177.
134. Armstrong RA. Review article What causes Alzheimer's disease? *Folia Neuropathol*. 2013;https://doi.org/10.5114/fn.2013.37702.
135. Hippius H, Neundörfer G. The discovery of Alzheimer's disease. *Dialogues Clin Neurosci*. 2003;5(1):101–108. http://www.ncbi.nlm.nih.gov/pubmed/22034141.
136. Minoshima S, Giordani B, Berent S, Frey KA, Foster NL, Kuhl DE. Metabolic reduction in the posterior cingulate cortex in very early Alzheimer's disease. *Ann Neurol*. 1997; https://doi.org/10.1002/ana.410420114.
137. Herholz K, Salmon E, Perani D, et al. Discrimination between Alzheimer dementia and controls by automated analysis of multicenter FDG PET. *NeuroImage*. 2002; https://doi.org/10.1006/nimg.2002.1208.
138. Mosconi L. Brain glucose metabolism in the early and specific diagnosis of Alzheimer's disease: FDG-PET studies in MCI and AD. *Eur J Nucl Med Mol Imaging*. 2005; https://doi.org/10.1007/s00259-005-1762-7.
139. Herholz K. Cerebral glucose metabolism in preclinical and prodromal Alzheimers disease. *Expert Rev Neurother*. 2010; https://doi.org/10.1586/ern.10.136.
140. de Leon MJ, Pirraglia E, De Santi S, et al. Maternal family history of Alzheimer's disease predisposes to reduced brain glucose metabolism. *Proc Natl Acad Sci*. 2007; https://doi.org/10.1073/pnas.0705036104.
141. Mosconi L. Glucose metabolism in normal aging and Alzheimer's disease: methodological and physiological considerations for PET studies. *Clin Transl Imaging*. 2013; https://doi.org/10.1007/s40336-013-0026-y.
142. Hardy J, Selkoe DJ. The amyloid hypothesis of Alzheimer's disease: progress and problems on the road to therapeutics. *Science*. 2002; https://doi.org/10.1126/science.1072994.
143. Pearson HA, Peers C. Physiological roles for amyloid beta peptides. *J Physiol*. 2006;575 (Pt 1):5–10. https://doi.org/10.1113/jphysiol.2006.111203.
144. Lowe VJ, Weigand SD, Senjem ML, et al. Association of hypometabolism and amyloid levels in aging, normal subjects. *Neurology*. 2014; https://doi.org/10.1212/WNL.0000000000000467.
145. Wolk DA, Nebes RD, Klunk WE, et al. Basal cerebral metabolism may modulate the cognitive effects of a in mild cognitive impairment: an example of brain reserve. *J Neurosci*. 2009; https://doi.org/10.1523/jneurosci.3669-09.2009.
146. Engler H, Forsberg A, Almkvist O, et al. Two-year follow-up of amyloid deposition in patients with Alzheimer's disease. *Brain*. 2006; https://doi.org/10.1093/brain/awl178.

147. Li Y, Rinne JO, Mosconi L, et al. Regional analysis of FDG and PIB-PET images in normal aging, mild cognitive impairment, and Alzheimer's disease. *Eur J Nucl Med Mol Imaging*. 2008; https://doi.org/10.1007/s00259-008-0833-y.
148. Rabinovici GD, Furst AJ, Alkalay A, et al. Increased metabolic vulnerability in early-onset Alzheimer's disease is not related to amyloid burden. *Brain*. 2010; https://doi.org/10.1093/brain/awp326.
149. Altmann A, Ng B, Landau SM, Jagust WJ, Greicius MD. Regional brain hypometabolism is unrelated to regional amyloid plaque burden. *Brain*. 2015; https://doi.org/10.1093/brain/awv278.
150. Pereira C, Santos MS, Oliveira C. Mitochondrial function impairment induced by amyloid beta-peptide on PC12 cells. *Neuroreport*. 1998;9(8):1749–1755. http://www.ncbi.nlm.nih.gov/pubmed/9665595.
151. Canevari L, Clark JB, Bates TE. β-Amyloid fragment 25–35 selectively decreases complex IV activity in isolated mitochondria. *FEBS Lett*. 1999; https://doi.org/10.1016/S0014-5793(99)01028-5.
152. Reddy PH. Amyloid beta, mitochondrial structural and functional dynamics in Alzheimer's disease. *Exp Neurol*. 2009; https://doi.org/10.1016/j.expneurol.2009.03.042.
153. Deshpande A. Different conformations of amyloid beta induce neurotoxicity by distinct mechanisms in human cortical neurons. *J Neurosci*. 2006; https://doi.org/10.1523/jneurosci.1189-06.2006.
154. Cabezas-Opazo FA, Vergara-Pulgar K, Pérez MJ, Jara C, Osorio-Fuentealba C, Quintanilla RA. Mitochondrial dysfunction contributes to the pathogenesis of Alzheimer's disease. *Oxidative Med Cell Longev*. 2015; https://doi.org/10.1155/2015/509654.
155. Wang H, Lim PJ, Karbowski M, Monteiro MJ. Effects of overexpression of huntingtin proteins on mitochondrial integrity. *Hum Mol Genet*. 2009; https://doi.org/10.1093/hmg/ddn404.
156. Barsoum MJ, Yuan H, Gerencser AA, et al. Nitric oxide-induced mitochondrial fission is regulated by dynamin-related GTPases in neurons. *EMBO J*. 2006; https://doi.org/10.1038/sj.emboj.7601253.
157. Hamblet NS, Ragland B, Ali M, Conyers B, Castora FJ. Mutations in mitochondrial-encoded cytochrome c oxidase subunits I,II, and III genes detected in Alzheimer's disease using single-strand conformation polymorphism. *Electrophoresis*. 2006; https://doi.org/10.1002/elps.200500420.
158. Angelova PR, Abramov AY. Role of mitochondrial ROS in the brain: from physiology to neurodegeneration. *FEBS Lett*. 2018;592(5):692–702. https://doi.org/10.1002/1873-3468.12964.
159. Lovell MA, Ehmann WD, Butler SM, Markesbery WR. Elevated thiobarbituric acid-reactive substances and antioxidant enzyme activity in the brain in Alzheimer's disease. *Neurology*. 1995; https://doi.org/10.1212/WNL.45.8.1594.
160. Montine KS, Reich E, Neely MD, et al. Distribution of reducible 4-hydroxynonenal adduct immunoreactivity in Alzheimer disease is associated with APOE genotype. *J Neuropathol Exp Neurol*. 1998; https://doi.org/10.1097/00005072-199805000-00005.
161. Anderson AJ, Su JH, Cotman CW. DNA damage and apoptosis in Alzheimer's disease: colocalization with c- Jun immunoreactivity, relationship to brain area, and effect of postmortem delay. *J Neurosci*. 1996; 16(5):1710–1719. https://doi.org/10.1523/JNEUROSCI.16-05-01710.1996.
162. Costantini LC, Barr LJ, Vogel JL, Henderson ST. Hypometabolism as a therapeutic target in Alzheimer's disease. *BMC Neurosci*. 2008; https://doi.org/10.1186/1471-2202-9-S2-S16.
163. Sorrentino V, Romani M, Mouchiroud L, et al. Enhancing mitochondrial proteostasis reduces amyloid-β proteotoxicity. *Nature*. 2017; https://doi.org/10.1038/nature25143.
164. Kim HJ, Khalimonchuk O, Smith PM, Winge DR. Structure, function, and assembly of heme centers in mitochondrial respiratory complexes. *Biochim Biophys Acta, Mol Cell Res*. 2012;1823(9):1604–1616. https://doi.org/10.1016/j.bbamcr.2012.04.008.

165. Kish SJ, Bergeron C, Rajput A, et al. Brain cytochrome oxidase in Alzheimer's disease. *J Neurochem*. 1992; https://doi.org/10.1111/j.1471-4159.1992.tb09439.x.
166. Ojaimi J, Masters CL, McLean C, Opeskin K, McKelvie P, Byrne E. Irregular distribution of cytochrome c oxidase protein subunits in aging and Alzheimer's disease. *Ann Neurol*. 1999; https://doi.org/10.1002/1531-8249(199910)46:4<656::AID-ANA16>3.0.CO;2-Q.
167. Mutisya EM, Bowling AC, Beal MF. Cortical cytochrome oxidase activity is reduced in Alzheimer's disease. *J Neurochem*. 1994;63:2179–2184.
168. Parker WD. Cytochrome oxidase deficiency in Alzheimer's disease. *Ann N Y Acad Sci*. 1991; https://doi.org/10.1111/j.1749-6632.1991.tb00191.x.
169. Cardoso SM, Proença MT, Santos S, Santana I, Oliveira CR. Cytochrome c oxidase is decreased in Alzheimer's disease platelets. *Neurobiol Aging*. 2004; https://doi.org/10.1016/S0197-4580(03)00033-2.
170. Asson-Batres M, Hares J. Effect of oxygen on the synthesis and assembly of mitochondrial encoded subunits of cytochrome oxidase and cytochrome b.cl in mouse embryo fibroblasts. *J Biol Chem*. 1991;266(15):9932–9938.
171. Maurer I, Zierz S, Möller HJ. A selective defect of cytochrome c oxidase is present in brain of Alzheimer disease patients. *Neurobiol Aging*. 2000;21(3):455–462. http://www.ncbi.nlm.nih.gov/pubmed/10858595.
172. Holper L, Ben-Shachar D, Mann JJ. Multivariate meta-analyses of mitochondrial complex I and IV in major depressive disorder, bipolar disorder, schizophrenia, Alzheimer disease, and Parkinson disease. *Neuropsychopharmacology*. 2018; https://doi.org/10.1038/s41386-018-0090-0.
173. Diaz F, Fukui H, Garcia S, Moraes CT. Cytochrome c oxidase is required for the assembly/stability of respiratory complex I in mouse fibroblasts. *Mol Cell Biol*. 2006;26(13): 4872–4881. https://doi.org/10.1128/mcb.01767-05.
174. Suthammarak W, Yang YY, Morgan PG, Sedensky MM. Complex I function is defective in complex IV-deficient Caenorhabditis elegans. *J Biol Chem*. 2009; https://doi.org/10.1074/jbc.M805733200.
175. Hinchliffe P, Sazanov LA. Biochemistry: organization of iron-sulfur clusters in respiratory complex I. *Science*. 2005;309(5735):771–774. https://doi.org/10.1126/science.1113988.
176. Sekar S, McDonald J, Cuyugan L, et al. Alzheimer's disease is associated with altered expression of genes involved in immune response and mitochondrial processes in astrocytes. *Neurobiol Aging*. 2015; https://doi.org/10.1016/j.neurobiolaging.2014.09.027.
177. Kim SH, Vlkolinsky R, Cairns N, Fountoulakis M, Lubec G. The reduction of NADH ubiquinone oxidoreductase 24- and 75-kDa subunits in brains of patients with down syndrome and Alzheimer's disease. *Life Sci*. 2001; https://doi.org/10.1016/S0024-3205(01)01074-8.
178. Coskun PE, Beal MF, Wallace DC. Alzheimer's brains harbor somatic mtDNA control-region mutations that suppress mitochondrial transcription and replication. *Proc Natl Acad Sci*. 2004; https://doi.org/10.1073/pnas.0403649101.
179. Rhein V, Song X, Wiesner A, et al. Amyloid-β and tau synergistically impair the oxidative phosphorylation system in triple transgenic Alzheimer's disease mice. *Proc Natl Acad Sci*. 2009; https://doi.org/10.1073/pnas.0905529106.
180. Zhang L, Zhang S, Maezawa I, et al. Modulation of mitochondrial complex I activity averts cognitive decline in multiple animal models of familial Alzheimer's disease. *EBioMedicine*. 2015; https://doi.org/10.1016/j.ebiom.2015.03.009.
181. Kim J, Yang G, Kim Y, Kim J, Ha J. AMPK activators: mechanisms of action and physiological activities. *Exp Mol Med*. 2016; https://doi.org/10.1038/emm.2016.16.
182. Arai H, Lee VM, Otvos L, et al. Defined neurofilament, t, and b-amyloid precursor protein epitopes distinguish Alzheimer from non-Alzheimer senile palques. *Proc Natl Acad Sci*. 1990;.

183. Ferrer I, López-González I, Carmona M, et al. Glial and neuronal tau pathology in tauopathies: characterization of disease-specific phenotypes and tau pathology progression. *J Neuropathol Exp Neurol.* 2014; https://doi.org/10.1097/NEN.0000000000000030.
184. Grundke-Iqbal I, Iqbal K, Tung YC, Quinlan M, Wisniewski HM, Binder LI. Abnormal phosphorylation of the microtubule-associated protein tau (tau) in Alzheimer cytoskeletal pathology. *Proc Natl Acad Sci U S A.* 1986;83(13):4913–4917.
185. Cai Z, Yan L-J, Li K, Quazi SH, Zhao B. Roles of AMP-activated protein kinase in Alzheimer's disease. *NeuroMolecular Med.* 2012;14(1):1–14. https://doi.org/10.1007/s12017-012-8173-2.
186. Cullen KM, Kócsi Z, Stone J. Pericapillary haem-rich deposits: evidence for microhaemorrhages in aging human cerebral cortex. *J Cereb Blood Flow Metab.* 2005; https://doi.org/10.1038/sj.jcbfm.9600155.
187. Ghosh C, Seal M, Mukherjee S, Ghosh DS. Alzheimer's disease: a heme-Aβ perspective. *Acc Chem Res.* 2015; https://doi.org/10.1021/acs.accounts.5b00102.
188. Perry G, Cash AD, Smith MA. Alzheimer disease and oxidative stress. *J Biomed Biotechnol.* 2003; https://doi.org/10.1155/S1110724302203010.
189. Hirose W, Ikematsu K, Tsuda R. Age-associated increases in heme oxygenase-1 and ferritin immunoreactivity in the autopsied brain. *Legal Med.* 2003; https://doi.org/10.1016/S1344-6223(02)00133-5.
190. Xing S, Shen D, Chen C, Wang J, Yu Z. Early induction of oxidative stress in a mouse model of Alzheimer's disease with heme oxygenase activity. *Mol Med Rep.* 2014; https://doi.org/10.3892/mmr.2014.2252.
191. Cataloglu B, Gultepe M, Cosar A, Muftuoglu T, Ipcioglu O. Decreased heme oxygenase activity in patients with Alzheimer's disease. *Dis Mol Med.* 2013;1(2):31. https://doi.org/10.5455/dmm.20130408021244.
192. Schipper HM, Chertkow H, Mehindate K, Frankel D, Melmed C, Bergman H. Evaluation of heme oxygenase-1 as a systemic biological marker of sporadic AD. *Neurology.* 2000; https://doi.org/10.1212/WNL.54.6.1297.
193. Praticò D. Oxidative stress hypothesis in Alzheimer's disease: a reappraisal. *Trends Pharmacol Sci.* 2008; https://doi.org/10.1016/j.tips.2008.09.001.
194. Padurariu M, Ciobica A, Lefter R, Serban IL, Stefanescu C, Chirita R. The oxidative stress hypothesis in Alzheimer's disease. *Psychiatr Danub.* 2013;25(4):401–409.
195. Nitti M, Piras S, Brondolo L, Marinari UM, Pronzato MA, Furfaro AL. Heme oxygenase 1 in the nervous system: does it favor neuronal cell survival or induce neurodegeneration? *Int J Mol Sci.* 2018; https://doi.org/10.3390/ijms19082260.
196. Chen J. Heme oxygenase in neuroprotection: from mechanisms to therapeutic implications. *Rev Neurosci.* 2014; https://doi.org/10.1515/revneuro-2013-0046.
197. Takeda A, Perry G, Abraham NG, et al. Overexpression of heme oxygenase in neuronal cells, the possible interaction with tau. *J Biol Chem.* 2000; https://doi.org/10.1074/jbc.275.8.5395.
198. Atamna H, Frey WH. A role for heme in Alzheimer's disease: heme binds amyloid and has altered metabolism. *Proc Natl Acad Sci.* 2004; https://doi.org/10.1073/pnas.0404349101.
199. Azimi S, Rauk A. The binding of Fe(II)-heme to the amyloid beta peptide of Alzheimer's disease: QM/MM investigations. *J Chem Theory Comput.* 2012;8(12):5150–5158. https://doi.org/10.1021/ct300716p.
200. Zhao LN, Mu Y, Chew LY. Heme prevents amyloid beta peptide aggregation through hydrophobic interaction based on molecular dynamics simulation. *Phys Chem Chem Phys.* 2013; https://doi.org/10.1039/c3cp52354c.
201. Raha AA, Vaishnav RA, Friedland RP, Bomford A, Raha-Chowdhury R. The systemic iron-regulatory proteins hepcidin and ferroportin are reduced in the brain in Alzheimer's disease. *Acta Neuropathol Commun.* 2013;1:55. https://doi.org/10.1186/2051-5960-1-55.

202. Atamna H, Boyle K. Amyloid-beta peptide binds with heme to form a peroxidase: relationship to the cytopathologies of Alzheimer's disease. *Proc Natl Acad Sci.* 2006; https://doi.org/10.1073/pnas.0600134103.
203. Mukherjee S, Seal M, Dey SG. Kinetics of serotonin oxidation by heme-Aβ relevant to Alzheimer's disease. *J Biol Inorg Chem.* 2014;19(8):1355–1365. https://doi.org/10.1007/s00775-014-1193-7.
204. Pramanik D, Mukherjee S, Dey SG. Apomyoglobin sequesters heme from heme bound Aβ peptides. *Inorg Chem.* 2013;52(19):10929–10935. https://doi.org/10.1021/ic401771j.
205. Seal M, Uppal S, Kundu S, Dey SG. Interaction of apoNeuroglobin with heme-Aβ complexes relevant to Alzheimer's disease. *J Biol Inorg Chem.* 2015;20(3):563–574. https://doi.org/10.1007/s00775-015-1241-y.
206. Lebouvier T, Chaumette T, Paillusson S, et al. The second brain and Parkinson's disease. *Eur J Neurosci.* 2009; https://doi.org/10.1111/j.1460-9568.2009.06873.x.
207. Damier P, Hirsch EC, Agid Y, Graybiel AM. The substantia nigra of the human brain: II. Patterns of loss of dopamine-containing neurons in Parkinson's disease. *Brain.* 1999; https://doi.org/10.1093/brain/122.8.1437.
208. Rodriguez-Oroz MC, Jahanshahi M, Krack P, et al. Initial clinical manifestations of Parkinson's disease: features and pathophysiological mechanisms. *Lancet Neurol.* 2009; https://doi.org/10.1016/S1474-4422(09)70293-5.
209. Gibb WRG, Lees AJ. The relevance of the Lewy body to the pathogenesis of idiopathic Parkinson's disease. *J Neurol Neurosurg Psychiatry.* 1988; https://doi.org/10.1136/jnnp.51.6.745.
210. Dias V, Junn E, Mouradian MM. The role of oxidative stress in Parkinson's disease. *J Park Dis.* 2013; https://doi.org/10.3233/JPD-130230.
211. Kroemer G, Blomgren K. Mitochondrial cell death control in familial Parkinson disease. *PLoS Biol.* 2007; https://doi.org/10.1371/journal.pbio.0050206.
212. Cardoso SM. The mitochondrial cascade hypothesis for Parkinsons disease. *Curr Pharm Des.* 2011; https://doi.org/10.2174/138161211798072508.
213. Langston J, Ballard P, Tetrud J, Irwin I. Chronic Parkinsonism in humans due to a product of meperidine-analog synthesis. *Science.* 2006; https://doi.org/10.1126/science.6823561.
214. Nicklas WJ, Vyas I, Heikkila RE. Inhibition of NADH-linked oxidation in brain mitochondria by 1-methyl-4-phenyl-pyridine, a metabolite of the neurotoxin, 1-methyl-4-phenyl-1,2,5,6-tetrahydropyridine. *Life Sci.* 1985; https://doi.org/10.1016/0024-3205(85)90146-8.
215. Vyas I, Heikkila RE, Nicklas WJ. Studies on the neurotoxicity of 1-methyl-4-phenyl-1,2,3,6-tetrahydropyridine: inhibition of NAD-linked substrate oxidation by its metabolite, 1-methyl-4-phenylpyridinium. *J Neurochem.* 1986; https://doi.org/10.1111/j.1471-4159.1986.tb01768.x.
216. Mizuno Y, Saitoh T, Sone N. Inhibition of mitochondrial NADH-ubiquinone oxidoreductase activity by 1-methyl-4-phenylpyridinium ion. *Biochem Biophys Res Commun.* 1987;143:294–299.
217. Dodson MW, Guo M. Pink1, Parkin, DJ-1 and mitochondrial dysfunction in Parkinson's disease. *Curr Opin Neurobiol.* 2007;17(3):331–337. https://doi.org/10.1016/j.conb.2007.04.010.
218. Narendra D, Walker JE, Youle R. Mitochondrial quality control mediated by PINK1 and Parkin: links to parkinsonism. *Cold Spring Harb Perspect Biol.* 2012; https://doi.org/10.1101/cshperspect.a011338.
219. Youle RJ, Narendra DP. Mechanisms of mitophagy. *Nat Rev Mol Cell Biol.* 2011;12(1):9–14. https://doi.org/10.1038/nrm3028.
220. Liu X, Hajnóczky G. Ca^{2+}-dependent regulation of mitochondrial dynamics by the Miro-Milton complex. *Int J Biochem Cell Biol.* 2009;41(10):1972–1976. https://doi.org/10.1016/j.biocel.2009.05.013.

221. Weihofen A, Thomas KJ, Ostaszewski BL, Cookson MR, Selkoe DJ. Pink1 forms a multiprotein complex with miro and milton, linking Pink1 function to mitochondrial trafficking. *Biochemistry*. 2009; https://doi.org/10.1021/bi8019178.
222. Liu S, Sawada T, Lee S, et al. Parkinson's disease-associated kinase PINK1 regulates Miro protein level and axonal transport of mitochondria. *PLoS Genet*. 2012;8(3):e1002537 https://doi.org/10.1371/journal.pgen.1002537.
223. Chan NC, Salazar AM, Pham AH, et al. Broad activation of the ubiquitin-proteasome system by Parkin is critical for mitophagy. *Hum Mol Genet*. 2011; https://doi.org/10.1093/hmg/ddr048.
224. Wang X, Winter D, Ashrafi G, et al. PINK1 and Parkin target miro for phosphorylation and degradation to arrest mitochondrial motility. *Cell*. 2011; https://doi.org/10.1016/j.cell.2011.10.018.
225. Schapira AHV, Mann VM, Cooper JM, et al. Anatomic and disease specificity of NADH CoQ1 reductase (Complex I) deficiency in Parkinson's disease. *J Neurochem*. 1990; https://doi.org/10.1111/j.1471-4159.1990.tb05809.x.
226. Dexter DT, Wells FR, Lee AJ, et al. Increased nigral iron content and alterations in other metal ions occurring in brain in Parkinson's disease. *J Neurochem*. 1989; https://doi.org/10.1111/j.1471-4159.1989.tb07264.x.
227. Mann VM, Cooper JM, Daniel SE, et al. Complex I, iron, and ferritin in Parkinson's disease substantia nigra. *Ann Neurol*. 1994; https://doi.org/10.1002/ana.410360612.
228. Parker WD, Parks JK, Swerdlow RH. Complex I deficiency in Parkinson's disease frontal cortex. *Brain Res*. 2008; https://doi.org/10.1016/j.brainres.2007.10.061.
229. Haas RH, Nasirian F, Nakano K, et al. Low platelet mitochondrial complex I and complex II/III activity in early untreated Parkinson's disease. *Ann Neurol*. 1995;37(6): 714–722. https://doi.org/10.1002/ana.410370604.
230. Gu M, Cooper JM, Taanman JW, Schapira AH. Mitochondrial DNA transmission of the mitochondrial defect in Parkinson's disease. *Ann Neurol*. 1998; https://doi.org/10.1002/ana.410440207.
231. Alam M, Schmidt WJ. Rotenone destroys dopaminergic neurons and induces parkinsonian symptoms in rats. *Behav Brain Res*. 2002; https://doi.org/10.1016/S0166-4328(02)00180-8.
232. Richardson JR, Caudle WM, Guillot TS, et al. Obligatory role for complex I inhibition in the dopaminergic neurotoxicity of 1-methyl-4-phenyl-1,2,3,6-tetrahydropyridine (MPTP). *Toxicol Sci*. 2007;95(1):196–204. https://doi.org/10.1093/toxsci/kfl133.
233. Fato R, Bergamini C, Bortolus M, et al. Differential effects of mitochondrial Complex I inhibitors on production of reactive oxygen species. *Biochim Biophys Acta Bioenerg*. 2009; https://doi.org/10.1016/j.bbabio.2008.11.003.
234. Ungerstedt U, Ljungberg T, Steg G. Behavioral, physiological, and neurochemical changes after 6-hydroxydopamine-induced degeneration of the nigro-striatal dopamine neurons. *Adv Neurol*. 1974;5:421–426. http://www.ncbi.nlm.nih.gov/pubmed/4531217.
235. Di Santo S, Skaper SD, Giusti P, Franceschini D, Dabbeni-Sala F. Melatonin protects against 6-OHDA-induced neurotoxicity in rats: a role for mitochondrial complex I activity. *FASEB J*. 2002; https://doi.org/10.1096/fj.00-0129com.
236. Matsui H, Gavinio R, Asano T, et al. PINK1 and parkin complementarily protect dopaminergic neurons in vertebrates. *Hum Mol Genet*. 2013; https://doi.org/10.1093/hmg/ddt095.
237. Marella M, Seo BB, Yagi T, Matsuno-Yagi A. Parkinson's disease and mitochondrial complex I: a perspective on the Ndi1 therapy. *J Bioenerg Biomembr*. 2009;41(6):493–497. https://doi.org/10.1007/s10863-009-9249-z.
238. Keeney PM, Xie J, Capaldi RA, Bennett JP. Parkinson's disease brain mitochondrial complex I has oxidatively damaged subunits and is functionally impaired and misassembled. *J Neurosci*. 2006;26(19):5256–5264. https://doi.org/10.1523/JNEUROSCI.0984-06.2006.

239. Lhermitte J, Kraus WM, McAlpine D. On the occurrence of abnormal deposits of iron in the brain in parkinsonism with special reference to its localisation. *J Neurol Neurosurg Psychiatry*. 1924; https://doi.org/10.1136/jnnp.s1-5.19.195.
240. Wang JY, Zhuang QQ, Zhu LB, et al. Meta-analysis of brain iron levels of Parkinson's disease patients determined by postmortem and MRI measurements. *Sci Rep*. 2016; https://doi.org/10.1038/srep36669.
241. Hirsch EC, Brandel J-P, Galle P, Javoy-Agid F, Agid Y. Iron and aluminum increase in the substantia nigra of patients with Parkinson's disease: an X-ray microanalysis. *J Neurochem*. 1991; https://doi.org/10.1111/j.1471-4159.1991.tb08170.x.
242. Drayer B, Burger P, Darwin R. Magnetic resonance imaging of brain iron. *Am J Neuroradiol*. 1986; https://doi.org/10.1016/S0022-510X(02)00431-8.
243. Sofic E, Riederer P, Heinsen H, et al. Increased iron (III) and total iron content in post mortem substantia nigra of parkinsonian brain. *J Neural Transm*. 1988; https://doi.org/10.1007/BF01244786.
244. Dexter DT, Jenner P, Schapira AHV, Marsden CD. Alterations in levels of iron, ferritin, and other trace metals in neurodegenerative diseases affecting the basal ganglia. *Ann Neurol*. 1992; https://doi.org/10.1002/ana.410320716.
245. Braak H, Ghebremedhin E, Rüb U, Bratzke H, Del Tredici K. Stages in the development of Parkinson's disease-related pathology. *Cell Tissue Res*. 2004; https://doi.org/10.1007/s00441-004-0956-9.
246. Medeiros MS, Schumacher-Schuh A, Cardoso AM, et al. Iron and oxidative stress in Parkinson's disease: an observational study of injury biomarkers. *PLoS ONE*. 2016; https://doi.org/10.1371/journal.pone.0146129.
247. Ben-Shachar D, Youdim MB. Intranigral iron injection induces behavioral and biochemical "parkinsonism" in rats. *J Neurochem*. 1991;57(6):2133–2135. http://www.ncbi.nlm.nih.gov/pubmed/1940919.
248. Jellinger K, Kienzl E, Rumpelmair G, et al. Iron-melanin complex in substantia nigra of parkinsonian brains: an X-ray microanalysis. *J Neurochem*. 1992; https://doi.org/10.1111/j.1471-4159.1992.tb08362.x.
249. Youdim MB, Riederer P. The role of iron in senescence of dopaminergic neurons in Parkinson's disease. *J Neural Transm Suppl*. 1993;40:57–67. http://www.ncbi.nlm.nih.gov/pubmed/8294901.
250. Song W, Kothari V, Velly AM, et al. Evaluation of salivary heme oxygenase-1 as a potential biomarker of early Parkinson's disease. *Mov Disord*. 2018; https://doi.org/10.1002/mds.27328.
251. Mateo I, Infante J, Sánchez-Juan P, et al. Serum heme oxygenase-1 levels are increased in Parkinson's disease but not in Alzheimer's disease. *Acta Neurol Scand*. 2010; https://doi.org/10.1111/j.1600-0404.2009.01261.x.
252. Song W, Zukor H, Lin SH, et al. Unregulated brain iron deposition in transgenic mice over-expressing HMOX1 in the astrocytic compartment. *J Neurochem*. 2012; https://doi.org/10.1111/j.1471-4159.2012.07914.x.
253. Cressatti M, Song W, Turk A, Galindez C, Schipper H. *Effects of Glial Heme Oxygenase-1 on Neuronal Alpha-Synuclein in the GFAP.HMOX1 Mouse Model of Parkinson's Disease*. 2018; Https://www.mdsabstracts.org/abstract/effects-of-glial-heme-oxygenase-1-on-neuronal-alpha-synuclein-in-the-gfap-hmox1-mouse-model-of-parkinsons-disease/.
254. Kragh CL, Romero-Ramos M, Halliday G, Jensen PH. Alpha synuclein in Parkinson's disease. In: *Handbook of Neurotoxicity*. 2014: https://doi.org/10.1007/978-1-4614-5836-4_14.
255. McGlashan TH. A selective review of recent North American long-term follow-up studies of schizophrenia. *Schizophr Bull*. 1988;14:515–542.
256. McGlashan TH, Hoffman RE. Schizophrenia as a disorder of developmentally reduced synaptic connectivity. *Arch Gen Psychiatry*. 2000;57(7):637–648.

257. van Os J, Rutten BP, Poulton R. Gene-environment interactions in schizophrenia: review of epidemiological findings and future directions. *Schizophr Bull*. 2008;34(6):1066–1082. https://doi.org/10.1093/schbul/sbn117.
258. Benes FM. Neural circuitry models of schizophrenia: is it dopamine, GABA, glutamate, or something else? *Biol Psychiatry*. 2009;65(12):1003–1005. https://doi.org/10.1016/j.biopsych.2009.04.006.
259. Roberts RC. Postmortem studies on mitochondria in schizophrenia. *Schizophr Res*. 2017;187:17–25. https://doi.org/10.1016/j.schres.2017.01.056.
260. Konradi C, Öngür D. Role of mitochondria and energy metabolism in schizophrenia and psychotic disorders. *Schizophr Res*. 2017;187:1–2. https://doi.org/10.1016/j.schres.2017.07.007.
261. Iwamoto K, Bundo M, Kato T. Altered expression of mitochondria-related genes in postmortem brains of patients with bipolar disorder or schizophrenia, as revealed by large-scale DNA microarray analysis. *Hum Mol Genet*. 2005;14(2):241–253. https://doi.org/10.1093/hmg/ddi022.
262. Ben-Shachar D. Mitochondrial multifaceted dysfunction in schizophrenia; complex I as a possible pathological target. *Schizophr Res*. 2017;187:3–10. https://doi.org/10.1016/j.schres.2016.10.022.
263. Robicsek O, Karry R, Petit I, et al. Abnormal neuronal differentiation and mitochondrial dysfunction in hair follicle-derived induced pluripotent stem cells of schizophrenia patients. *Mol Psychiatry*. 2013;18(10):1067–1076. https://doi.org/10.1038/mp.2013.67mp201367.
264. Robicsek O, Ene HM, Karry R, et al. Isolated mitochondria transfer improves neuronal differentiation of schizophrenia-derived induced pluripotent stem cells and rescues deficits in a rat model of the disorder. *Schizophr Bull*. 2018;44(2):432–442. https://doi.org/10.1093/schbul/sbx077.
265. Wirth C, Brandt U, Hunte C, Zickermann V. Structure and function of mitochondrial complex I. *Biochim Biophys Acta Bioenerg*. 2016;1857(7):902–914. https://doi.org/10.1016/j.bbabio.2016.02.013.
266. Washizuka S, Kametani M, Sasaki T, et al. Association of mitochondrial complex I subunit gene NDUFV2 at 18p11 with schizophrenia in the Japanese population. *Am J Med Genet B Neuropsychiatr Genet*. 2006;141 B(3):301–304. https://doi.org/10.1002/ajmg.b.30285.
267. Martorell L, Segues T, Folch G, et al. New variants in the mitochondrial genomes of schizophrenic patients. *Eur J Hum Genet*. 2006;14(5):520–528. http://www.ncbi.nlm.nih.gov/entrez/query.fcgi?cmd=Retrieve&db=PubMed&dopt=Citation&list_uids=16538224.
268. Dhar S, Martin ET, Lephart PR, et al. Risk factors and outcomes for carbapenem-resistant *Klebsiella pneumoniae* isolation, stratified by its multilocus sequence typing: ST258 versus non-ST258. *Open Forum Infect Dis*. 2016;3(1):ofv213. https://doi.org/10.1093/ofid/ofv213ofv213.
269. Burkhardt C, Kelly JP, Lim YH, Filley CM, Parker WD. Neuroleptic medications inhibit complex I of the electron transport chain. *Ann Neurol*. 1993;33(5):512–517. https://doi.org/10.1002/ana.410330516.
270. Bergman O, Karry R, Milhem J, Ben-Shachar D. NDUFV2 pseudogene (NDUFV2P1) contributes to mitochondrial complex I deficits in schizophrenia. *Mol Psychiatry*. 2018; https://doi.org/10.1038/s41380-018-0309-9.
271. Maurer I, Zierz S, Moller H, Möller H-J. Evidence for a mitochondrial oxidative phosphorylation defect in brains from patients with schizophrenia. *Schizophr Res*. 2001;48(1):125–136. https://doi.org/10.1016/S0920-9964(00)00075-X.
272. Rosenfeld M, Brenner-Lavie H, Ari SG, Kavushansky A, Ben-Shachar D. Perturbation in mitochondrial network dynamics and in complex I dependent cellular respiration in schizophrenia. *Biol Psychiatry*. 2011;69(10):980–988. https://doi.org/10.1016/j.biopsych.2011.01.010 [pii:S0006–3223(11)00056–4].

273. Petit I, Kesner NS, Karry R, et al. Induced pluripotent stem cells from hair follicles as a cellular model for neurodevelopmental disorders. *Stem Cell Res*. 2012;8(1):134–140. https://doi.org/10.1016/j.scr.2011.09.003 [pii:S1873–5061(11)00117–6].
274. Weiser M, Levkowitch Y, Neuman M, Yehuda S. Decrease of serum iron in acutely psychotic schizophrenic patients. *Int J Neurosci*. 1994;78(1–2):49–52. http://www.ncbi.nlm.nih.gov/pubmed/7829291.
275. Insel BJ, Schaefer CA, McKeague IW, Susser ES, Brown AS. Maternal iron deficiency and the risk of schizophrenia in offspring. *Arch Gen Psychiatry*. 2008;65(10):1136–1144. https://doi.org/10.1001/archpsyc.65.10.1136.
276. Sørensen HJ, Nielsen PR, Pedersen CB, Mortensen PB. Association between prepartum maternal iron deficiency and offspring risk of schizophrenia: population-based cohort study with linkage of danish national registers. *Schizophr Bull*. 2011;37(5):982–987. https://doi.org/10.1093/schbul/sbp167.
277. Ifhar L, Ene H, Ben-Shachar D. Impaired heme metabolism in schizophrenia-derived cell lines and in a rat model of the disorder: possible involvement of mitochondrial complex I. *Eur Neuropsychopharmacol*. 2019; https://doi.org/10.1016/j.euroneuro.2019.03.011.
278. English JA, Fan Y, Föcking M, et al. Reduced protein synthesis in schizophrenia patient-derived olfactory cells. *Transl Psychiatry*. 2015;5(10):e663. https://doi.org/10.1038/tp.2015.119.
279. Trinh MA, Kaphzan H, Wek RC, Pierre P, Cavener DR, Klann E. Brain-specific disruption of the eIF2α kinase PERK decreases ATF4 expression and impairs behavioral flexibility. *Cell Rep*. 2012; https://doi.org/10.1016/j.celrep.2012.04.010.
280. Homedan C, Schmitt C, Laafi J, et al. Mitochondrial energetic defects in muscle and brain of a Hmbs−/− mouse model of acute intermittent porphyria. *Hum Mol Genet*. 2015;24(17):5015–5023. https://doi.org/10.1093/hmg/ddv222.
281. Gielisch I, Meierhofer D. Metabolome and proteome profiling of complex i deficiency induced by rotenone. *J Proteome Res*. 2015;14(1):224–235. https://doi.org/10.1021/pr500894v.
282. Sharma LK, Lu J, Bai Y. Mitochondrial respiratory complex I: structure, function and implication in human diseases. *Curr Med Chem*. 2009;16(10):1266–1277. https://doi.org/10.2174/092986709787846578.
283. Lenaz G. The mitochondrial production of reactive oxygen species: mechanisms and implications in human pathology. *IUBMB Life*. 2002; https://doi.org/10.1080/15216540152845957.
284. Sanz A, Stefanatos R, McIlroy G. Production of reactive oxygen species by the mitochondrial electron transport chain in *Drosophila melanogaster*. *J Bioenerg Biomembr*. 2010; https://doi.org/10.1007/s10863-010-9281-z.
285. Tretter L, Sipos I, Adam-Vizi V. Initiation of neuronal damage by complex I deficiency and oxidative stress in Parkinson's disease. *Neurochem Res*. 2004;29(3):569–577. https://doi.org/10.1023/B:NERE.0000014827.94562.4b.
286. Heinz S, Freyberger A, Lawrenz B, Schladt L, Schmuck G, Ellinger-Ziegelbauer H. Mechanistic investigations of the mitochondrial complex I inhibitor rotenone in the context of pharmacological and safety evaluation. *Sci Rep*. 2017; https://doi.org/10.1038/srep45465.
287. Scialo F, Mallikarjun V, Stefanatos R, Sanz A. Regulation of lifespan by the mitochondrial electron transport chain: reactive oxygen species-dependent and reactive oxygen species-independent mechanisms. *Antioxid Redox Signal*. 2012; https://doi.org/10.1089/ars.2012.4900.
288. Onyango IG, Dennis J, Khan SM. Mitochondrial dysfunction in Alzheimer's disease and the rationale for bioenergetics based therapies. *Aging Dis*. 2016; https://doi.org/10.14336/ad.2015.1007.

289. Federico A, Cardaioli E, Da Pozzo P, Formichi P, Gallus GN, Radi E. Mitochondria, oxidative stress and neurodegeneration. *J Neurol Sci*. 2012;322(1–2):254–262. https://doi.org/10.1016/j.jns.2012.05.030.
290. Atamna H. Heme, iron, and the mitochondrial decay of ageing. *Ageing Res Rev*. 2004;3(3):303–318. https://doi.org/10.1016/j.arr.2004.02.002.
291. Lovell MA, Robertson JD, Teesdale WJ, Campbell JL, Markesbery WR. Copper, iron and zinc in Alzheimer's disease senile plaques. *J Neurol Sci*. 1998; https://doi.org/10.1016/S0022-510X(98)00092-6.
292. Schipper HM, Cissé S, Stopa EG. Expression of heme oxygenase-1 in the senescent and Alzheimer-diseased brain. *Ann Neurol*. 1995; https://doi.org/10.1002/ana.410370609.
293. Ben-Shachar D, Ene HM. Mitochondrial targeted therapies: where do we stand in mental disorders? *Biol Psychiatry*. 2017; https://doi.org/10.1016/j.biopsych.2017.08.007.
294. Kerr DS. Treatment of mitochondrial electron transport chain disorders: a review of clinical trials over the past decade. *Mol Genet Metab*. 2010;99(3):246–255. https://doi.org/10.1016/j.ymgme.2009.11.005.
295. Parikh S, Saneto R, Falk MJ, Anselm I, Cohen BH, Haas R. A modern approach to the treatment of mitochondrial disease. *Curr Treat Options Neurol*. 2009;11(6):414–430. https://doi.org/10.1007/s11940-009-0046-0.
296. Du J, Zhu M, Bao H, et al. The role of nutrients in protecting mitochondrial function and neurotransmitter signaling: implications for the treatment of depression, PTSD, and suicidal behaviors. *Crit Rev Food Sci Nutr*. 2016;56(15):2560–2578. https://doi.org/10.1080/10408398.2013.876960.
297. Cheng Y, Rong J. Therapeutic potential of heme oxygenase-1/carbon monoxide system against ischemia-reperfusion injury. *Curr Pharm Des*. 2017;23(26). https://doi.org/10.2174/1381612823666170413122439.
298. Fernandez-Marcos PJ, Auwerx J. Regulation of PGC-1α, a nodal regulator of mitochondrial biogenesis. *Am J Clin Nutr*. 2011;93. https://doi.org/10.3945/ajcn.110.001917.
299. Itoh K, Ye P, Matsumiya T, Tanji K, Ozaki T. Emerging functional cross-talk between the Keap1-Nrf2 system and mitochondria. *J Clin Biochem Nutr*. 2015;56(2):91–97. https://doi.org/10.3164/jcbn.14-134.

CHAPTER

9

Neuroprotective effects of lithium in neuropsychiatric disorders

Galila Agam[a], *Joseph Levine*[b]

[a]Department of Clinical Biochemistry and Pharmacology and Psychiatry Research Unit, Faculty of Health Sciences, Ben-Gurion University of the Negev and Beer-Sheva Mental Health Center, Beer-Sheva, Israel, [b]Ben Gurion University of the Negev, Beer-Sheva, Israel

Introduction

Lithium salts (lithium) have been introduced by John Cade into the arsenal of psychiatric treatments in 1948 following the successful treatment of subjects with manic episodes.[1,2] Nowadays, about 70 years later, lithium is still considered a first-line treatment for bipolar disorders with antimanic, antisuicidal, and prophylactic effects and, in addition, as adjunctive treatment to antidepressants in major depression.[3–5] It is of note that lithium, a monovalent cation, may also demonstrate beneficial effects in schizoaffective disorder.[6,7] Furthermore, some evidence suggests that lithium is beneficial as augmentation in sporadic cases of schizophrenia[8–10] and in autism spectrum disorders (ASD).[11–13]

Contrarily to other ions that are major constituents of the body, that is, chloride, protons, sodium, potassium, calcium, and magnesium, the concentration of which is highly regulated by selective channels and active transporters, lithium lacks biological regulation in body fluids/tissues via selective transports or binding sites.[14] Nonetheless, it does have an array of important biochemical and physiological effects in mammals, in general, and in the human body, in particular, mainly by competing with other cations having similar ionic radius as the divalent cation magnesium and the monovalent cation sodium, on numerous molecular targets.[14,15]

https://doi.org/10.1016/B978-0-12-814037-6.00006-9

Several key cell signaling-related enzymes were found to be lithium inhibitable due to the ion's competition with magnesium ions including glycogen synthase kinase 3 beta,[16, 17] inositol monophosphatase,[18] phosphoadenosylphosphate phosphatase,[19] and Akt/beta-arrestin 2 (Akt).[20] Processes associated with these enzymes include autophagy, neurotrophic factor release, neuroinflammation and cytokine release, mitochondrial energy metabolism, and oxidative stress.[21–31] The involvement of these processes in a variety of brain disorders beyond bipolar disorders, such as Parkinson's disease, Alzheimer's disease (AD), ALS, Huntington's disease (HD), ASD, and traumatic brain injury (TBI), raised, in the last decade, the interest in exploring lithium's potential neuroprotective property not only in psychiatric disorders but also in neurodegenerative and neurodevelopmental disorders.[21, 32–37]

The first to use the term "lithium-induced neuroprotection" were Chen et al.[38] when reporting that chronic administration of lithium and valproate (VPA, another mood stabilizer) increased the levels of Bcl-2, an antiapoptotic factor, in the frontal cortex. In the following pages, we will first discuss lithium's effects contributing to the drug's neuroprotection such as autophagy enhancement, neurotrophic factor release, attenuation of neuroinflammation and regulation of cytokine release as well as boosting mitochondrial function and attenuating apoptosis. We then review animal model studies and clinical trials with lithium in several neurodegenerative, neurodevelopmental, and some other neurological disorders.

Lithium's effect on autophagy, BDNF, inflammation, mitochondrial function and apoptosis

Autophagy

Autophagy (macroautophagy) is the process that degrades unnecessary or dysfunctional cellular components (for review, see Ravikumar et al.[38] and Scrivo et al.[39]). Two main routes enhance autophagy: (a) inhibition of the mammalian target of rapamycin (mTOR) (Dunlop et al.[40]), a negative regulator of autophagy, and (b) inhibition of inositol monophosphatase (IMPase).[41] Lithium has long been recognized as an inhibitor of IMPase at therapeutically relevant concentrations.[18] Hence, lithium enhances autophagy as first shown by Sarkar et al.[41] and thereby enhances the clearance of autophagy substrates. This effect is not mediated by glycogen synthase kinase 3 beta inhibition (via the mTOR pathway), another signaling enzyme claimed to be inhibited by lithium.[41]

IMPase catalyzes the hydrolysis of inositol monophosphate (IP_1) into free inositol required for the phosphatidylinositol (PI) signaling pathway.[42] Simplistically, lithium induces free inositol depletion, which, in turn, decreases myo-inositol-1,4,5-trisphosphate (IP_3) levels. However, we have found that lithium treatment increased frontal cortex and hippocampal phosphoinositols (which include IP_3) in the frontal cortex of wild-type mice.[43] Using cells in culture, Sarkar et al.[41] reported that increased inositol or IP_3 levels inhibit autophagy and reverse lithium's effect; Criollo et al.[44] showed that IP_3 and the stimulation of its receptor(s) suppress autophagy. On the other hand, we found[43] that intracerbroventricular administration of IP_3 trapped in liposomes increased hippocampal messenger RNA levels of beclin 1 (required for autophagy execution) and hippocampal and frontal cortex protein levels ratio of beclin 1/p62 (p62 is degraded by autophagy), indicative of enhanced autophagy. Further studies are required to unequivocally reveal the role of IP_3 in autophagy.

Inositol depletion is a common suggested mechanism of the mood-stabilizing therapeutic effects of lithium, carbamazepine (CBZ), and valproic acid (VPA).[45] This is consistent with the finding that CBZ and VPA also enhanced the clearance of aggregate-prone proteins suggesting enhanced autophagy.[41] Lithium is also a direct[46] and indirect[47] inhibitor of GSK-3β. However, several studies demonstrate that the drug's autophagy-enhancing effect cannot be attributed to GSK-3β inhibition. In this regard a specific GSK-3β inhibitor, SB21672, attenuated the clearance of aggregate-prone huntingtin and α-synuclein and impaired autophagosome synthesis.[48] The ambiguity of the question whether GSK-3 inhibition accelerates or attenuates autophagy is further reflected in the following reports. In an AD model a substrate-competitive GSK-3 inhibitor, L803-mts, restored the activity of mTOR, inhibited autophagy, and ameliorated the Alzheimer-like pathology.[49] Zhang et al.[50] reported, as expected, that lithium treatment upregulated the inactivate form of GSK-3β but, unexpectedly, attenuated the autophagy activation in an AD mouse model. Rapamycin, an mTOR inhibitor, while enhancing autophagy, was reported to elevate the levels of the inactive form of GSK-3β (Phospho-Ser) in mutant P301S tau transgenic mice.[51] It may be summarized that there is an unequivocal consensus that lithium's inhibition of IMPase plays a role in autophagy enhancement, as shown in the clearance of huntingtin,[41, 48, 52] α-synuclein[53–55] and ceroid lipofuscinosis.[56] On the other hand, it is yet unclear whether GSK-3β is an autophagy positive or negative regulator. If this is a matter of the status of autophagy dysregulation than lithium's inhibition of GSK-3 may not be responsible for the amelioration of the aberration.

BDNF

Brain-derived neurotrophic factor (BDNF)[57] is a member of the neurotrophin family of growth factors. BDNF acts on certain neurons in the brain supporting their survival and affecting the growth and differentiation of new neurons and synapses.[58, 59] BDNF binds to the tropomyosin receptor kinase B (TrkB) receptor, the activation of which is pivotal for neuronal growth. Chronic lithium treatment increases cortical neuron BDNF mRNA and protein levels in culture and in the cortex, hippocampus, and amygdala in rodents.[60–63] It has been suggested that chronic lithium treatment elicits its antimanic effects via BDNF-TrkB-dependent synaptic downscaling[64] even at subtherapeutic concentrations[28] implicating the involvement of BDNF in the neuroprotective effect of lithium.[63, 65, 66] These findings in cells in culture and in rodents were further established in patients with bipolar mania.[67] Furthermore, good response to lithium was associated with improved cognitive function, higher serum BDNF levels, and genotype of the BDNF gene,[68–71] implying the involvement of lithium-BDNF interaction in lithium's neuroprotective effect and prevention of cellular degeneration.[72] As Won and Kim[73] recently summarized "Numerous studies have reported decreased BDNF levels in patients with bipolar depression and mania, along with low levels of BDNF to be correlated with severity of depression and mania symptoms (Cunha et al[74]; Machado-Vieira et al[75])." Nevertheless, there is still a debate whether lithium exerts both its antimanic and antidepressant effects via the BDNF-TrkB signaling pathway. Gideons et al.[64] recently reported that this signaling is required for the antimanic-like but not the antidepressant-like response to lithium. In this respect, it is of note that previous studies showed that BDNF is required for the antidepressant response to several antidepressants[76–78] including ketamine.[79, 80]

Inflammation

The earliest finding associating lithium with the immune/inflammatory system was that lithium administration elevates leukocyte count.[81] Following inflammatory stimuli, microglia and astrocytes undergo morphological changes and secrete numerous inflammatory mediators, which facilitate inflammatory processes in the CNS and regulate production of neurotransmitters. The effects of lithium on glial secretion of pro- and anti-inflammatory mediators have been examined extensively.[82–91] The drug affects both pro- and anti-inflammatory mediators both in the blood and in the brain.

Regarding proinflammatory mediators:

(A) In vivo chronic treatment attenuated brain cyclooxygenase (COX)-2 expression and prostaglandin (PG)E2 production.[92, 93]

(B) The reports of lithium's effect on tumor necrosis factor (TNF)-α are inconsistent. While some found that lithium reduces this cytokine levels (reviewed in Nassar and Azab[29]), others report the opposite.[94] The differences may stem from different experimental conditions.

(C) Interleukin-1β (IL-1β) production was mostly reported to be inhibited by lithium.[86, 88, 94–103]

(D) The levels of interleukin-4 (IL-4) were reported to be increased by lithium[104–106] although others reported no effect or reduction.[100, 107]

(E) In multiple studies, interleukin-6 (IL-6) production was found to be attenuated by lithium,[89, 99, 108–112] while others reported it enhances secretion[84, 100, 103, 104, 113–116] or a lack of effect.[109, 110]

(F) Reports related to interferon-γ (IFN-γ) are also inconclusive, but most of them found that lithium inhibits INF-γ synthesis[82, 98, 99, 110, 111, 117] attesting for a possible antiinflammatory effect.

As for the antiinflammatory cytokines:

(A) Numerous studies demonstrated that lithium increases IL-2 secretion,[103, 107, 118–120] but others demonstrated that the drug either does not alter this cytokine's levels[100, 104] or decreases its production.[106, 111, 121]

(B) Most of the studies that investigated the effect of lithium on IL-10 found that the drug increases this cytokine's production[86, 87, 98, 106, 110, 115, 122–125] supporting a strong antiinflammatory action of the drug.

Although not unequivocally the data summarized imply that lithium induces an antiinflammatory effect. According to Nassar and Azab[29] "GSK-3β plays a pivotal role in the mechanism underlying the effects of lithium on inflammation" but GSK-3β-independent mechanisms may also be involved. Indeed, Damri et al.[126] have recently reported that the mechanism mediating the effect of lithium on brain cytokine levels differs among brain regions. Unlike the effect of the drug in the hippocampus, which favors the involvement of the PI cycle, the effect in the hypothalamus did not support this mechanism, and the lack of effect in the frontal cortex was uninformative. As summarized by the authors "The discrepancy among our results in the three brain regions studied corroborates the notion that different brain regions play different and specific roles in the disorder and its treatment. This area-specific response demonstrates the intricate effects of lithium, in general, and on the inflammatory system, in particular, mediated by different signaling pathways which may lead to a common outcome."

Mitochondria

Multiple studies have shown that the enhancement of mitochondrial function plays a role in lithium's neuroprotective characteristic. Thus Chen et al.[38] reported that lithium acts on intracellular calcium signaling by inhibiting IMPase and upregulating mitochondrial outer membrane Bcl-2. Scola et al.[127] demonstrated in rat primary cortical neurons that both mitochondrial complex I dysfunction and the consequent increase in DNA methylation are prevented by lithium. Investigating the effects of lithium-induced autophagy upregulating in *Caenorhabditis elegans*, Tam et al.[128] found that lithium treatment increased both the life span and health span of the nematode accompanied by improved mitochondrial energetic function. Shalbuyeva et al.[129] reported that lithium desensitizes mitochondria to elevated calcium and diminishes cytochrome *c* release from brain mitochondria by antagonizing the Ca^{2+}-induced mitochondrial permeability transition. Interpretation of the latter study should be considered with caution concerning potential therapeutic effects of lithium since the findings were obtained using 125 mM LiCl, a concentration not compatible with survival of mammals (lithium is already toxic above 1.5 mM). These properties of lithium could be useful to prevent and/or counteract dysfunctions underlying brain aging and neurodegenerative diseases, such as a defect in glucose transport and metabolism, excessive reactive oxygen species (ROS) production, and neuronal loss. Another well-established characteristic of mitochondrial function is the production of ROS, which is enhanced by mitochondria-directed toxins (e.g., rotenone and paraquant). Lithium alleviates this consequence.[130] As summarized by Nciri et al.[131]: "These properties of lithium could be useful to prevent and/or counteract dysfunctions underlying brain ageing and neurodegenerative diseases, such as a defect in glucose transport and metabolism, excessive reactive oxygen species (ROS) production and neuronal loss."

Maurer et al.[132] reported that the activities of mitochondrial complexes I+III and II+III were elevated by lithium dose-dependently and suggested that lithium stimulates mitochondrial respiratory chain enzyme activities at clinically relevant concentrations. However, assessing the effect of 6-week lithium treatment of bipolar depressed patients on their leukocyte electron transport chain (ETC) complexes I–IV activities, de Sousa et al.[25] found that only mitochondrial complex I activity was significantly increased with no changes in other complexes. Elucidating the possible involvement of glutathione in lithium's ability to decrease protein carbonylation and nitration produced by complex I inhibition, Nascimento et al.[133] found that glutathione is essential for lithium's amelioration of carbonylation but not nitration.

Omic studies have further provided compelling evidence that lithium positively regulates mitochondrial function via its effect on inositol metabolism. A microarray gene expression study found that the phosphatidylinositol metabolism is affected by lithium, that mRNA expression changes were observed in genes encoding for ATP binding- or ATPase activity-related proteins, and that the drug significantly affected mitochondia-related genes.[134] Genomic and proteomic studies of our group in the frontal cortex of mice treated chronically with lithium and in inositol metabolism-related knockout mice indicated upregulation of mitochondria-related genes and proteins.[135, 136] In studies investigating the effects of lithium on D-amphetamine- and methamphetamine-induced mitochondrial respiratory chain activity dysfunction in rat brain, lithium's preventive effect varied depending on the brain region and the treatment regimen.[137–139] Similar results of the rescue by lithium were obtained with chronic NMDA-induced mitochondrial damage.[140] To sum up, there is a consensus that lithium's neuroprotective effect involves amelioration of mitochondrial dysfunction. Interestingly, as for the molecular mechanism mediating this effect, as far as our literature search could allocate, only IMPase inhibition has been raised.

Apoptosis

Lithium was consistently shown to prevent apoptotic processes,[141] possibly through the inhibition of GSK-3β.[142–144] The GSK-3 signaling pathway modulates apoptosis.[145] Ngok-Ngam et al.[146] demonstrated that GSK-3 promotes the mitochondrial translocation of p53, enabling its interaction with B-cell lymphoma 2 (Bcl-2) to allow Bax oligomerization and the subsequent release of cytochrome *c*. This leads to caspase activation in the mitochondrial pathway of intrinsic apoptotic signaling. Studies have shown that by inhibiting GSK-3 lithium is able to (a) decrease proapoptotic factors such as p53 and bax[143]; (b) normalize intracellular calcium levels and regulate ATP production, which could antagonize the activation of apoptotic machinery[147]; and (c) normalize Bcl-2;[148] (d) Attenuate serum deprivation-induced apoptosis in PC12 cells through the regulation of the Akt/FoxO1 signaling pathways.[149]

Finally, Aminzadeh et al.[150] examined the effects of lithium on high glucose-induced neurotoxicity. They reported that lithium pretreatment protected PC12 cells against mitochondrial apoptotic-induced cell death. It reduced ROS generation, Bax/Bcl-2 ratio, caspase-3 activation and JNK, and P38 MAPK phosphorylation.

Lithium and neuropsychiatric disorders

Preface

As Mattson[151] eloquently summarized "neuronal death underlies the symptoms of many human neurological disorders, including Alzheimer's, Parkinson's and Huntington's diseases, stroke, and amyotrophic lateral sclerosis. Many signals can initiate apoptosis in neurons, including lack of neurotrophic factor support, overactivation of glutamate receptors (leading to calcium influx), increased oxidative stress and metabolic stress."

As mentioned earlier, cumulative data suggest that lithium induces a variety of neuroprotective effects[152] including increasing levels of the neuroprotrophic factor BDNF and the antiapoptotic factor Bcl-2 along with decreasing proapoptotic factors, enhancing mitochondrial function, decreasing activity of key proinflammatory cytokines, affecting autophagy, and attenuating apoptosis. In addition, lithium has been reported to enhance neurogenesis and gliogenesis[153]; reduce the lesions size in animal models of hemorrhagic stroke, HD, AD, ALS, fragile X syndrome (FXS), PD, and multiple sclerosis (MS)[22]; and decrease the prevalence of several neurological disorders.[154] This data along with additional evidence presented in the succeeding text opened new avenues for lithium as a candidate treatment in these debilitating neurological disorders.

Stroke

Stroke occurs when there is poor blood flow to the brain, which may result in cell death. There are two main types of stroke: ischemic, due to occlusion of a brain blood vessel, and hemorrhagic, due to intracerebral bleeding. Stroke is reported to be associated with dysfunctional autophagy,[155] neuroinflammation,[156–158] decreased BDNF levels,[159] aberrant mitochondrial function,[160] and neuronal apoptosis.[161]

Animal studies

Neuroprotective effects of lithium against cerebral ischemia were first demonstrated in animal models. Long-term pretreatment with lithium has been reported to decrease infarct volume and reduce neurological deficits in a model induced by permanent middle cerebral artery occlusion.[162] Similarly, Xu et al.[163] and Silachev et al.[164] found a beneficial effect of lithium in transient occlusion models followed by reperfusion. In a rat model of middle cerebral artery occlusion, postinsult lithium administration, a better approximate of the clinical management of stroke, reduced brain infarct volume. This was achieved using therapeutically-relevant doses (in bipolar disorder) and was accompanied by facilitated neurological recovery.[165] These studies of prestroke and post stroke lithium administration encouraged human studies of lithium in stroke.

Human studies

To the best of our knowledge, only two human studies dealt with lithium in stroke, both suggesting that lithium treatment may be beneficial for patients following stroke. Lan et al.[166] conducted a population-based retrospective study in lithium-treated bipolar patients. They report significantly reduced risk of stroke associated both with the highest cumulative lithium dose and the longest cumulative exposure period. Mohammadianinejad et al.[167] conducted a double-blind, placebo-controlled, randomized clinical trial of lithium treatment on early motor recovery of patients following ischemic stroke. Treatment was initiated 48h after stroke and continued for 30days. Motor impairment was evaluated on the 5th and 30th day of treatment. Patients on lithium improved significantly better compared with the placebo group. These preliminary results call for further research of lithium treatment in stroke.

Huntington's disease

HD, a devastating inherited, autosomal dominant, neurodegenerative disease, is caused by a mutant huntingtin protein due to an expanded polyglutamine (polyQ) cytosine-adenine-guanine (CAG) repeat sequence at exon 1 of the gene huntingtin.[168] The mutant huntingtin accumulates in intraneuronal aggregates[169] leading to the loss of neurons in key brain areas.[170] The symptoms of HD include motor dysfunction characterized by chorea, psychiatric disturbances, and cognitive decline that might lead to dementia. There is no treatment to reverse the deteriorating course of the disease.

HD is reported to demonstrate dysfunctional autophagy,[171] deficient BDNF[172] and mitochondrial function,[173–175] as well as neuroinflammtion[176, 177] and neuronal apoptosis,[178, 179] all suggested to be alleviated by lithium.

Nonhuman studies

HD models exist in cells, flies (Drosophila) and rodents (mice and rats). In a cell and in a Drosophila model, lithium reduced polyQ toxicity.[180, 181] In a mouse model the drug significantly improved motor performance.[182] In an excitotoxic rat model induced by quinolinic acid presenting hyperactivation of NMDA receptors—which appears to contribute to the pathophysiology of HD[183]—lithium treatment reduced the loss of striatal medium-sized neurons, thereby suppressing striatal lesions, reducing neurodegeneration and stimulating cell proliferation.[184, 185] Three different studies explored the effects of a combination of lithium with a second drug in HD animal models: the first studied lithium with rapamycin in a fly model and the second and third studied lithium with valproate in mouse models. The first two demonstrated synergistic beneficial effect of the combination on neurodegeneration and motor deficits compared with either agent alone.[48, 186] It is conceivable that combining lithium that activates autophagy in an mTOR-independent pathway with rapamycin that activates autophagy through mTOR-dependent pathway results in stronger induction of autophagy.[187] The third study showed that preconditioning of bone marrow-derived mesenchymal stem cells with the presence, compared with the absence, of combined lithium and valproic acid enhanced therapeutic efficacy in a mouse model of HD.[188] It should, nevertheless, be mentioned that Wood and Morton[182] reported that chronic lithium treatment had variable effects on motor behavior and survival of mice transgenic for the Huntington's disease mutation. Overall the data support the notion that lithium administration, either by itself or in combination with other drugs, may be beneficial in the treatment of HD patients.

Human studies

Regrettably, only few human clinical studies were conducted in the 1970s. These small studies explored lithium treatment for several weeks in patients with HD. Overall, no improvement in neurological symptomatology was noted suggesting that lithium should not be considered as a potential treatment in HD.[189–191] In contrast, three case reports in which low doses of lithium were administered over extended periods of 2–4 years showed no further progression of chorea over this period.[192] Additional solid data are required to reappraise long-term lithium treatment in HD as an agent slowing the progression of the disease.

Alzheimer's disease

AD is the most common neurodegenerative form of dementia characterized clinically by deteriorated cognition and pathophysiologically by brain atrophy accompanied by the finding of neurofibrillary tangles and amyloid-β (Aβ) plaques in brain tissue. Deteriorated cognition is the hallmark of AD. Hence, it is of note that among bipolar patients, lithium-treated subjects showed better scores than nonlithium subjects in tests measuring a variety of memory functions.[193, 194] Nevertheless, reviewing the literature on lithium treatment and cognition in bipolar subjects, Paterson and Parker[195] suggested that "any impact of lithium on memory in patients with bipolar disorders is unclear as the literature is contradictory and any such effect may be overshadowed by the greater impact of residual mood symptoms."

AD is associated with dysfunctional autophagy,[196] neuroinflammation,[177, 197, 198] deteriorated BDNF signaling,[199, 200] aberrant mitochondrial function[174, 175, 197] and neuronal apoptosis,[178, 179] all suggested to be modified by lithium.

Tissue and animal studies

Since abnormal levels of GSK-3 were reported to be associated with pathogenesis and neuronal death in subjects with AD,[201] lithium, a GSK-3 inhibitor,[16, 202] was studied for its effect on cellular and animal models of AD. In this regard, lithium inhibits the phosphorylation of glycogen synthase kinase 3α and 3β, which are related to amyloid precursor protein (APP) processing and tau hyperphosphorylation in pathological conditions, respectively. Thus, several studies claimed that lithium treatment induces a significant reduction of GSK-3 activity associated with decreased tau phosphorylation, insoluble aggregated tau accumulation and axonal degeneration.[203–206] Using double transgenic mice (AβPPSwe/PS1A246E) displaying amyloid deposits, Zhang et al.[50] demonstrated reduced γ-cleavage of APP followed by lowering of amyloid-β plaque formation. These changes were accompanied by improved spatial learning and memory abilities along with increased autophagy. Hence, nonhuman studies suggest beneficial effects of lithium on the pivotal pathology of AD (neurofibrillary tangles and amyloid-β plaques) and on cognitive dysfunction, paving the way to human studies.

Risk of Alzheimer's disease in lithium users

Several studies explored whether lithium use may reduce risk of dementia, in general, and AD, in particular. Terao et al.[207] compared nondemented subjects over 60 years old who received or were currently receiving lithium with a control group. The results showed that lithium-receiving (past and/or present) subjects had better Mini-Mental Status Examination (MMSE, a test that measures cognition) scores. The authors suggest that their results provide some evidence to support the contention that lithium could offer hope as a preventive treatment for Alzheimer's disease. Nunes et al.[208] and Kessing et al.[209] further demonstrated that the clinical use of lithium in subjects with bipolar disorder reduced the incidence of dementia in old age. In a population-based, nested, case-control study of above 60,000 cases of Alzheimer's disease and above 120,000 matched controls, Cheng et al.[210] have recently assessed whether the risk of Alzheimer's disease is associated with the use of lithium. In contrast with Nunes et al and Kessing et al, their findings did not support an increased or decreased risk of Alzheimer's disease among lithium users when taking into account the indication for lithium use as a potential confounding factor. Furthermore, using data from the General Practice Research Database in the United Kingdom, Dunn et al.[211] found that patients who received lithium treatment had a higher risk of diagnosis of dementia, believed to correlate with higher doses of the drug.[211] Interestingly, studying the association between groundwater lithium with the diagnosis of bipolar disorder and dementia in the United States, Parker et al.[212] have recently reported that "the purported association of high-lithium concentrations in drinking water with mental health disorders (including dementia) is driven by unaccounted variation in demographics, health care resources, and diagnosis practices."

Human clinical studies

Theorizing that GSK-3 inhibition by lithium may reduce the pathological brain changes associated with AD and possibly lead to beneficial effects on its symptomatology led clinicians to carry out clinical trials of lithium in AD. In the succeeding text we summarize them in a chronological manner according to the year of publication.

First, Macdonald et al.[213] conducted an open-labeled feasibility and tolerability study of low-dose lithium administered to AD patients for 1 year along with a comparison group of AD patients matched for cognition and age not receiving lithium therapy. No beneficial effect of lithium was observed according to MMSE and the lack of difference in deaths. Lithium treatment had relatively few mild and reversible side effects.

Second, Hampel et al.[214] conducted a placebo-controlled, randomized, single-blind, multicenter study in which lithium was given to 71 AD patients over a period of 10 weeks. Cerebrospinal fluid (CSF) and plasma biomarkers (total tau, phosphorylated tau, and Aβ42) were monitored during that time period, and there was no significant difference in the biomarkers or in cognitive performance as compared with patients receiving placebo. However, further different analyses of a subset of Humple et al.'s study reported, in two additional papers, positive results. (1) Leyhe et al.[215] found a significant increase in serum BDNF levels and a significant decrease in the Alzheimer's Disease Assessment Scale-Cognitive Subscale (ADAS-Cog subscale) score. Diminution of cognitive impairment was inversely correlated with serum lithium concentrations. The authors suggest that elevation of BDNF might be part of a neuroprotective effect of lithium in AD patients. (2) Based on their previous finding that CSF glial cell line-derived neurotrophic factor (GDNF) levels were elevated in early AD patients, Straten et al.[216] assessed the influence of lithium treatment on serum and CSF GDNF concentrations in a subset of Humple et al.'s study. They found a significant negative correlation of serum lithium concentration both with CSF GDNF concentration at the end of treatment and with the difference in CSF GDNF concentration before and after treatment. However, there was no difference in serum or CSF GDNF levels between the patients after the treatment with lithium or placebo.

Third, Forlenza et al.[217] conducted a long-term (2 years) lithium treatment versus placebo in patients with mild cognitive impairment (MCI). The researchers measured CSF Aβ42, phosphorylated tau and total tau, as well as the subjects' cognitive performance. They found reduced levels of CSF phosphorylated tau in the lithium-treated patients and nonsignificant fewer conversions from mild cognitive impairment to AD. In elderly women with mild cognitive impairment small lithium doses were more potent than placebo in slowing the rate of cognitive deterioration over a year. The authors suggested that lithium may slow the disease progression from cognitive impairment to dementia.[217]

Fourth, Nunes et al.[218] evaluated the effect of a microdose of 300 μg lithium (as compared with 150–600 mg used in other AD clinical trials) administered once daily in AD patients for 15 months. The treated group showed no decreased performance in the MMSE test, while lower scores were observed for the control group.

Finally a recent metaanalysis by Matsunaga et al.[219] of 3 randomized placebo-controlled studies including 232 patients with AD and individuals with MCI indicated that lithium treatment may have beneficial effects on cognitive performance in these disorders.

To sum up, in their review of the possible role of lithium in AD, Morris and Berk[220] conclude that "data regarding efficacy in human trials and animal models of AD are mixed, but recent data using 'micro-dose' lithium in MCI is encouraging, hence lithium could be a putative multi target treatment in these patients."

Parkinson's disease

PD is characterized by neuronal brain Lewy bodies that are aggregates containing mutant protein, mainly α-synuclein.[221, 222] PD is manifested by resting tremor, muscular rigidity, bradykinesia and postural instability. The disease may also be accompanied by psychiatric comorbidity and the development of dementia. It is associated with dysfunctional autophagy,[223–226] neuroinflammation,[177, 227] aberrant mitochondrial function,[174, 175, 228] increased risk of dyskinesia in certain genetic variants of the BDNF receptor[229] and neuronal apoptosis,[178, 179] all suggested to be alleviated by lithium.

Studies in neuronal cells in culture

1-methyl-4-phenylpyridinium (MPP) and rotenone, inhibitors of mitochondrial respiration complex I, are frequently used to model PD characteristics in cells and animals.[230] Applying these agents to human neuroblastoma SH-SY5Y cells King et al.[142] observed time- and dose-dependent caspase-3 (the apoptosis-associated cysteine protease) activation and morphological changes characteristic of apoptosis. Lithium attenuated MPP- and rotenone-induced caspase-3 activation. Based on reports that lithium inhibits GSK-3 the authors conclude that "overall, these results indicate that inhibition of GSK3beta provides protection against the toxic effects of agents such as MPP and rotenone that impair mitochondrial function." In this regard, it was also demonstrated that lithium administered to cultured neurons prevents 6-hydroxydopamine-induced neuronal death suggested to be mediated by its inhibiting of GSK-3β.

Soleimani and Ghasemi[231] investigated the effect of lithium on dopaminergic differentiation of human immortalized RenVm cells aiming to obtain a population of fully differentiated cells for transplantation in Parkinson's disease. Their results indicated that LiCl can promote dopaminergic differentiation of RenVm cells in a dose-dependent manner.

Animal studies

Animal models of PD include the "classic" models based on neurotoxins selectively targeting catecholaminergic neurons (i.e., 6-hydroxydopamine [6-OHDA], 1-methyl-1,2,3,6-tetrahydropiridine [MPTP], agricultural pesticides), along with models that introduce genetic mutations similar to those found in familial cases of PD (α-synuclein, DJ-1, PINK1 and Parkin) or models that selectively disrupt nigrostriatal neurons (MitoPark, Pitx3, and Nurr1).[232] Some of these animal models were applied to explore the effects of lithium administration as follows.

In the MPTP-induced neurotoxicity PD mice model, Youdim and Arraf[233] reported that chronic lithium treatment normalized the MPTP-induced downregulation and upregulation of striatal Bcl-2 and Bax, respectively. Lithium at therapeutically relevant concentrations has been shown to facilitate clearance of the mutant form of α-synuclein, an autophagy substrate in the "classic" animal models of PD induced by neurotoxins such as rotenone, 6-OHDA, 1-methyl-4-phenylpyridinium (MPPM+), and the MPP precursor MPTP.[41] Evaluating the effect of lithium treatment in an aged mouse model expressing a Parkin mutation within dopaminergic neurons Lieu et al.[234] found that low-dose lithium treatment prevented motor impairment, striatal dopaminergic degeneration, and Parkin-induced striatal astrogliosis and microglial activation. Their findings provide further validation that lithium could be repurposed as a therapy for PD and suggest that the drug's antiinflammatory effects may contribute to its neuroprotective mechanisms. It is noticeable that Lei et al.[235] hypothesized that since lithium lowers neuronal tau, the drug endangers substantia nigra (SN) iron accumulation (a robust feature of PD), thereby inducing parkinsonian neurodegeneration. Indeed, using magnetic resonance imaging of lithium-treated subjects they found iron levels elevation in the SN. In mice, they found that lithium treatment lowered brain tau levels and increased nigral and cortical iron elevation, and in neuronal cultures, that lithium attenuated iron efflux by lowering tau protein that traffics APP to facilitate iron efflux. They also found that tau- and APP-knockout mice were protected against lithium-induced iron elevation and neurotoxicity. They conclude that their findings "challenge the appropriateness of lithium as a potential treatment for disorders where brain iron is elevated."

Human studies

Lithium's effect to stabilize the increased and decreased behavioral manifestations of bipolar disorder ignited the imagination of clinicians to examine the drug's efficacy in normalizing the increased and decreased behavioral manifestations of the levodopa-induced on-off phenomenon in PD patients. Following a dispute concerning the therapeutic value of lithium in modifying PD manifestations and diminishing levodopa-induced dyskinesias, McCaul and Stern[236] ran a single-blinded clinical trial of up to 6-week lithium in 21 PD patients including eight patients on levodopa without drug-induced movements, seven on levodopa with an acceptable degree of dyskinesia, four who were untreated with levodopa and two who were unable to tolerate levodopa. The only possible beneficial effect of lithium was the alleviation of painful muscle spasms and cramps in three patients with no anti-Parkinsonian effect. A decade later, Coffey et al.[237] reported that lithium treatment of five patients with the "on-off" phenomenon of parkinsonism at doses resulting in therapeutic serum concentrations (according to bipolar disorder) induced dyskinesias associated with reduction in akinesia. Recently, Guttuso[238] reported that in a case series of PD patients, low-dose lithium adjunct therapy reduced offtime and did not increase dyskinesias calling for placebo-controlled trials to assess low-dose lithium's effects on offtime in PD subjects. Nevertheless, it should be noted that even the low lithium dose was poorly tolerated in two PD patients with dementia.

Fragile X syndrome

Fragile X syndrome (FXS) is an inherited disorder caused by abnormal expansion of the trinucleotide CGG repeat in the promoter region of the fragile X mental retardation-1 (FMR1) gene leading to various degrees of epigenetic transcriptional silencing of this gene and its protein product, fragile X mental retardation protein (FMRP).[239, 240] Since this gene is required for the formation and function of synapses and development of neural circuits, it is clinically characterized by intellectual disability, ASD, hyperactivity, sensory hypersensitivity, hyperarousal and anxiety. Some of the cellular processes affected by lithium have been implicated in FXS including autophagy,[241] BDNF signaling,[242] mitochondrial function and ROS.[243]

Animal studies

Lithium treatment has been studied extensively in both mouse and fruit fly models of FXS, and it has been shown to reverse numerous behavioral, physiological, cellular and molecular phenotypes[13] as follows.

Drosophila

In a Drosophila FXS model Choi et al.[244] showed that lithium or metabotropic glutamate receptor (mGluR) antagonists administration rescued age-related cognitive impairments, and Mcbride et al.[245] showed that lithium treatment of these flies at adulthood increased naive courtship and restored short-term memory.

Rodents

Several models of FXS have been developed, including the *Fmr1* knockout mouse,[246] the conditional knockout mice and knockout rats displaying certain FXS behaviors including autism-relevant behavioral phenotypes.[247] Min et al.[248] reported that in the FVB/NJ FMR1 knockout mice the inhibitory serine phosphorylation of GSK-3 is impaired, corroborating with Yuskaitis et al.'s[249] and Liu et al.'s[115] findings that lithium ameliorated impaired learning ability, dysfunctional social interactions and anxiety levels as well as blocked aberrant dendritic spine morphology in several FXS mouse models. Similarly, King and Jope[142] explored if lithium treatment can ameliorate impairments in four cognitive tasks in Fmr1 KO mice reporting that (a) Fmr1 KO mice exhibit severe impairments in these cognitive tasks; (b) that lithium is effective in normalizing cognition in these tasks; (c) that such effectiveness can be seen in young or adult mice; and (d) that for the cognitive improvements to be sustained lithium administration has to be continued. It may be summarized that lithium administration in FXS animal models provides evidence that lithium administration may be beneficial for FXS patients.

Human studies

The use of the earlier models led to potential targets for pharmaceutical treatment; however, although treatments were shown to be efficient in preclinical studies, they failed thus far to turn into therapy.[247] As for lithium, only Berry-Kravis et al.[250] conducted a pilot clinical trial of lithium (titrated to serum levels of 0.8–1.2 mEq/L) in 15 individuals with FXS aged 6–23 years. They reported improvements in behavior and function following 2 months of treatment.

To sum up regarding lithium's role in FXS, Liu et al.[13] commented that "Lithium treatment has been studied extensively in both mouse and fruit fly models of FXS, and it has been shown to reverse numerous behavioral, physiological, cellular, and molecular phenotypes…(and results) of a pilot clinical trial…indicated that…a double-blind clinical trial of lithium treatment in FXS patients is now needed."

Amyotrophic lateral sclerosis

ALS or motor neuron disease, also known as Lou Gehrig's disease, is an adult-onset devastating neurodegenerative disease with, up to date, no effective treatment, characterized by progressive loss of brain and spinal cord motor neurons, resulting in generalized weakness, muscle atrophy, paralysis, and, eventually, death within 3–5 years of disease onset.[251, 252] ALS is a familial disease in about 10% of patients, with the remaining 90% developing sporadic ALS.[253] ALS is associated with dysfunctional autophagy,[223, 254] neuroinflammation,[177, 255] aberrant mitochondrial function[174, 175, 256] and enhanced apoptosis,[178, 179] all known to be alleviated by lithium.

Nonhuman studies

Certain aspects of ALS have been modeled, guided by recent insights into the disease's genetic and molecular basis, to investigate the complex process of motorneuron degeneration. The models vary from in vitro biochemical systems to cell culture systems, invertebrates and nonmammalian vertebrates, extend to rodent models and, recently, to human patient-derived stem cell models.[257] Thus, rodent models for ALS-associated mutations in Cu/Zn superoxide dismutase-1 (*SOD1*), transactive response DNA-binding protein (*TARDBP*), fused in sarcoma (*FUS*), and chromosome 9 open reading frame 72 (*C9orf72*) were developed. These animal models provided important insights into motorneuron degeneration pathways (Ferraiuolo et al.[258]). In particular, mouse models in which human genomic mutant *SOD1* is overexpressed have been used intensively.[257, 259] The model demonstrates ALS-like phenotypes such as formation of intracellular aggregates of SOD1 in the brain and spinal cord, a variety of behavioral abnormalities and premature death.

Both positive and negative results of lithium in ALS animal models have been reported. Thus, Pizzasegola et al.[260] and Gill et al.[261] reported that lithium treatment did not improve disease progression in different strains of SOD1 mutant mice. Contrarily, several groups reported that treatment with lithium either alone or in combination with an antioxidant or with valproate, another mood stabilizer, improved motor function and slowed disease progression in a mouse model of ALS.[262–265] The earlier results, although mixed, led some authors to suggest that lithium may offer some promise as a treatment for human patients affected by ALS.

Human studies

Clinical trials of lithium in ALS patients were carried out either in combination with other drugs (mainly riluzole, the only FDA-approved drug for ALS) or on its own.

Lithium combined with another drug

Fornai et al.[263] carried out a 15-month randomized pilot clinical trial in ALS patients. They reported that the combination of lithium and riluzole resulted in reduced mortality compared with patients treated with riluzole alone. Aggarwal et al.[266] conducted a double-blind, placebo-controlled trial. Patients who were receiving riluzole were assigned to either lithium or placebo. In an interim analysis, decreased functioning or death was not different between the two groups, results that led to halt the study. Verstraete et al.[267] aimed to determine the safety and efficacy of lithium in ALS. Over 100 patients were randomized to receive lithium or placebo as add-on treatment with riluzole. Similarly to Aggarwal et al.'s results, no beneficial effect of lithium addition on survival or functional decline was discerned. In this regard, Yáñez et al.[268] reported that lithium-induced neuroprotection may be antagonized by riluzole, perhaps suggesting that lithium should not be researched in ALS as addition to riluzole.

Lithium by its own

Overall, clinical trials of lithium on its own were negative. In a multicenter consortium study, Chiò et al.[269] reported that lithium treatment was not effective in ameliorating ALS symptoms. Miller et al.[270] found similar results also reporting that the lithium group was more prone to adverse events. In a large cohort, the UK Motor Neurone Disease (UKMND)-lithium carbonate in ALS (LiCALS) Study Group[271] conducted a double-blind, placebo-controlled trial of lithium administered daily for 18 months. Two hundred fourteen patients received either lithium or placebo. Again, no evidence for benefit of lithium on survival was noted, but there were no safety concerns. Despite the previously mentioned, discussing the methodological problems and equivocal results, Agam and Isrealson[272] concluded: "We believe that the potential beneficial effect of lithium for neurodegenerative disorders deserves serious reconsideration" and that "lithium studies for ALS treatment should not be halted prematurely." Indeed, in a metaanalysis of pharmacogenetic interactions, van Eijk et al.[273] investigated whether genetic subgroups in ALS trials did respond to lithium treatment. When clinical data were matched with the UNC13A (*The UNC13A protein is involved in synaptic vesicle maturation and neuronal outgrowth; see Broeke et al.*[274]) and C9orf72 genotypes, the results showed that 12-month survival probability for UNC13A carriers on lithium improved, suggesting that the patient's genotype should be considered in clinical trials.

Multiple sclerosis

MS is an inflammatory demyelinating disease manifested predominantly by demyelinated lesions in the CNS white matter.[274] The disease course is characterized by relapsing-remitting nature and secondary progressive disease development over time. Signs and symptoms depend on the amount of nerve damage. Unfortunately the therapeutic effects of the existing drugs are limited, in general, and on long-term outcome, in particular.[206] Neuropathology and neuroprotective mechanisms associated with MS and its treatment seem to involve autophagy,[275] BDNF signaling,[276, 277] neuroinflammation,[177, 278, 279] mitochondrial function[280] and necroptosis (an alternative mode of regulated cell death mimicking features of apoptosis and necrosis),[281] and, as mentioned earlier, lithium affects all these biological processes.

Animal studies

The three most characterized animal models of MS are (a) experimental autoimmune/allergic encephalomyelitis (EAE); (b) a virally induced chronic demyelinating disease, the Theiler's murine encephalomyelitis virus (TMEV) infection; and (c) a toxin-induced demyelination. Mainly, EAE is a model that closely mimics the autoimmune pathogenesis in MS, frequently applied to explore potential novel treatments. The two other models mimic axonal injury and remyelination occurring in MS.[282, 283]

De Sarno et al.[89] demonstrated that pretreatment with lithium at a therapeutically relevant dose suppressed the onset of experimental EAE in mice and reduced demyelination, microglia activation, and leukocyte infiltration in the spinal cord. When administered postimmunization (after the onset of EAE), lithium led to reduced disease severity and facilitated partial recovery. In the same line, Ahn et al.[284] recently evaluated the potential role of lithium in a rat EAE model. Lithium significantly delayed the onset of EAE paralysis, ameliorated its severity, reduced serum level of the proinflammatory tumor necrosis factor but not of IL10, suppressed the activation of ionized calcium binding protein-1-positive microglial cells and vascular cell adhesion molecule-1 expression in the spinal cords.

Human studies

Rinker et al.[285] conducted a retrospective chart review of the safety and tolerability of lithium among 21,847 US veterans with MS of whom 101 were treated by lithium for at least 6months. Eighteen percent of these veterans experienced lithium-associated adverse events associated with later age of MS onset. Lithium did not increase the risk of brain lesions, but relapse rates per year were higher on lithium. In addition, Expanded Disability Status Scale score change was greater in the off-lithium period than the on-lithium period. The authors conclude that no consistent effect of lithium on MS disease activity was apparent.

Summary

In this chapter, we aimed to review studies reporting the effect of lithium treatment on autophagy, BDNF signaling, neuroinflammation, mitochondrial function, and apoptosis. Since these processes are involved in a variety of brain diseases beyond bipolar disorders including stroke, HD, AD, PD, FXS, ALS, and MS, we further reviewed lithium's role in models of these diseases and in patients. Based on data coming from cellular and animal models, human clinical trials were held in neurodegenerative and neurodevelopmental diseases suggesting the following: (1) Preliminary results call for further research of lithium treatment in stroke.

(2) Albeit negative results, additional solid data are required to reappraise long-term (years) lithium treatment in HD subjects as an agent slowing the progression of the disease. (3) Data regarding efficacy of lithium in AD subjects are mixed, but recent data using "microdose" lithium in MCI and AD seem encouraging. (4) Based on small studies, lithium seems not to be effective in PD patients; however, it may still be that low-dose lithium adjunct therapy will reduce offtime while not increasing dyskinesias. (5) Preliminary human data suggest the need for double-blind clinical trial of lithium treatment in FXS patients. (6) Lithium does not seem to be effective in ALS, although it may still be of benefit in a certain genetic subgroups of ALS patients yet to be defined. (7) Lithium does not seem to be effective in MS.

References

1. Cade JF. Lithium salts in the treatment of psychotic excitement. 1949. *Bull World Health Organ*. 2000;78(4):518–520.
2. Schioldann J. John Cade's seminal lithium paper turns fifty. *Acta Psychiatr Scand*. 1999;100(6):403–405.
3. Belmaker RH. Bipolar disorder. *N Engl J Med*. 2004;351(5):476–486.
4. Muller-Oerlinghausen B, Lewitzka U. Lithium reduces pathological aggression and suicidality: a mini-review. *Neuropsychobiology*. 2010;62(1):43–49.
5. Fawcett JA. Lithium combinations in acute and maintenance treatment of unipolar and bipolar depression. *J Clin Psychiatry*. 2003;64(Suppl 5):32–37.
6. Baethge C, Gruschka P, Berghofer A, et al. Prophylaxis of schizoaffective disorder with lithium or carbamazepine: outcome after long-term follow-up. *J Affect Disord*. 2004;79(1–3):43–50.
7. Maj M. Lithium prophylaxis of schizoaffective disorders: a prospective study. *J Affect Disord*. 1988;14(2):129–135.
8. Leucht S, Helfer B, Dold M, Kissling W, McGrath JJ. Lithium for schizophrenia. *Cochrane Database Syst Rev*. 2015;(10):CD003834.
9. Citrome L. Adjunctive lithium and anticonvulsants for the treatment of schizophrenia: what is the evidence? *Expert Rev Neurother*. 2009;9(1):55–71.
10. Leucht S, Kissling W, McGrath J. Lithium for schizophrenia. *Cochrane Database Syst Rev*. 2007;(3):CD003834.
11. Serret S, Thummler S, Dor E, Vesperini S, Santos A, Askenazy F. Lithium as a rescue therapy for regression and catatonia features in two SHANK3 patients with autism spectrum disorder: case reports. *BMC Psychiatry*. 2015;15:107.
12. Siegel M, Beresford CA, Bunker M, et al. Preliminary investigation of lithium for mood disorder symptoms in children and adolescents with autism spectrum disorder. *J Child Adolesc Psychopharmacol*. 2014;24(7):399–402.
13. Liu Z, Smith CB. Lithium: a promising treatment for fragile X syndrome. *ACS Chem Neurosci*. 2014;5(6):477–483.
14. Jakobsson E, Arguello-Miranda O, Chiu SW, et al. Towards a unified understanding of lithium action in basic biology and its significance for applied biology. *J Membr Biol*. 2017;250(6):587–604.
15. Roux M, Dosseto A. From direct to indirect lithium targets: a comprehensive review of omics data. *Metallomics*. 2017;9(10):1326–1351.
16. Klein PS, Melton DA. A molecular mechanism for the effect of lithium on development. *Proc Natl Acad Sci U S A*. 1996;93(16):8455–8459.

17. Gould TD, Manji HK. The Wnt signaling pathway in bipolar disorder. *Neuroscientist.* 2002;8(5):497–511.
18. Hallcher LM, Sherman WR. The effects of lithium ion and other agents on the activity of myo-inositol-1-phosphatase from bovine brain. *J Biol Chem.* 1980;255(22): 10896–10901.
19. York JD, Guo S, Odom AR, Spiegelberg BD, Stolz LE. An expanded view of inositol signaling. *Adv Enzym Regul.* 2001;41:57–71.
20. Beaulieu JM, Marion S, Rodriguiz RM, et al. A beta-arrestin 2 signaling complex mediates lithium action on behavior. *Cell.* 2008;132(1):125–136.
21. Dell'Osso L, Del Grande C, Gesi C, Carmassi C, Musetti L. A new look at an old drug: neuroprotective effects and therapeutic potentials of lithium salts. *Neuropsychiatr Dis Treat.* 2016;12:1687–1703.
22. Chiu CT, Wang Z, Hunsberger JG, Chuang DM. Therapeutic potential of mood stabilizers lithium and valproic acid: beyond bipolar disorder. *Pharmacol Rev.* 2013;65(1):105–142.
23. Hantson P. Mechanisms of toxic cardiomyopathy. *Clin Toxicol (Phila).* 2019;57(1):1–9.
24. Yang K, Chen Z, Gao J, et al. The key roles of GSK-3beta in regulating mitochondrial activity. *Cell Physiol Biochem.* 2017;44(4):1445–1459.
25. de Sousa RT, Streck EL, Zanetti MV, et al. Lithium increases leukocyte mitochondrial complex I activity in bipolar disorder during depressive episodes. *Psychopharmacology.* 2015;232(1):245–250.
26. Motoi Y, Shimada K, Ishiguro K, Hattori N. Lithium and autophagy. *ACS Chem Neurosci.* 2014;5(6):434–442.
27. van den Ameele S, van Diermen L, Staels W, et al. The effect of mood-stabilizing drugs on cytokine levels in bipolar disorder: a systematic review. *J Affect Disord.* 2016;203:364–373.
28. De-Paula VJ, Kerr DS, Scola G, Gattaz WF, Forlenza OV. Lithium distinctly modulates the secretion of pro- and anti-inflammatory interleukins in co-cultures of neurons and glial cells at therapeutic and sub-therapeutic concentrations. *Curr Alzheimer Res.* 2016;13(8):848–852.
29. Nassar A, Azab AN. Effects of lithium on inflammation. *ACS Chem Neurosci.* 2014;5(6):451–458.
30. Amare AT, Schubert KO, Baune BT. Pharmacogenomics in the treatment of mood disorders: strategies and opportunities for personalized psychiatry. *EPMA J.* 2017;8(3):211–227.
31. Dwivedi T, Zhang H. Lithium-induced neuroprotection is associated with epigenetic modification of specific BDNF gene promoter and altered expression of apoptotic-regulatory proteins. *Front Neurosci.* 2014;8:457.
32. Lazzara CA, Kim YH. Potential application of lithium in Parkinson's and other neurodegenerative diseases. *Front Neurosci.* 2015;9:403.
33. Forlenza OV, De-Paula VJ, Diniz BS. Neuroprotective effects of lithium: implications for the treatment of Alzheimer's disease and related neurodegenerative disorders. *ACS Chem Neurosci.* 2014;5(6):443–450.
34. Morlet E, Hozer F, Costemale-Lacoste JF. Neuroprotective effects of lithium: what are the implications in humans with neurodegenerative disorders? *Geriatr Psychol Neuropsychiatr Vieil.* 2018;16(1):78–86.
35. Leeds PR, Yu F, Wang Z, et al. A new avenue for lithium: intervention in traumatic brain injury. *ACS Chem Neurosci.* 2014;5(6):422–433.
36. Fornai F, Longone P, Ferrucci M, et al. Autophagy and amyotrophic lateral sclerosis: the multiple roles of lithium. *Autophagy.* 2008;4(4):527–530.
37. Scheuing L, Chiu CT, Liao HM, Linares GR, Chuang DM. Preclinical and clinical investigations of mood stabilizers for Huntington's disease: what have we learned? *Int J Biol Sci.* 2014;10(9):1024–1038.

38. Chen G, Zeng WZ, Yuan PX, et al. The mood-stabilizing agents lithium and valproate robustly increase the levels of the neuroprotective protein bcl-2 in the CNS. *J Neurochem*. 1999;72(2):879–882.
39. Scrivo A, Bourdenx M, Pampliega O, Cuervo AM. Selective autophagy as a potential therapeutic target for neurodegenerative disorders. *Lancet Neurol*. 2018;17(9):802–815.
40. Dunlop EA, Tee AR. mTOR and autophagy: a dynamic relationship governed by nutrients and energy. *Semin Cell Dev Biol*. 2014;36:121–129.
41. Sarkar S, Floto RA, Berger Z, et al. Lithium induces autophagy by inhibiting inositol monophosphatase. *J Cell Biol*. 2005;170(7):1101–1111.
42. Atack JR. Inositol monophosphatase, the putative therapeutic target for lithium. *Brain Res Brain Res Rev*. 1996;22(2):183–190.
43. Sade Y, Toker L, Kara NZ, et al. IP3 accumulation and/or inositol depletion: two downstream lithium's effects that may mediate its behavioral and cellular changes. *Transl Psychiatry*. 2016;6(12):e968.
44. Criollo A, Vicencio JM, Tasdemir E, Maiuri MC, Lavandero S, Kroemer G. The inositol trisphosphate receptor in the control of autophagy. *Autophagy*. 2007;3(4):350–353.
45. Williams RS, Cheng L, Mudge AW, Harwood AJ. A common mechanism of action for three mood-stabilizing drugs. *Nature*. 2002;417(6886):292–295.
46. Ryves WJ, Harwood AJ. Lithium inhibits glycogen synthase kinase-3 by competition for magnesium. *Biochem Biophys Res Commun*. 2001;280(3):720–725.
47. Chalecka-Franaszek E, Chuang DM. Lithium activates the serine/threonine kinase Akt-1 and suppresses glutamate-induced inhibition of Akt-1 activity in neurons. *Proc Natl Acad Sci U S A*. 1999;96(15):8745–8750.
48. Sarkar S, Krishna G, Imarisio S, Saiki S, O'Kane CJ, Rubinsztein DC. A rational mechanism for combination treatment of Huntington's disease using lithium and rapamycin. *Hum Mol Genet*. 2008;17(2):170–178.
49. Avrahami L, Farfara D, Shaham-Kol M, Vassar R, Frenkel D, Eldar-Finkelman H. Inhibition of glycogen synthase kinase-3 ameliorates beta-amyloid pathology and restores lysosomal acidification and mammalian target of rapamycin activity in the Alzheimer disease mouse model: in vivo and in vitro studies. *J Biol Chem*. 2013;288(2):1295–1306.
50. Zhang X, Heng X, Li T, et al. Long-term treatment with lithium alleviates memory deficits and reduces amyloid-beta production in an aged Alzheimer's disease transgenic mouse model. *J Alzheimers Dis*. 2011;24(4):739–749.
51. Caccamo A, Magri A, Medina DX, et al. mTOR regulates tau phosphorylation and degradation: implications for Alzheimer's disease and other tauopathies. *Aging Cell*. 2013;12(3):370–380.
52. Sarkar S, Rubinsztein DC. Inositol and IP3 levels regulate autophagy: biology and therapeutic speculations. *Autophagy*. 2006;2(2):132–134.
53. Li XZ, Chen XP, Zhao K, Bai LM, Zhang H, Zhou XP. Therapeutic effects of valproate combined with lithium carbonate on MPTP-induced parkinsonism in mice: possible mediation through enhanced autophagy. *Int J Neurosci*. 2013;123(2):73–79.
54. Cotman SL, Vrbanac V, Lebel LA, et al. Cln3(Deltaex7/8) knock-in mice with the common JNCL mutation exhibit progressive neurologic disease that begins before birth. *Hum Mol Genet*. 2002;11(22):2709–2721.
55. Kohlschutter A, Gardiner RM, Goebel HH. Human forms of neuronal ceroid-lipofuscinosis (Batten disease): consensus on diagnostic criteria, Hamburg 1992. *J Inherit Metab Dis*. 1993;16(2):241–244.
56. Chang JW, Choi H, Cotman SL, Jung YK. Lithium rescues the impaired autophagy process in CbCln3(Deltaex7/8/Deltaex7/8) cerebellar cells and reduces neuronal vulnerability to cell death via IMPase inhibition. *J Neurochem*. 2011;116(4):659–668.
57. Binder DK, Scharfman HE. Brain-derived neurotrophic factor. *Growth Factors*. 2004;22(3):123–131.

58. Acheson A, Conover JC, Fandl JP, et al. A BDNF autocrine loop in adult sensory neurons prevents cell death. *Nature*. 1995;374(6521):450–453.
59. Huang EJ, Reichardt LF. Neurotrophins: roles in neuronal development and function. *Annu Rev Neurosci*. 2001;24:677–736.
60. Garcia del Campo AL, Huecas V, Fernandez A, Puerta ML. Hematology and blood chemistry of macaws, Ara rubrogenys. *Comp Biochem Physiol A Comp Physiol*. 1991;100(4):943–944.
61. Jornada LK, Moretti M, Valvassori SS, et al. Effects of mood stabilizers on hippocampus and amygdala BDNF levels in an animal model of mania induced by ouabain. *J Psychiatr Res*. 2010;44(8):506–510.
62. Yasuda S, Liang MH, Marinova Z, Yahyavi A, Chuang DM. The mood stabilizers lithium and valproate selectively activate the promoter IV of brain-derived neurotrophic factor in neurons. *Mol Psychiatry*. 2009;14(1):51–59.
63. Hashimoto R, Takei N, Shimazu K, Christ L, Lu B, Chuang DM. Lithium induces brain-derived neurotrophic factor and activates TrkB in rodent cortical neurons: an essential step for neuroprotection against glutamate excitotoxicity. *Neuropharmacology*. 2002;43(7):1173–1179.
64. Gideons ES, Lin PY, Mahgoub M, Kavalali ET, Monteggia LM. Chronic lithium treatment elicits its antimanic effects via BDNF-TrkB dependent synaptic downscaling. *elife*. 2017;6:e25480. https://doi.org/10.7554/eLife.25480.
65. Fan M, Jin W, Zhao H, et al. Lithium chloride administration prevents spatial learning and memory impairment in repeated cerebral ischemia-reperfusion mice by depressing apoptosis and increasing BDNF expression in hippocampus. *Behav Brain Res*. 2015;291:399–406.
66. Motaghinejad M, Seyedjavadein Z, Motevalian M, Asadi M. The neuroprotective effect of lithium against high dose methylphenidate: possible role of BDNF. *Neurotoxicology*. 2016;56:40–54.
67. de Sousa RT, van de Bilt MT, Diniz BS, et al. Lithium increases plasma brain-derived neurotrophic factor in acute bipolar mania: a preliminary 4-week study. *Neurosci Lett*. 2011;494(1):54–56.
68. Vincze I, Perroud N, Buresi C, et al. Association between brain-derived neurotrophic factor gene and a severe form of bipolar disorder, but no interaction with the serotonin transporter gene. *Bipolar Disord*. 2008;10(5):580–587.
69. Post RM. Role of BDNF in bipolar and unipolar disorder: clinical and theoretical implications. *J Psychiatr Res*. 2007;41(12):979–990.
70. Rybakowski JK, Suwalska A. Excellent lithium responders have normal cognitive functions and plasma BDNF levels. *Int J Neuropsychopharmacol*. 2010;13(5):617–622.
71. Rybakowski JK. Response to lithium in bipolar disorder: clinical and genetic findings. *ACS Chem Neurosci*. 2014;5(6):413–421.
72. Emamghoreishi M, Keshavarz M, Nekooeian AA. Acute and chronic effects of lithium on BDNF and GDNF mRNA and protein levels in rat primary neuronal, astroglial and neuroastroglia cultures. *Iran J Basic Med Sci*. 2015;18(3):240–246.
73. Won E, Kim YK. An oldie but goodie: lithium in the treatment of bipolar disorder through neuroprotective and neurotrophic mechanisms. *Int J Mol Sci*. 2017;18(12).
74. Cunha AB, Frey BN, Andreazza AC, et al. Serum brain-derived neurotrophic factor is decreased in bipolar disorder during depressive and manic episodes. *Neurosci Lett*. 2006;398(3):215–219.
75. Machado-Vieira R, Dietrich MO, Leke R, et al. Decreased plasma brain derived neurotrophic factor levels in unmedicated bipolar patients during manic episode. *Biol Psychiatry*. 2007;61(2):142–144.
76. Adachi M, Barrot M, Autry AE, Theobald D, Monteggia LM. Selective loss of brain-derived neurotrophic factor in the dentate gyrus attenuates antidepressant efficacy. *Biol Psychiatry*. 2008;63(7):642–649.

77. Ibarguen-Vargas Y, Surget A, Vourc'h P, et al. Deficit in BDNF does not increase vulnerability to stress but dampens antidepressant-like effects in the unpredictable chronic mild stress. *Behav Brain Res*. 2009;202(2):245–251.
78. Monteggia LM, Barrot M, Powell CM, et al. Essential role of brain-derived neurotrophic factor in adult hippocampal function. *Proc Natl Acad Sci U S A*. 2004;101(29):10827–10832.
79. Autry AE, Adachi M, Nosyreva E, et al. NMDA receptor blockade at rest triggers rapid behavioural antidepressant responses. *Nature*. 2011;475(7354):91–95.
80. Lepack AE, Fuchikami M, Dwyer JM, Banasr M, Duman RS. BDNF release is required for the behavioral actions of ketamine. *Int J Neuropsychopharmacol*. 2014;18:1–6.
81. Murphy DL, Goodwin FK, Bunney Jr. WE. Leukocytosis during lithium treatment. *Am J Psychiatry*. 1971;127(11):1559–1561.
82. Wang HM, Zhang T, Li Q, Huang JK, Chen RF, Sun XJ. Inhibition of glycogen synthase kinase-3beta by lithium chloride suppresses 6-hydroxydopamine-induced inflammatory response in primary cultured astrocytes. *Neurochem Int*. 2013;63(5):345–353.
83. Yuskaitis CJ, Jope RS. Glycogen synthase kinase-3 regulates microglial migration, inflammation, and inflammation-induced neurotoxicity. *Cell Signal*. 2009;21(2):264–273.
84. Hashioka S, Klegeris A, Monji A, et al. Antidepressants inhibit interferon-gamma-induced microglial production of IL-6 and nitric oxide. *Exp Neurol*. 2007;206(1):33–42.
85. Koriyama Y, Nakayama Y, Matsugo S, et al. Anti-inflammatory effects of lipoic acid through inhibition of GSK-3beta in lipopolysaccharide-induced BV-2 microglial cells. *Neurosci Res*. 2013;77(1–2):87–96.
86. Green HF, Nolan YM. GSK-3 mediates the release of IL-1beta, TNF-alpha and IL-10 from cortical glia. *Neurochem Int*. 2012;61(5):666–671.
87. Kang K, Kim YJ, Kim YH, et al. Lithium pretreatment reduces brain injury after intracerebral hemorrhage in rats. *Neurol Res*. 2012;34(5):447–454.
88. Li H, Li Q, Du X, et al. Lithium-mediated long-term neuroprotection in neonatal rat hypoxia-ischemia is associated with antiinflammatory effects and enhanced proliferation and survival of neural stem/progenitor cells. *J Cereb Blood Flow Metab*. 2011;31(10):2106–2115.
89. De Sarno P, Axtell RC, Raman C, Roth KA, Alessi DR, Jope RS. Lithium prevents and ameliorates experimental autoimmune encephalomyelitis. *J Immunol*. 2008;181(1):338–345.
90. Yu F, Wang Z, Tanaka M, et al. Posttrauma cotreatment with lithium and valproate: reduction of lesion volume, attenuation of blood-brain barrier disruption, and improvement in motor coordination in mice with traumatic brain injury. *J Neurosurg*. 2013;119(3):766–773.
91. Gao M, Yan X, Weng HR. Inhibition of glycogen synthase kinase 3beta activity with lithium prevents and attenuates paclitaxel-induced neuropathic pain. *Neuroscience*. 2013;254:301–311.
92. Nahman S, Belmaker RH, Azab AN. Effects of lithium on lipopolysaccharide-induced inflammation in rat primary glia cells. *Innate Immun*. 2012;18(3):447–458.
93. Bosetti F, Rintala J, Seemann R, et al. Chronic lithium downregulates cyclooxygenase-2 activity and prostaglandin E(2) concentration in rat brain. *Mol Psychiatry*. 2002;7(8):845–850.
94. Hull M, Lee E, Lee T, Anand N, LaLone V, Parameswaran N. Lithium chloride induces TNFalpha in mouse macrophages via MEK-ERK-dependent pathway. *J Cell Biochem*. 2014;115(1):71–80.
95. Bazinet RP, Rao JS, Chang L, Rapoport SI, Lee HJ. Chronic carbamazepine decreases the incorporation rate and turnover of arachidonic acid but not docosahexaenoic acid in brain phospholipids of the unanesthetized rat: relevance to bipolar disorder. *Biol Psychiatry*. 2006;59(5):401–407.
96. Park HJ, Kim HJ, Bae GS, et al. Selective GSK-3beta inhibitors attenuate the cisplatin-induced cytotoxicity of auditory cells. *Hear Res*. 2009;257(1–2):53–62.

97. Tan WF, Cao XZ, Wang JK, Lv HW, Wu BY, Ma H. Protective effects of lithium treatment for spatial memory deficits induced by tau hyperphosphorylation in splenectomized rats. *Clin Exp Pharmacol Physiol*. 2010;37(10):1010–1015.
98. Tay TF, Maheran M, Too SL, Hasidah MS, Ismail G, Embi N. Glycogen synthase kinase-3beta inhibition improved survivability of mice infected with Burkholderia pseudomallei. *Trop Biomed*. 2012;29(4):551–567.
99. Kwon YJ, Yoon CH, Lee SW, Park YB, Lee SK, Park MC. Inhibition of glycogen synthase kinase-3beta suppresses inflammatory responses in rheumatoid arthritis fibroblast-like synoviocytes and collagen-induced arthritis. *Joint Bone Spine*. 2014;81(3):240–246.
100. Himmerich H, Bartsch S, Hamer H, et al. Impact of mood stabilizers and antiepileptic drugs on cytokine production in-vitro. *J Psychiatr Res*. 2013;47(11):1751–1759.
101. Zhu ZF, Wang QG, Han BJ, William CP. Neuroprotective effect and cognitive outcome of chronic lithium on traumatic brain injury in mice. *Brain Res Bull*. 2010;83(5):272–277.
102. Zhao L, Wang F, Gui B, Hua F, Qian Y. Prophylactic lithium alleviates postoperative cognition impairment by phosphorylating hippocampal glycogen synthase kinase-3beta (Ser9) in aged rats. *Exp Gerontol*. 2011;46(12):1031–1036.
103. Knijff EM, Breunis MN, Kupka RW, et al. An imbalance in the production of IL-1beta and IL-6 by monocytes of bipolar patients: restoration by lithium treatment. *Bipolar Disord*. 2007;9(7):743–753.
104. Guloksuz S, Cetin EA, Cetin T, Deniz G, Oral ET, Nutt DJ. Cytokine levels in euthymic bipolar patients. *J Affect Disord*. 2010;126(3):458–462.
105. Al-Azemi M, Omu FE, Kehinde EO, Anim JT, Oriowo MA, Omu AE. Lithium protects against toxic effects of cadmium in the rat testes. *J Assist Reprod Genet*. 2010;27(8):469–476.
106. Rapaport MH, Manji HK. The effects of lithium on ex vivo cytokine production. *Biol Psychiatry*. 2001;50(3):217–224.
107. Ohteki T, Parsons M, Zakarian A, et al. Negative regulation of T cell proliferation and interleukin 2 production by the serine threonine kinase GSK-3. *J Exp Med*. 2000;192(1):99–104.
108. Albayrak A, Halici Z, Polat B, et al. Protective effects of lithium: a new look at an old drug with potential antioxidative and anti-inflammatory effects in an animal model of sepsis. *Int Immunopharmacol*. 2013;16(1):35–40.
109. Chen K, Wu Y, Zhu M, et al. Lithium chloride promotes host resistance against Pseudomonas aeruginosa keratitis. *Mol Vis*. 2013;19:1502–1514.
110. Agrawal S, Gollapudi S, Gupta S, Agrawal A. Dendritic cells from the elderly display an intrinsic defect in the production of IL-10 in response to lithium chloride. *Exp Gerontol*. 2013;48(11):1285–1292.
111. Boufidou F, Nikolaou C, Alevizos B, Liappas IA, Christodoulou GN. Cytokine production in bipolar affective disorder patients under lithium treatment. *J Affect Disord*. 2004;82(2):309–313.
112. Natsume H, Tokuda H, Adachi S, et al. Wnt3a regulates tumor necrosis factor-alpha-stimulated interleukin-6 release in osteoblasts. *Mol Cell Endocrinol*. 2011;331(1):66–72.
113. Victoratos P, Yiangou M, Avramidis N, Hadjipetrou L. Regulation of cytokine gene expression by adjuvants in vivo. *Clin Exp Immunol*. 1997;109(3):569–578.
114. Ballanger F, Tenaud I, Volteau C, Khammari A, Dreno B. Anti-inflammatory effects of lithium gluconate on keratinocytes: a possible explanation for efficiency in seborrhoeic dermatitis. *Arch Dermatol Res*. 2008;300(5):215–223.
115. Liu KJ, Lee YL, Yang YY, et al. Modulation of the development of human monocyte-derived dendritic cells by lithium chloride. *J Cell Physiol*. 2011;226(2):424–433.
116. Vines A, Cahoon S, Goldberg I, Saxena U, Pillarisetti S. Novel anti-inflammatory role for glycogen synthase kinase-3beta in the inhibition of tumor necrosis factor-alpha- and interleukin-1beta-induced inflammatory gene expression. *J Biol Chem*. 2006;281(25):16985–16990.

117. Rowse AL, Naves R, Cashman KS, et al. Lithium controls central nervous system autoimmunity through modulation of IFN-gamma signaling. *PLoS ONE*. 2012;7(12):e52658.
118. Wu YY, Yang XH. Enhancement of interleukin 2 production in human and Gibbon T cells after in vitro treatment with lithium. *Proc Soc Exp Biol Med*. 1991;198(1):620–624.
119. Kucharz EJ, Sierakowski SJ, Goodwin JS. Lithium in vitro enhances interleukin-2 production by T cells from patients with systemic lupus erythematosus. *Immunopharmacol Immunotoxicol*. 1993;15(5):515–523.
120. Szuster-Ciesielska A, Tustanowska-Stachura A, Slotwinska M, Marmurowska-Michalowska H, Kandefer-Szerszen M. In vitro immunoregulatory effects of antidepressants in healthy volunteers. *Pol J Pharmacol*. 2003;55(3):353–362.
121. Matsebatlela T, Gallicchio V, Becker R. Lithium modulates cancer cell growth, apoptosis, gene expression and cytokine production in HL-60 promyelocytic leukaemia cells and their drug-resistant sub-clones. *Biol Trace Elem Res*. 2012;149(3):323–330.
122. Martin M, Rehani K, Jope RS, Michalek SM. Toll-like receptor-mediated cytokine production is differentially regulated by glycogen synthase kinase 3. *Nat Immunol*. 2005;6(8):777–784.
123. Zhang P, Katz J, Michalek SM. Glycogen synthase kinase-3beta (GSK3beta) inhibition suppresses the inflammatory response to Francisella infection and protects against tularemia in mice. *Mol Immunol*. 2009;46(4):677–687.
124. Coant N, Simon-Rudler M, Gustot T, et al. Glycogen synthase kinase 3 involvement in the excessive proinflammatory response to LPS in patients with decompensated cirrhosis. *J Hepatol*. 2011;55(4):784–793.
125. Baban B, Liu JY, Mozaffari MS. Pressure overload regulates expression of cytokines, gammaH2AX, and growth arrest- and DNA-damage inducible protein 153 via glycogen synthase kinase-3beta in ischemic-reperfused hearts. *Hypertension*. 2013;61(1):95–104.
126. Damri O, Sade Y, Toker L, et al. Molecular effects of lithium are partially mimicked by inositol-monophosphatase (IMPA)1 knockout mice in a brain region-dependent manner. *Eur Neuropsychopharmacol*. 2015;25(3):425–434.
127. Scola G, Kim HK, Young LT, Salvador M, Andreazza AC. Lithium reduces the effects of rotenone-induced complex I dysfunction on DNA methylation and hydroxymethylation in rat cortical primary neurons. *Psychopharmacology*. 2014;231(21):4189–4198.
128. Tam ZY, Gruber J, Ng LF, Halliwell B, Gunawan R. Effects of lithium on age-related decline in mitochondrial turnover and function in Caenorhabditis elegans. *J Gerontol A Biol Sci Med Sci*. 2014;69(7):810–820.
129. Shalbuyeva N, Brustovetsky T, Brustovetsky N. Lithium desensitizes brain mitochondria to calcium, antagonizes permeability transition, and diminishes cytochrome C release. *J Biol Chem*. 2007;282(25):18057–18068.
130. Alural B, Ozerdem A, Allmer J, Genc K, Genc S. Lithium protects against paraquat neurotoxicity by NRF2 activation and miR-34a inhibition in SH-SY5Y cells. *Front Cell Neurosci*. 2015;9:209.
131. Nciri R, Desmoulin F, Allagui MS, et al. Neuroprotective effects of chronic exposure of SH-SY5Y to low lithium concentration involve glycolysis stimulation, extracellular pyruvate accumulation and resistance to oxidative stress. *Int J Neuropsychopharmacol*. 2013;16(2):365–376.
132. Maurer IC, Schippel P, Volz HP. Lithium-induced enhancement of mitochondrial oxidative phosphorylation in human brain tissue. *Bipolar Disord*. 2009;11(5):515–522.
133. Nascimento C, Kim HK, Young LT, et al. Glutathione-mediated effects of lithium in decreasing protein oxidation induced by mitochondrial complex I dysfunction. *J Neural Transm*. 2015;122(6):741–746.
134. McQuillin A, Rizig M, Gurling HM. A microarray gene expression study of the molecular pharmacology of lithium carbonate on mouse brain mRNA to understand the neurobiology of mood stabilization and treatment of bipolar affective disorder. *Pharmacogenet Genomics*. 2007;17(8):605–617.

135. Toker L, Bersudsky Y, Plaschkes I, et al. Inositol-related gene knockouts mimic lithium's effect on mitochondrial function. *Neuropsychopharmacology*. 2014;39(2):319–328.
136. Toker L, Agam G. Lithium, inositol and mitochondria. *ACS Chem Neurosci*. 2014;5(6):411–412.
137. Valvassori SS, Rezin GT, Ferreira CL, et al. Effects of mood stabilizers on mitochondrial respiratory chain activity in brain of rats treated with d-amphetamine. *J Psychiatr Res*. 2010;44(14):903–909.
138. Feier G, Valvassori SS, Varela RB, et al. Lithium and valproate modulate energy metabolism in an animal model of mania induced by methamphetamine. *Pharmacol Biochem Behav*. 2013;103(3):589–596.
139. Bachmann RF, Wang Y, Yuan P, et al. Common effects of lithium and valproate on mitochondrial functions: protection against methamphetamine-induced mitochondrial damage. *Int J Neuropsychopharmacol*. 2009;12(6):805–822.
140. Kim HK, Isaacs-Trepanier C, Elmi N, Rapoport SI, Andreazza AC. Mitochondrial dysfunction and lipid peroxidation in rat frontal cortex by chronic NMDA administration can be partially prevented by lithium treatment. *J Psychiatr Res*. 2016;76:59–65.
141. Zhou X, da Li W, Yuan BL, et al. Lithium treatment prevents apoptosis in neonatal rat hippocampus resulting from sevoflurane exposure. *Neurochem Res*. 2016;41(8): 1993–2005.
142. King TD, Bijur GN, Jope RS. Caspase-3 activation induced by inhibition of mitochondrial complex I is facilitated by glycogen synthase kinase-3beta and attenuated by lithium. *Brain Res*. 2001;919(1):106–114.
143. Quiroz JA, Gray NA, Kato T, Manji HK. Mitochondrially mediated plasticity in the pathophysiology and treatment of bipolar disorder. *Neuropsychopharmacology*. 2008;33(11):2551–2565.
144. Tanno M, Kuno A, Ishikawa S, et al. Translocation of glycogen synthase kinase-3beta (GSK-3beta), a trigger of permeability transition, is kinase activity-dependent and mediated by interaction with voltage-dependent anion channel 2 (VDAC2). *J Biol Chem*. 2014;289(42):29285–29296.
145. Carter CJ. Multiple genes and factors associated with bipolar disorder converge on growth factor and stress activated kinase pathways controlling translation initiation: implications for oligodendrocyte viability. *Neurochem Int*. 2007;50(3):461–490.
146. Ngok-Ngam P, Watcharasit P, Thiantanawat A, Satayavivad J. Pharmacological inhibition of GSK3 attenuates DNA damage-induced apoptosis via reduction of p53 mitochondrial translocation and Bax oligomerization in neuroblastoma SH-SY5Y cells. *Cell Mol Biol Lett*. 2013;18(1):58–74.
147. Li Q, Li H, Roughton K, et al. Lithium reduces apoptosis and autophagy after neonatal hypoxia-ischemia. *Cell Death Dis*. 2010;1:e56.
148. Keshavarz M, Emamghoreishi M, Nekooeian AA, J JW, Zare HR. Increased bcl-2 protein levels in rat primary astrocyte culture following chronic lithium treatment. *Iran J Med Sci*. 2013;38(3):255–262.
149. Zeng Z, Wang H, Shang F, et al. Lithium ions attenuate serum-deprivation-induced apoptosis in PC12 cells through regulation of the Akt/FoxO1 signaling pathways. *Psychopharmacology*. 2016;233(5):785–794.
150. Aminzadeh A, Dehpour AR, Safa M, Mirzamohammadi S, Sharifi AM. Investigating the protective effect of lithium against high glucose-induced neurotoxicity in PC12 cells: involvements of ROS, JNK and P38 MAPKs, and apoptotic mitochondria pathway. *Cell Mol Neurobiol*. 2014;34(8):1143–1150.
151. Mattson MP. Apoptosis in neurodegenerative disorders. *Nat Rev Mol Cell Biol*. 2000;1(2):120–129.
152. Jope RS, Nemeroff CB. Lithium to the rescue. *Cerebrum*. 2016;cer-02-16:1–11.
153. Rowe MK, Chuang DM. Lithium neuroprotection: molecular mechanisms and clinical implications. *Expert Rev Mol Med*. 2004;6(21):1–18.

154. Prosser JM, Fieve RR. Patients receiving lithium therapy have a reduced prevalence of neurological and cardiovascular disorders. *Prog Neuro-Psychopharmacol Biol Psychiatry*. 2016;71:39–44.
155. Wang Y, Zhang Q, Wang B, Li P, Liu P. LiCl treatment induces programmed cell death of schwannoma cells through AKT- and MTOR-mediated necroptosis. *Neurochem Res*. 2017;42(8):2363–2371.
156. Danton GH, Dietrich WD. Inflammatory mechanisms after ischemia and stroke. *J Neuropathol Exp Neurol*. 2003;62(2):127–136.
157. Iadecola C, Anrather J. The immunology of stroke: from mechanisms to translation. *Nat Med*. 2011;17(7):796–808.
158. McColl BW, Rothwell NJ, Allan SM. Systemic inflammatory stimulus potentiates the acute phase and CXC chemokine responses to experimental stroke and exacerbates brain damage via interleukin-1- and neutrophil-dependent mechanisms. *J Neurosci Off J Soc Neurosci*. 2007;27(16):4403–4412.
159. Santos GL, Alcantara CC, Silva-Couto MA, Garcia-Salazar LF, Russo TL. Decreased brain-derived neurotrophic factor serum concentrations in chronic post-stroke subjects. *J Stroke Cerebrovasc Dis*. 2016;25(12):2968–2974.
160. Yang JL, Mukda S, Chen SD. Diverse roles of mitochondria in ischemic stroke. *Redox Biol*. 2018;16:263–275.
161. Radak D, Katsiki N, Resanovic I, et al. Apoptosis and acute brain ischemia in ischemic stroke. *Curr Vasc Pharmacol*. 2017;15(2):115–122.
162. Nonaka S, Chuang DM. Neuroprotective effects of chronic lithium on focal cerebral ischemia in rats. *Neuroreport*. 1998;9(9):2081–2084.
163. Xu J, Culman J, Blume A, Brecht S, Gohlke P. Chronic treatment with a low dose of lithium protects the brain against ischemic injury by reducing apoptotic death. *Stroke*. 2003;34(5):1287–1292.
164. Silachev DN, Gulyaev MV, Zorova LD, et al. Magnetic resonance spectroscopy of the ischemic brain under lithium treatment. Link to mitochondrial disorders under stroke. *Chem Biol Interact*. 2015;237:175–182.
165. Ren M, Senatorov VV, Chen RW, Chuang DM. Postinsult treatment with lithium reduces brain damage and facilitates neurological recovery in a rat ischemia/reperfusion model. *Proc Natl Acad Sci U S A*. 2003;100(10):6210–6215.
166. Hollenbach K, Sleet DA. Resource guide to occupant protection programs and materials. *Health Educ Q*. 1984;11(2):215–220.
167. Mohammadianinejad SE, Majdinasab N, Sajedi SA, Abdollahi F, Moqaddam MM, Sadr F. The effect of lithium in post-stroke motor recovery: a double-blind, placebo-controlled, randomized clinical trial. *Clin Neuropharmacol*. 2014;37(3):73–78.
168. A novel gene containing a trinucleotide repeat that is expanded and unstable on Huntington's disease chromosomes. The Huntington's Disease Collaborative Research Group. *Cell*. 1993;72(6):971–983.
169. Martin JB, Gusella JF. Huntington's disease. Pathogenesis and management. *N Engl J Med*. 1986;315(20):1267–1276.
170. Friedlander RM. Apoptosis and caspases in neurodegenerative diseases. *N Engl J Med*. 2003;348(14):1365–1375.
171. Martin DD, Ladha S, Ehrnhoefer DE, Hayden MR. Autophagy in Huntington disease and huntingtin in autophagy. *Trends Neurosci*. 2015;38(1):26–35.
172. Zuccato C, Cattaneo E. Huntington's disease. *Handb Exp Pharmacol*. 2014;220:357–409.
173. Carmo C, Naia L, Lopes C, Rego AC. Mitochondrial dysfunction in Huntington's disease. *Adv Exp Med Biol*. 2018;1049:59–83.
174. Martin LJ. Mitochondrial pathobiology in Parkinson's disease and amyotrophic lateral sclerosis. *J Alzheimers Dis*. 2010;20(Suppl 2):S335–S356.

175. Elliott R, Jiang X, Head J. Mitochondria and neurodegeneration "could mitochondrial organelle transfer be a cellular biotherapy for neurodegenerative diseases?". *SOJ Biochem*. 2016;2(1):5.
176. Crotti A, Glass CK. The choreography of neuroinflammation in Huntington's disease. *Trends Immunol*. 2015;36(6):364–373.
177. Frank-Cannon TC, Alto LT, McAlpine FE, Tansey MG. Does neuroinflammation fan the flame in neurodegenerative diseases? *Mol Neurodegener*. 2009;4:47.
178. Radi E, Formichi P, Battisti C, Federico A. Apoptosis and oxidative stress in neurodegenerative diseases. *J Alzheimers Dis*. 2014;42(Suppl 3):S125–S152.
179. Ghavami S, Shojaei S, Yeganeh B, et al. Autophagy and apoptosis dysfunction in neurodegenerative disorders. *Prog Neurobiol*. 2014;112:24–49.
180. Carmichael J, Sugars KL, Bao YP, Rubinsztein DC. Glycogen synthase kinase-3beta inhibitors prevent cellular polyglutamine toxicity caused by the Huntington's disease mutation. *J Biol Chem*. 2002;277(37):33791–33798.
181. Berger Z, Ttofi EK, Michel CH, et al. Lithium rescues toxicity of aggregate-prone proteins in Drosophila by perturbing Wnt pathway. *Hum Mol Genet*. 2005;14(20):3003–3011.
182. Wood NI, Morton AJ. Chronic lithium chloride treatment has variable effects on motor behaviour and survival of mice transgenic for the Huntington's disease mutation. *Brain Res Bull*. 2003;61(4):375–383.
183. Taylor-Robinson SD, Weeks RA, Bryant DJ, et al. Proton magnetic resonance spectroscopy in Huntington's disease: evidence in favour of the glutamate excitotoxic theory. *Mov Disord*. 1996;11(2):167–173.
184. Wei H, Qin ZH, Senatorov VV, et al. Lithium suppresses excitotoxicity-induced striatal lesions in a rat model of Huntington's disease. *Neuroscience*. 2001;106(3):603–612.
185. Senatorov VV, Ren M, Kanai H, Wei H, Chuang DM. Short-term lithium treatment promotes neuronal survival and proliferation in rat striatum infused with quinolinic acid, an excitotoxic model of Huntington's disease. *Mol Psychiatry*. 2004;9(4):371–385.
186. Chiu CT, Liu G, Leeds P, Chuang DM. Combined treatment with the mood stabilizers lithium and valproate produces multiple beneficial effects in transgenic mouse models of Huntington's disease. *Neuropsychopharmacology*. 2011;36(12):2406–2421.
187. Sarkar S, Ravikumar B, Floto RA, Rubinsztein DC. Rapamycin and mTOR-independent autophagy inducers ameliorate toxicity of polyglutamine-expanded huntingtin and related proteinopathies. *Cell Death Differ*. 2009;16(1):46–56.
188. Linares GR, Chiu CT, Scheuing L, et al. Preconditioning mesenchymal stem cells with the mood stabilizers lithium and valproic acid enhances therapeutic efficacy in a mouse model of Huntington's disease. *Exp Neurol*. 2016;281:81–92.
189. Aminoff MJ, Marshall J. Treatment of Huntington's chorea with lithium carbonate. A double-blind trial. *Lancet*. 1974;1(7848):107–109.
190. Leonard DP, Kidson MA, Brown JG, Shannon PJ, Taryan S. A double blind trial of lithium carbonate and haloperidol in Huntington's chorea. *Aust N Z J Psychiatry*. 1975;9(2):115–118.
191. Vestergaard P, Baastrup PC, Petersson H. Lithium treatment of Huntington's chorea. A placebo-controlled clinical trial. *Acta Psychiatr Scand*. 1977;56(3):183–188.
192. Danivas V, Moily NS, Thimmaiah R, et al. Off label use of lithium in the treatment of Huntington's disease: a case series. *Indian J Psychiatry*. 2013;55(1):81–83.
193. Bersani G, Quartini A, Zullo D, Iannitelli A. Potential neuroprotective effect of lithium in bipolar patients evaluated by neuropsychological assessment: preliminary results. *Hum Psychopharmacol*. 2016;31(1):19–28.
194. Quartini A, Iannitelli A, Bersani G. Lithium: from mood stabilizer to putative cognitive enhancer. *Neural Regen Res*. 2016;11(8):1234–1235.
195. Paterson A, Parker G. Lithium and cognition in those with bipolar disorder. *Int Clin Psychopharmacol*. 2017;32(2):57–62.

196. Li Q, Liu Y, Sun M. Autophagy and Alzheimer's disease. *Cell Mol Neurobiol.* 2017;37(3):377–388.
197. Stapinski A, Popielarski M. Participation of district dermato-venereological clinics in preventing HIV infection in Poland. *Przegl Dermatol.* 1989;76(4):345–347.
198. Calsolaro V, Edison P. Neuroinflammation in Alzheimer's disease: current evidence and future directions. *Alzheimers Dement.* 2016;12(6):719–732.
199. Bothwell M. Recent advances in understanding neurotrophin signaling. *F1000Research* 2016;5(F1000 Faculty Rev):1885.
200. Tanila H. The role of BDNF in Alzheimer's disease. *Neurobiol Dis.* 2017;97(Pt B):114–118.
201. Bhat RV, Budd Haeberlein SL, Avila J. Glycogen synthase kinase 3: a drug target for CNS therapies. *J Neurochem.* 2004;89(6):1313–1317.
202. Stambolic V, Ruel L, Woodgett JR. Lithium inhibits glycogen synthase kinase-3 activity and mimics wingless signalling in intact cells. *Curr Biol.* 1996;6(12):1664–1668.
203. Hong M, Chen DC, Klein PS, Lee VM. Lithium reduces tau phosphorylation by inhibition of glycogen synthase kinase-3. *J Biol Chem.* 1997;272(40):25326–25332.
204. Munoz-Montano JR, Moreno FJ, Avila J, Diaz-Nido J. Lithium inhibits Alzheimer's disease-like tau protein phosphorylation in neurons. *FEBS Lett.* 1997;411(2–3):183–188.
205. Alvarez G, Munoz-Montano JR, Satrustegui J, Avila J, Bogonez E, Diaz-Nido J. Lithium protects cultured neurons against beta-amyloid-induced neurodegeneration. *FEBS Lett.* 1999;453(3):260–264.
206. Noble W, Planel E, Zehr C, et al. Inhibition of glycogen synthase kinase-3 by lithium correlates with reduced tauopathy and degeneration in vivo. *Proc Natl Acad Sci U S A.* 2005;102(19):6990–6995.
207. Terao T, Nakano H, Inoue Y, Okamoto T, Nakamura J, Iwata N. Lithium and dementia: a preliminary study. *Prog Neuro-Psychopharmacol Biol Psychiatry.* 2006;30(6):1125–1128.
208. Nunes PV, Forlenza OV, Gattaz WF. Lithium and risk for Alzheimer's disease in elderly patients with bipolar disorder. *Br J Psychiatry J Ment Sci.* 2007;190:359–360.
209. Kessing LV, Forman JL, Andersen PK. Does lithium protect against dementia? *Bipolar Disord.* 2010;12(1):87–94.
210. Cheng C, Zandi P, Stuart E, et al. Association between lithium use and risk of Alzheimer's disease. *J Clin Psychiatry.* 2017;78(2):e139–e145.
211. Dunn N, Holmes C, Mullee M. Does lithium therapy protect against the onset of dementia? *Alzheimer Dis Assoc Disord.* 2005;19(1):20–22.
212. Parker WF, Gorges RJ, Gao YN, Zhang Y, Hur K, Gibbons RD. Association between groundwater lithium and the diagnosis of bipolar disorder and dementia in the United States. *JAMA Psychiat.* 2018;75(7):751–754.
213. Macdonald A, Briggs K, Poppe M, Higgins A, Velayudhan L, Lovestone S. A feasibility and tolerability study of lithium in Alzheimer's disease. *Int J Geriatr Psychiatry.* 2008;23(7):704–711.
214. Hampel H, Ewers M, Burger K, et al. Lithium trial in Alzheimer's disease: a randomized, single-blind, placebo-controlled, multicenter 10-week study. *J Clin Psychiatry.* 2009;70(6):922–931.
215. Leyhe T, Eschweiler GW, Stransky E, et al. Increase of BDNF serum concentration in lithium treated patients with early Alzheimer's disease. *J Alzheimers Dis.* 2009;16(3):649–656.
216. Straten G, Saur R, Laske C, et al. Influence of lithium treatment on GDNF serum and CSF concentrations in patients with early Alzheimer's disease. *Curr Alzheimer Res.* 2011;8(8):853–859.
217. Forlenza OV, Diniz BS, Radanovic M, Santos FS, Talib LL, Gattaz WF. Disease-modifying properties of long-term lithium treatment for amnestic mild cognitive impairment: randomised controlled trial. *Br J Psychiatry J Ment Sci.* 2011;198(5):351–356.
218. Nunes MA, Viel TA, Buck HS. Microdose lithium treatment stabilized cognitive impairment in patients with Alzheimer's disease. *Curr Alzheimer Res.* 2013;10(1):104–107.

219. Matsunaga S, Kishi T, Annas P, Basun H, Hampel H, Iwata N. Lithium as a treatment for Alzheimer's disease: a systematic review and meta-analysis. *J Alzheimers Dis*. 2015;48(2):403–410.
220. Morris G, Berk M. The putative use of lithium in Alzheimer's disease. *Curr Alzheimer Res*. 2016;13(8):853–861.
221. Kruger R, Kuhn W, Muller T, et al. Ala30Pro mutation in the gene encoding alpha-synuclein in Parkinson's disease. *Nat Genet*. 1998;18(2):106–108.
222. Polymeropoulos MH, Lavedan C, Leroy E, et al. Mutation in the alpha-synuclein gene identified in families with Parkinson's disease. *Science*. 1997;276(5321):2045–2047.
223. Son JH, Shim JH, Kim KH, Ha JY, Han JY. Neuronal autophagy and neurodegenerative diseases. *Exp Mol Med*. 2012;44(2):89–98.
224. Karabiyik C, Lee MJ, Rubinsztein DC. Autophagy impairment in Parkinson's disease. *Essays Biochem*. 2017;61(6):711–720.
225. Xilouri M, Brekk OR, Stefanis L. Autophagy and alpha-synuclein: relevance to Parkinson's disease and related synucleopathies. *Mov Disord*. 2016;31(2):178–192.
226. Chang D, Nalls MA, Hallgrimsdottir IB, et al. A meta-analysis of genome-wide association studies identifies 17 new Parkinson's disease risk loci. *Nat Genet*. 2017;49(10):1511–1516.
227. Vivekanantham S, Shah S, Dewji R, Dewji A, Khatri C, Ologunde R. Neuroinflammation in Parkinson's disease: role in neurodegeneration and tissue repair. *Int J Neurosci*. 2015;125(10):717–725.
228. Liu XL, Wang YD, Yu XM, Li DW, Li GR. Mitochondria-mediated damage to dopaminergic neurons in Parkinson's disease (review). *Int J Mol Med*. 2018;41(2):615–623.
229. Kusters CDJ, Paul KC, Guella I, et al. Dopamine receptors and BDNF-haplotypes predict dyskinesia in Parkinson's disease. *Parkinsonism Relat Disord*. 2018;47:39–44.
230. Chen G, Bower KA, Ma C, Fang S, Thiele CJ, Luo J. Glycogen synthase kinase 3beta (GSK3beta) mediates 6-hydroxydopamine-induced neuronal death. *FASEB J*. 2004;18(10):1162–1164.
231. Soleimani M, Ghasemi N. Lithium chloride can induce differentiation of human immortalized RenVm cells into dopaminergic neurons. *Avicenna J Med Biotechnol*. 2017;9(4):176–180.
232. Gubellini P, Kachidian P. Animal models of Parkinson's disease: an updated overview. *Rev Neurol*. 2015;171(11):750–761.
233. Youdim MB, Arraf Z. Prevention of MPTP (N-methyl-4-phenyl-1,2,3,6-tetrahydropyridine) dopaminergic neurotoxicity in mice by chronic lithium: involvements of Bcl-2 and Bax. *Neuropharmacology*. 2004;46(8):1130–1140.
234. Lieu CA, Dewey CM, Chinta SJ, et al. Lithium prevents parkinsonian behavioral and striatal phenotypes in an aged parkin mutant transgenic mouse model. *Brain Res*. 2014;1591:111–117.
235. Lei P, Ayton S, Appukuttan AT, et al. Lithium suppression of tau induces brain iron accumulation and neurodegeneration. *Mol Psychiatry*. 2017;22(3):396–406.
236. McCaul JA, Letter SGM. Lithium in Parkinson's disease. *Lancet*. 1974;1(7866):1117.
237. Coffey CE, Ross DR, Massey EW, Olanow CW. Dyskinesias associated with lithium therapy in parkinsonism. *Clin Neuropharmacol*. 1984;7(3):223–229.
238. Guttuso Jr. T. Low-dose lithium adjunct therapy associated with reduced off-time in Parkinson's disease: a case series. *J Neurol Sci*. 2016;368:221–222.
239. Crawford DC, Meadows KL, Newman JL, et al. Prevalence of the fragile X syndrome in African-Americans. *Am J Med Genet*. 2002;110(3):226–233.
240. Garber K, Smith KT, Reines D, Warren ST. Transcription, translation and fragile X syndrome. *Curr Opin Genet Dev*. 2006;16(3):270–275.
241. Yan J, Porch MW, Court-Vazquez B, Bennett MVL, Zukin RS. Activation of autophagy rescues synaptic and cognitive deficits in fragile X mice. *Proc Natl Acad Sci U S A*. 2018;115(41):E9707–E9716.

242. Castren ML, Castren E. BDNF in fragile X syndrome. *Neuropharmacology*. 2014;76(Pt C):729–736.
243. Valenti D, de Bari L, De Filippis B, Henrion-Caude A, Vacca RA. Mitochondrial dysfunction as a central actor in intellectual disability-related diseases: an overview of Down syndrome, autism, Fragile X and Rett syndrome. *Neurosci Biobehav Rev*. 2014;46(Pt 2):202–217.
244. Choi CH, McBride SM, Schoenfeld BP, et al. Age-dependent cognitive impairment in a Drosophila fragile X model and its pharmacological rescue. *Biogerontology*. 2010;11(3):347–362.
245. McBride SM, Choi CH, Wang Y, et al. Pharmacological rescue of synaptic plasticity, courtship behavior, and mushroom body defects in a Drosophila model of fragile X syndrome. *Neuron*. 2005;45(5):753–764.
246. Fmr1 knockout mice: a model to study fragile X mental retardation. The Dutch-Belgian Fragile X Consortium. *Cell*. 1994;78(1):23–33.
247. Dahlhaus R. Of men and mice: modeling the Fragile X syndrome. *Front Mol Neurosci*. 2018;11:41.
248. Min WW, Yuskaitis CJ, Yan Q, et al. Elevated glycogen synthase kinase-3 activity in Fragile X mice: key metabolic regulator with evidence for treatment potential. *Neuropharmacology*. 2009;56(2):463–472.
249. Yuskaitis CJ, Mines MA, King MK, Sweatt JD, Miller CA, Jope RS. Lithium ameliorates altered glycogen synthase kinase-3 and behavior in a mouse model of fragile X syndrome. *Biochem Pharmacol*. 2010;79(4):632–646.
250. Berry-Kravis E, Sumis A, Hervey C, et al. Open-label treatment trial of lithium to target the underlying defect in fragile X syndrome. *J Dev Behav Pediatr*. 2008;29(4):293–302.
251. Rowland LP. Amyotrophic lateral sclerosis: human challenge for neuroscience. *Proc Natl Acad Sci U S A*. 1995;92(5):1251–1253.
252. Kiernan MC, Vucic S, Cheah BC, et al. Amyotrophic lateral sclerosis. *Lancet*. 2011;377(9769):942–955.
253. Al-Chalabi A, van den Berg LH, Veldink J. Gene discovery in amyotrophic lateral sclerosis: implications for clinical management. *Nat Rev Neurol*. 2017;13(2):96–104.
254. Ramesh N, Pandey UB. Autophagy dysregulation in ALS: when protein aggregates get out of hand. *Front Mol Neurosci*. 2017;10:263.
255. Liu J, Wang F. Role of neuroinflammation in amyotrophic lateral sclerosis: cellular mechanisms and therapeutic implications. *Front Immunol*. 2017;8:1005.
256. Smith EF, Shaw PJ, De Vos KJ. The role of mitochondria in amyotrophic lateral sclerosis. *Neurosci Lett*. 2017. https://doi.org/10.1016/j.neulet.2017.06.052.
257. Van Damme P, Robberecht W, Van Den Bosch L. Modelling amyotrophic lateral sclerosis: progress and possibilities. *Dis Model Mech*. 2017;10(5):537–549.
258. Ferraiuolo L, Kirby J, Grierson AJ, Sendtner M, Shaw PJ. Molecular pathways of motor neuron injury in amyotrophic lateral sclerosis. *Nat Rev Neurol*. 2011;7(11):616–630.
259. Gurney ME. Transgenic-mouse model of amyotrophic lateral sclerosis. *N Engl J Med*. 1994;331(25):1721–1722.
260. Pizzasegola C, Caron I, Daleno C, et al. Treatment with lithium carbonate does not improve disease progression in two different strains of SOD1 mutant mice. *Amyotroph Lateral Scler*. 2009;10(4):221–228.
261. Gill A, Kidd J, Vieira F, Thompson K, Perrin S. No benefit from chronic lithium dosing in a sibling-matched, gender balanced, investigator-blinded trial using a standard mouse model of familial ALS. *PLoS ONE*. 2009;4(8):e6489.
262. Shin JH, Cho SI, Lim HR, et al. Concurrent administration of Neu2000 and lithium produces marked improvement of motor neuron survival, motor function, and mortality in a mouse model of amyotrophic lateral sclerosis. *Mol Pharmacol*. 2007;71(4):965–975.
263. Fornai F, Longone P, Cafaro L, et al. Lithium delays progression of amyotrophic lateral sclerosis. *Proc Natl Acad Sci U S A*. 2008;105(6):2052–2057.
264. Ferrucci M, Spalloni A, Bartalucci A, et al. A systematic study of brainstem motor nuclei in a mouse model of ALS, the effects of lithium. *Neurobiol Dis*. 2010;37(2):370–383.

265. Feng HL, Leng Y, Ma CH, Zhang J, Ren M, Chuang DM. Combined lithium and valproate treatment delays disease onset, reduces neurological deficits and prolongs survival in an amyotrophic lateral sclerosis mouse model. *Neuroscience*. 2008;155(3):567–572.
266. Aggarwal SP, Zinman L, Simpson E, et al. Safety and efficacy of lithium in combination with riluzole for treatment of amyotrophic lateral sclerosis: a randomised, double-blind, placebo-controlled trial. *Lancet Neurol*. 2010;9(5):481–488.
267. Verstraete E, Veldink JH, Huisman MH, et al. Lithium lacks effect on survival in amyotrophic lateral sclerosis: a phase IIb randomised sequential trial. *J Neurol Neurosurg Psychiatry*. 2012;83(5):557–564.
268. Yanez M, Matias-Guiu J, Arranz-Tagarro JA, et al. The neuroprotection exerted by memantine, minocycline and lithium, against neurotoxicity of CSF from patients with amyotrophic lateral sclerosis, is antagonized by riluzole. *Neurodegener Dis*. 2014;13(2–3):171–179.
269. Chio A, Borghero G, Calvo A, et al. Lithium carbonate in amyotrophic lateral sclerosis: lack of efficacy in a dose-finding trial. *Neurology*. 2010;75(7):619–625.
270. Miller RG, Moore DH, Forshew DA, et al. Phase II screening trial of lithium carbonate in amyotrophic lateral sclerosis: examining a more efficient trial design. *Neurology*. 2011;77(10):973–979.
271. UKMND-LiCALS Study Group, Morrison KE, Dhariwal S, et al. Lithium in patients with amyotrophic lateral sclerosis (LiCALS): a phase 3 multicentre, randomised, double-blind, placebo-controlled trial. *Lancet Neurol*. 2013;12(4):339–345.
272. Agam G, Israelson A. Why lithium studies for ALS treatment should not be halted prematurely. *Front Neurosci*. 2014;8:267.
273. van Eijk RPA, Jones AR, Sproviero W, et al. Meta-analysis of pharmacogenetic interactions in amyotrophic lateral sclerosis clinical trials. *Neurology*. 2017;89(18):1915–1922.
274. Broeke JH, Roelandse M, Luteijn MJ, et al. Munc18 and Munc13 regulate early neurite outgrowth. *Biol Cell*. 2010;102(8):479–488.
275. Liang P, Le W. Role of autophagy in the pathogenesis of multiple sclerosis. *Neurosci Bull*. 2015;31(4):435–444.
276. Frota ER, Rodrigues DH, Donadi EA, Brum DG, Maciel DR, Teixeira AL. Increased plasma levels of brain derived neurotrophic factor (BDNF) after multiple sclerosis relapse. *Neurosci Lett*. 2009;460(2):130–132.
277. Al-Temaimi R, AbuBaker J, Al-Khairi I, Alroughani R. Remyelination modulators in multiple sclerosis patients. *Exp Mol Pathol*. 2017;103(3):237–241.
278. Datta G, Colasanti A, Rabiner EA, et al. Neuroinflammation and its relationship to changes in brain volume and white matter lesions in multiple sclerosis. *Brain J Neurol*. 2017;140(11):2927–2938.
279. Mahad DH, Trapp BD, Lassmann H. Pathological mechanisms in progressive multiple sclerosis. *Lancet Neurol*. 2015;14(2):183–193.
280. Patergnani S, Fossati V, Bonora M, et al. Mitochondria in multiple sclerosis: molecular mechanisms of pathogenesis. *Int Rev Cell Mol Biol*. 2017;328:49–103.
281. Ofengeim D, Ito Y, Najafov A, et al. Activation of necroptosis in multiple sclerosis. *Cell Rep*. 2015;10(11):1836–1849.
282. Procaccini C, De Rosa V, Pucino V, Formisano L, Matarese G. Animal models of multiple sclerosis. *Eur J Pharmacol*. 2015;759:182–191.
283. Noseworthy JH, Lucchinetti C, Rodriguez M, Weinshenker BG. Multiple sclerosis. *N Engl J Med*. 2000;343(13):938–952.
284. Ahn M, Kim J, Park C, et al. Potential involvement of glycogen synthase kinase (GSK)-3beta in a rat model of multiple sclerosis: evidenced by lithium treatment. *Anat Cell Biol*. 2017;50(1):48–59.
285. Rinker 2nd JR, Cossey TC, Cutter GR, Culpepper WJ. A retrospective review of lithium usage in veterans with multiple sclerosis. *Mult Scler Relat Disord*. 2013;2(4):327–333.

SECTION IV

Alzheimer and neurodegenerative diseases

CHAPTER

10

Tau-based therapies for Alzheimer's disease: Promising novel neuroprotective approaches

Carmen Laura Sayas

Institute of Biomedical Technologies (ITB), University of La Laguna (ULL), Tenerife, Spain

Introduction

Alzheimer's disease (AD) is the most prevalent neurodegenerative disorder, accounting for 60%–80% of all dementia. AD currently affects tens of millions worldwide, but as population ages, the number of affected people is expected to increase dramatically, making AD a seriously growing global health problem.[1] Moreover, AD diagnosis is usually made too late, when patients have an extensive neuronal death, and brain damage is irreversible. As a consequence, currently available therapies are symptomatic, neither blocking nor reducing disease progression. Among these treatments against AD symptoms are the approved acetylcholinesterase inhibitors (AchE) and memantine.[2,3] AchE inhibitors increase acetylcholine levels, which have been shown to be reduced in AD.[4,5] Memantine, a noncompetitive NMDA antagonist, restores glutamatergic homeostasis, which is severely disrupted in patients with AD.[3] These drugs slightly enhance the cognitive abilities of AD patients, improving temporarily their quality of life, but they neither stop nor reduce the progression of neurodegeneration.[6] Anyway, it is important to develop new symptomatic treatments for those who have already progressed too far in the disease process. Actually, since the last symptomatic AD drug was approved around 14 years ago, hundreds of clinical trials have failed, and no new drug to treat AD has been approved.[7] For all these reasons, there is an

https://doi.org/10.1016/B978-0-12-814037-6.00005-7

urgent unmet need for new disease-modifying therapies that prevent or delay the onset of AD in subjects who are at risk for the disease or slow the progression in those with mild cognitive impairment.

The main hallmarks of AD are amyloid plaques and neurofibrillary tangles (NFTs) in the brain, accompanied by synapse loss and neuron progressive death, leading to cognitive impairment and, ultimately, dementia.[1] Plaques are composed by amyloid-β (Aβ) peptide, and NFTs are mostly made of tau protein.[8] Thus, amyloid-β (Aβ) and tau have been positioned as central players in AD pathogenesis and progression. However, the precise connection between them and the nature of their toxic forms are still unclear. In the early 1990s, Aβ peptide was postulated to be the causative agent of AD pathology in "the Aβ hypothesis".[9] Since then, the development of therapies for AD has primarily focused on Aβ; however, most Aβ-based therapies have failed in clinical trials.[10]

Recent findings in the field show that toxic aggregates of conformationally misfolded tau spread in a prion-like manner, contributing to the propagation of AD.[11,12] Of note, tau pathology has been shown to correlate better with the progression of AD cognitive impairment than Aβ lesions do, suggesting that tau pathology may induce AD cognitive deficits.[13,14] As a consequence of this significant correlation and the overall failure of Aβ-based therapies for AD, the focus of AD drug development has been switched toward tau in the last decade. At first, tau-based potential therapies were mainly based on inhibition of tau kinases or tau aggregation or on microtubule (MT) stabilization, but most of these approaches have been discontinued because of toxicity and/or lack of efficacy.[11] Since trans-neuronal spread of tau appears to play a key role in AD progression, the search for new therapeutic approaches is currently mainly directed to target extracellular tau by blocking the neuronal uptake of tau aggregates and/or promoting their clearance by tau-directed immunotherapy.[11,12] Hence, some of the most promising recent potential tau-based therapies are either passive or active immunotherapies (anti-tau antibodies or vaccines, respectively), which target extracellular tau, and have proven to exert beneficial effects in several preclinical studies in mice.[15]

In this chapter, the current state of tau-targeting therapies for AD is summarized, including: agents that modulate posttranslational modifications of tau, such as phosphorylation, acetylation, or *O*-GlcNAcylation; MT stabilizers; inhibitors of tau aggregation and activators of tau degradation by autophagy; immunotherapy (passive immunotherapy with anti-tau antibodies and active immunotherapy with tau vaccines); and agents that reduce tau expression (antisense oligonucleotides (ASOs); for a schematic view, see Fig. 10.1). The focus has been placed on novel tau-based approaches that have reached clinical trials, briefly describing the current state of tau-directed AD drug development pipeline (see Tables 10.1 and 10.2).

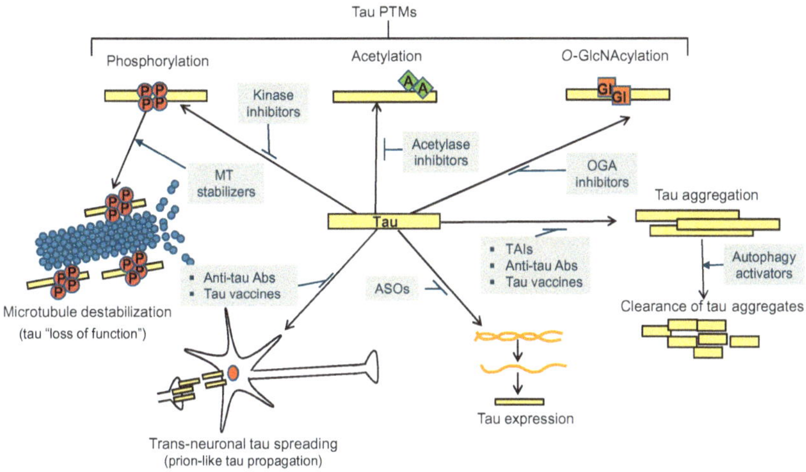

FIG. 10.1 Tau-targeting approaches as novel potential therapies for AD. Scheme representing different aspects of tau biology as targets for the development of novel potential disease-modifying therapies for AD. Different therapeutic approaches are directed to modulate tau posttranslational modifications (PTMs) that play key roles in tau pathology, including phosphorylation (kinase inhibitors), acetylation (acetylase inhibitors), or O-GlcNAcylation (OGA inhibitors). Hyperphosphorylated tau detaches from MTs, destabilizing them and inducing several deleterious effects in cells. MT stabilizers emerged as potential therapeutic agents to counteract tau "loss of function." Tau aggregation gives rise to the formation of toxic forms of tau, a phenomenon that is targeted by tau aggregation inhibitors (TAIs). Immunotherapy, through the use of anti-tau antibodies (Abs) (passive immunotherapy) or vaccines (active immunotherapy), is a promising approach to target extracellular tau protein, blocking tau trans-neuronal spreading and prion-like propagation of tau pathology, and tau aggregation. Autophagy activators induce degradation of toxic tau aggregates, being thereby possible therapeutic agents for AD. Specific antisense oligonucleotides (ASOs) that target tau mRNA and inhibit tau translation are also potential AD therapies. All the mentioned approaches have reached clinical trials in humans at different phases, with some of them currently ongoing.

Modulators of tau posttranslational modifications

Tau is subject of numerous posttranslational modifications (PTMs) that influence different features of the protein such as its turnover, assembly, or function.[16] The best-studied tau PTM is phosphorylation, but it also undergoes acetylation, glycosylation, methylation, nitration, and sumoylation.[17] Several of these PTMs are under intensive research as potential therapeutic targets for AD. The most successful ones to date, which have reached clinical trials, namely, phosphorylation, acetylation, and O-GlcNAcylation, are briefly described in the succeeding text.

TABLE 10.1 Tau-based therapeutic approaches tested in clinical trials

Type	Mechanism of action	Target	Name	Effect on tau	Phase cl. trial/ stage	Clinical trial n.	Therapeutic effects in humans
Modulators of tau PTMs	Inhibitors of tau kinases	GSK-3	Lithium	Inhibit tau Ser/Thr phosphorylation	II (Discontinued)	NCT00088387 SRCTN72046462	Mild improvement of cognitive performance
			Tideglusib		II (Discontinued)	NCT00948259, IIa NCT01350362. IIb	Overall, no clear therapeutic effects
		Fyn	Saracatinib	Inhibits tau Tyr phosphorylation	II (Discontinued)	NCT01864655, IIb NCT02167256, IIa	Failed to slow AD progression
	Inhibitor of tau acetylation	p300	Salsalate	Clears tau aggregates	I (Ongoing)	NCT03277573, Ib	To be reported
	Inhibits O-GlcNAc removal from tau	OGA	ASN120290	Reduces tau aggregates	Not used yet in AD clinical trials but in Ia and II PSP trials FDA-designated as orphan drug for PSP treatment		
MT stabilizers	Stabilize MTs	MTs	Epothilone D	N.r.	I (Discontinued)	NCT01492374	Not reported (probable lack of positive results)
			Abeotaxane	N.r.	I (Discontinued)	NCT02133846	No significant improvement
		MTs tau/EBs	NAP	Interacts with tau and MTs; protects MTs from destabilization	Tested in phase II clinical trials in MCI and phase II/III in PSP; not tested yet in AD patients		Improved cognition in patients with MCI Failed to improve PSP
TAIs	Inhibit tau aggregation	Tau	Methylene blue	Inhibit tau-tau interaction	II (Discontinued)	NCT00515333 NCT00684944	Beneficial effects at low doses, no effects at high doses
			LMTX		III (Ongoing)	NCT01689246 NCT01689233 NCT02245568 NCT03446001	Significant improvement in cognitive and functional outcomes
		c-Abl	Nilotinib	Autophagic clearance of tau	II (Ongoing)	NCT02947893	To be reported

Different potential therapeutic approaches that target distinct aspects of tau biology and have reached human clinical trials are detailed in this table. General mechanisms of action, specific targets, and effects of *Modulators of PTMs*, *MT stabilizers*, and *TAIs* are briefly summarized. The table details the phase, stage, and number of clinical trials reached by each therapy and whether they have been discontinued or are currently ongoing, as well as their therapeutic effects in humans, if reported.

TABLE 10.2 Tau-based therapeutic approaches tested in clinical trials

Type	Mechanism of action	Target	Name	Effect on tau	Phase c. trial/ stage	Clinical trial n.	Therapeutic effects in humans
Immunotherapy	Passive immunotherapy: anti-tau antibodies	Tau N-t (extracellular tau)	C2N-8E12	Interact with pathological extracellular tau	II (Ongoing)	NCT02880956	To be reported in AD, orphan drug for PSP (FDA and EMA)
			RO 7105705	Inhibit tau spreading and/or promote tau clearance	I II (Ongoing)	NCT02820896, I NCT03289143, II NCT03828747, II	To be reported
		Tau mid-region	JNJ-63733657		I (Ongoing)	NCT03375697	To be reported
			UCB0107		I (Ongoing)	NCT03464227	To be reported
		Pathological conformation of tau	LY3303560		I (Ongoing) II (Ongoing)	NCT02754830, I NCT03019536, I NCT03518073, II	To be reported
		Monomeric and fibrillar tau	BIIB076	Not reported	I (Ongoing)	NCT03056729	To be reported
	Active immunotherapy: tau vaccines	Pathological truncated tau	AADvac1	Trigger immune response to produce antibodies against pathological tau	I II (Ongoing)	Eu-CT 2012-003916-29 NCT01850238, I Eu-CT 2013-004499-36 NCT02031198, I NCT02579252, II	Reduction of hippocampal atrophy and cognitive decline (at high titers, ph. I) To be reported (ph. II)
		Phosphorylated pathological tau	ACI-35	Inhibit tau spreading and/or promote tau clearance	Ib	ISRCTN13033912	Not reported
ASOs	Reduction of tau expression	Tau mRNA	BIIB080	Inhibits tau translation	I (Ongoing)	NCT03186989	Not reported

Different potential therapeutic approaches that target distinct aspects of tau biology and have reached human clinical trials are detailed in this table. General mechanisms of action, specific targets, and effects of tau-based immunotherapy (passive and active) and *ASOs* are briefly summarized. The table details the phase, stage, and number of clinical trials reached by each therapy and whether they have been discontinued or are currently ongoing, as well as their therapeutic effects in humans, if reported.

Inhibitors of kinases that phosphorylate tau

Hyperphosphorylated tau is the main constituent of the toxic forms of tau present in AD, such as aggregates, fibrils, or NFTs.[18–20] Since tau hyperphosphorylation is one of the hallmarks of AD,[18–20] kinases involved in tau phosphorylation have arisen as potential therapeutic targets for AD.[21] Several kinases are known to phosphorylate tau, some of which have been involved in the pathogenesis of AD, such as the proline-directed serine/threonine kinases, glycogen synthase kinase-3 (GSK-3)[22] and cyclin-dependent kinase 5 (CDK5)[23] or the tyrosine kinase Fyn.[24]

The constitutively active serine/threonine kinase glycogen synthase kinase-3 beta (GSK-3β) is one of the main players involved in tau phosphorylation.[25] GSK-3β is widely expressed in the brain and phosphorylates tau in serines 199, 396, and 413.[25] Hence, over the last years, a plethora of small molecules that inhibit GSK-3β have been developed in an attempt to obtain potential therapeutic agents for AD treatment.[26] GSK-3 inhibitors have been classified into three different classes: metal cations, ATP competitive, and non-ATP competitive.[27] Different methodologies have been followed to design these inhibitory molecules, ranging from structure-based to computer-based approaches, including pharmacophore map generation, molecular field methods, and quantitative structure activity relationship (QSAR) studies.[28,29] Some compounds isolated from natural sources have also proven to inhibit GSK-3.[30]

Lithium is a metal cation that inhibits GSK-3 through a direct competition with magnesium[31] and indirectly by increasing GSK-3 inhibitory serine phosphorylation downstream of Akt or PKCα.[32,33] Lithium reduces tau phosphorylation by inhibition of GSK-3[34] and diminishes tau pathology and neurodegeneration in transgenic mouse models overexpressing human tau.[35] Lithium has been shown to be neuroprotective against a number of neuronal insults.[36] Although lithium has a long-standing use in the treatment of neurological disorders (e.g., bipolar disorders[37]), its use as a therapeutic agent in AD is still controversial. Different phase II clinical trials gave rise to mixed results (NCT00088387 and SRCTN72046462[38]). Some of these studies reported no effects, while others yielded only mild effects of lithium on AD symptoms.[39,40] The later studies include beneficial effects on cognitive performance in patients with AD dementia, even with doses lower than those used for mood stabilization. However, lithium studies with AD patients were interrupted since toxic side effects were found in some of them.[41] Moreover, since lithium also inhibits the activity of inositol monophosphatase, its effects might all not be due to GSK-3 inhibition.[42]

ATP-competitive GSK-3β inhibitors interact specifically with the amino acids in the ATP-binding site. Among them are thiazoles, indirubins, maleimides, paullones, aloisines, and oxadiazoles.[43] Non-ATP competitive GSK-3 inhibitors include compounds from the thiazolidinedione family

(TDZDs), halomethylketones (HMKs), pyrimidinones, and compounds extracted from marine organisms.[43]

Despite the high number of developed and/or discovered ATP-competitive and non-ATP competitive GSK-3 inhibitors, only a few have reached clinical trials in humans, such as Tideglusib. Tideglusib, also known as NP031112/NP-12, is an irreversible non-ATP competitive GSK-3 inhibitor that belongs to the TDZD family.[44] In preclinical studies in transgenic mice, Tideglusib acts as a neuroprotective drug, reducing tau phosphorylation and amyloid deposition, gliosis, and neuron loss in the hippocampus and mouse entorhinal cortex and reversing spatial memory deficit in transgenic mice.[45] In phase II clinical trials, this compound has shown to be generally well tolerated and safe (NCT00948259 IIa, moderate AD; NCT01350362: IIb, mild-to-moderate AD). A particular concentration of Tideglusib treatment in patients with mild-to-moderate AD previously treated for months with cholinesterase inhibitors showed a significant improvement in cognitive function. However, overall, clinical trials showed no clear therapeutic effects.[46]

GSK-3 has numerous targets and participates in diverse signaling pathways involved in proliferation, apoptosis, metabolism, and differentiation. Thus, its complete inhibition may be toxic and induce tumor progression or hypoglycemia. Since GSK-3 is overactive in AD, only weak inhibitors that diminish partially GSK-3 activity (~25%), like lithium, down to basal physiological levels, may be safe effective therapeutic agents.[47]

The serine/threonine kinase CDK5 becomes deregulated in AD, leading to tau hyperphosphorylation and neuronal cell death.[48] Inhibition of CDK5 has been shown to be neuroprotective under a number of different neuronal insults in cell culture and animal model studies.[49] CDK5 is thus a potential target in the treatment of AD.[50] However, CDK5 is expressed in several cell types and tissues and is involved in many cellular processes, and thus, targeting it is challenging. Moreover, most known CDK inhibitors are not specific for a particular CDK but inhibit more than one, interfering with several cell pathways and producing unwanted side effects.[51] CDK5 is overactive not only in AD and other neurodegenerative disorders[52,53] but also in a number of cancers.[54] Hence, some of the currently available CDK inhibitors, including those that inhibit CDK5 successfully in preclinical studies, are used in clinical trials for different cancers (e.g., leukemia, lymphoma, or multiple myeloma).[55] Remarkably, tamoxifen, a drug widely used for the treatment of breast cancer, has recently been reported to inhibit CDK5 and reduce tau phosphorylation, making it a potential treatment for AD.[56] However, while some studies indicate that long-term use of tamoxifen diminishes the risk of dementia,[57] others point to an increased risk of AD in aged women that had been treated with tamoxifen against breast cancer, most likely because of their longer survival.[58]

The tyrosine kinase Fyn phosphorylates tau in tyrosine and is present in NFTs.[59] Of note, Fyn regulates local translation of tau downstream of Aβ peptide, contributing to the somatodendritic accumulation of tau typical from AD.[60] Fyn inhibition by the compound saracatinib (AZD0530) is neuroprotective in a mouse model of AD, as it rescues synapse loss and improves memory.[61] This Fyn inhibitor, which was originally developed to treat cancer, has been repurposed as a potential therapy for AD. Results of a phase Ib clinical trial indicate that it can be administered safely and is well tolerated in patients with mild-to-moderate AD (NCT01864655). A multicenter phase IIa clinical trial (NCT02167256) yielded negative results as saracatinib failed to slow AD progression (as reported in the Clinical Trials of Alzheimer's Diseases 2018 meeting).

The development of tau-kinase inhibitors as potential treatment compounds for AD faces important challenges. Most tau kinases are key players in various important cellular processes (e.g., GSK3β is involved in glycogen metabolism, CDK5 is crucial for neuronal development, and Fyn is a tumor suppressor), raising the possibility of a number of off-target side effects. Moreover, despite the successful use of several kinase inhibitors to treat different cancers, it still has to be determined whether these compounds could be safely used as a chronic treatment in AD patients. Additionally, any compound intended to be used in AD treatment has to be deliverable to the brain and subsequently distributed to specific brain regions. These drugs have to be lipophilic to cross blood-brain barrier (BBB) to exert their actions. Provided that oral administration is the preferred way to treat AD, new compounds have to be also somewhat hydrophilic. Thus, equilibrium between both chemical features and suitable brain-to-plasma ratios may be determinant for a successful delivery of these compounds. Active research is underway to obtain safe and effective kinase inhibitors that pass the BBB and present tolerable side effects so that they can be clinically used as potential therapies in AD. Novel compounds that act as dual inhibitors of both CDK5 and GSK-3β are promising multifunctional drugs to treat AD.[62–64]

Besides hyperphosphorylation, other PTMs, such as acetylation and glycosylation, have been shown to modulate tau function, contributing to the initiation of tau aggregation,[65] thus serving as targets for potential therapies for AD.

Inhibitors of tau acetylation

Tau protein undergoes extensive lysine acetylation, mostly in the MT-binding region (MBR).[66] Tau acetylation is catalyzed by two highly homologous acetylases: p300 and histone acetyltransferase cAMP-response element–binding protein (CREB)–binding protein (CBP).[67,68] Remarkably, tau has intrinsic acetyltransferase enzymatic activity, thus being able to get self-acetylated.[69] On the other hand, tau deacetylation is catalyzed by histone deacetylase 6 (HDAC6) and NAD+-dependent sirtuin 1 (SIRT1),[67,70,71] which may act on different Lys residues leading to different pathological outcomes. Acetylation prevents tau binding to MTs and promotes its pathological aggregation.[72] Both acetylation and ubiquitination occur at the same Lys residues of Tau. Thus, if tau becomes acetylated, it does not get ubiquitinated, and as a consequence, tau degradation by the ubiquitin-proteasome (UPS) system is reduced, this favoring the formation of tau toxic aggregates.[16] It has been shown that under pathological conditions, tau acetylation is enhanced, leading to a decrease in tau degradation and a subsequent increase in total tau levels and phosphorylated tau.[67] In addition, elevated tau acetylation inhibits synaptic plasticity and induces memory deficits associated with AD dementia.[73] Importantly, tau acetylation is an early event in AD brain that initiates neuronal loss by promoting tau hyperphosphorylation.[67,74] For all these reasons, inhibition of tau acetylation was hypothesized to clear pathogenic forms of tau and decrease neurodegeneration, emerging as a novel potential therapeutic target for AD.

Salsalate, a nonsteroidal antiinflammatory drug (NSAID) that has been used for decades to treat rheumatoid arthritis and related disorders[75] and diabetes type 2,[76] has been shown to inhibit p300 acetyltransferase.[76,77] Salsalate proved to have neuroprotective effects in a mouse model of tauopathy, rescuing memory deficits and reducing hippocampal atrophy.[67] A phase Ib clinical trial is currently ongoing to test salsalate safety and tolerability in patients with mild-to-moderate AD (NCT03277573, estimated completion; October 2019). This study will analyze cerebrospinal fluid (CSF) proteins, brain magnetic resonance imaging (MRI), and cognitive performance and abilities.

Modulators of *O*-GlcNAcylation

Another PTM that regulates tau function in the human brain is *O*-GlcNAcylation, a noncanonical glycosylation that implicates the attachment of *O*-linked *N*-acetylglucosamine (*O*-GlcNAc) moieties to tau residues of serine and threonine.[78] Thus, *O*-GlcNAcylation and tau phosphorylation in serine/threonine are related processes, with hyperphosphorylated tau presenting fourfold less attached *O*-GlcNAc.[79] In line with this, in short-term fasted mice, *O*-GlcNAc levels are reduced, and tau phosphorylation levels are increased.[80] Notably, the levels of tau *O*-GlcNAcylation are substantially reduced in AD brains in correlation with the decay in brain glucose metabolism, insulin signaling, and glucose transporters, pointing to the metabolic nature of this PTM.[79,80] These reports on tau *O*-GlcNAcylation reinforce the association between glucose hypometabolism and tau pathology in AD.[79,80]

Two enzymes have been shown to regulate *O*-GlcNAcylation, a reversible process: *O*-GlcNAc transferase (OGT) that catalyzes the transfer of GlcNAc to proteins and *O*-GlcNAcase (OGA) (also known as *N*-acetylglucosaminidase) that removes GlcNAc from proteins.[81] Targeting *O*-GlcNAcylation enzymes has emerged as a promising pharmacological therapy in AD. Thiamet G is an OGA inhibitor that has been shown to be neuroprotective in preclinical studies, reducing tau phosphorylation levels[82] and preventing tangle formation and neuronal loss in tau transgenic mice.[82] However, since this compound failed to enter the brain efficiently, it was chemically improved, giving rise to ASN-561, now named ASN120290, a drug that crosses the BBB and inhibits strongly OGA. ASN120290 showed its therapeutic potential in preclinical studies with a wide reduction in the accumulation of toxic tau aggregates into neurofibrillary tangles.[83] Notably, ASN120290 demonstrated safety and excellent tolerability in a phase 1b clinical trial in healthy volunteers as novel therapy for progressive supranuclear palsy (PSP) and related tauopathies.[84] This compound received orphan drug designation by the FDA for PSP. ASN120290 has recently initiated a PET neuroimaging trial to get a guide dose selection for a planned clinical efficacy trial in PSP, and a phase II proof-of-concept study for PSP treatment is planned to be developed soon. The use of this compound of other OGA inhibitors as potential therapeutic agents for AD remains under study.

MT stabilizers

Tau best known function is its role as a classical MT-associated protein (MAP) that binds along MTs and stabilizes them, mostly in neurons.[85] This contributes to the maintenance of neuronal architecture and to proper axonal transport.[86] Tau hyperphosphorylation that occurs in AD promotes tau detachment from MTs.[19] This tau "loss of function" has been proposed to induce MT destabilization and impaired axonal transport as well as synaptic dysfunction.[87] As a consequence, MT stabilizers have arisen as promising therapeutic drugs for AD.[88] Among them are epothilone D, abeotaxane, and the ADNP-derived peptide, NAP.

While the MT stabilizer paclitaxel (taxol) does not cross the BBB, taxol-derived epothilone D (Epo D), also known as BMS-241027, crosses readily the BBB.[89] Epo D treatment in different tau transgenic mice reduced axonal dystrophy, improved fast axonal transport and cognitive performance, and protected hippocampal neurons.[90–92] Epothilone D was used in a phase I clinical study (NCT01492374) in subjects with mild AD to test safety, tolerability, and effects on CSF biomarkers. However, although the study ended in 2013, no results have been reported, and the assessment of Epo D in AD has been discontinued, most likely for its lack of efficacy.

Abeotaxane or TP1287, another taxol-derived molecule,[93] was more recently assessed in a phase I clinical trial in AD patients (NCT02133846) to test safety and tolerability. AD patients tolerated the drug poorly, and the treatment had to be discontinued in some of them. No significant improvement was found upon the analysis of the exploratory cognitive end points. These results were presented in Clinical Trials on Alzheimer's Disease (CTAD) meeting in Barcelona in December 2017. Researchers in charge of the NCT02133846 clinical trial confirmed their plan to interrupt abeotaxane assessment in AD patients.

NAP (also known as davunetide or AL-108) is an eight amino acid peptide derived from the activity-dependent neuroprotective protein (ADNP)[94] (for further details see chapter by I. Gozes in this book). NAP has shown wide neuroprotective actions upon a number of neuronal insults in mice.[95] NAP protects MTs through its direct interaction with tau and with the microtubular proteins end-binding proteins 1 and 3 (EB1 and 3).[96] NAP has been tested in four phase II and III clinical trials, using a nasal spray formulation. The effect of NAP was initially assessed in a phase II trial in subjects with mild cognitive impairment (MCI), frequently a prodromal phase of AD, proving to be, in general, safe and well tolerated by intranasal administration and to improve cognition.[97] NAP was tested further, with PSP patients (NCT01110720)—not in AD—getting through successful phase II clinical trials but failing in a late phase II/phase III, in which it missed all the end points. NAP neuroprotective actions in AD are subject to intensive research efforts (for more detailed information of NAP, see I. Gozes' chapter in this book).

Changes in MT dynamics and stability that occur in AD are common to other neurodegenerative disorders and seem to be due to more indirect effects of tau "loss of function" than originally expected. In addition, it appears that clinical trials for MT stabilizers should be reassessed to get better results (e.g., lower doses at longer treatment times).

Inhibitors of tau aggregation

Under pathological conditions, tau becomes misfolded and forms aggregates of different sizes (monomers, oligomers, prefilaments, granules, fibrils, and insoluble aggregates).[98] Tau aggregates have been proposed to act as the toxic species of tau protein, involved in AD pathogenesis.[98] Therefore, tau aggregation inhibitors (TAIs) have recently emerged as potential therapeutic agents for AD. Intensive cell-based and/or in vitro screening assays have yielded a number of novel compounds that inhibit tau aggregation, such as *N*-phenylamines, flavonoids, polyphenols, anthraquinones, rhodanines, phenylthiazolyl hydrazides, (thio)carbocyanines, and phenothiazines.[99]

The efficacy of most of these TAIs still has to be demonstrated in vivo. At the moment, the TAIs that are more advanced in terms of clinical development for AD therapy are methylene blue (Rember)—chemically methylthioninium chloride (MTC)—and derived compounds. Methylene blue, the first synthetic drug, has been used for more than 120 years as a staining reagent and to treat malaria and other disorders.[100] More than 20 years ago, methylene blue was shown to inhibit tau aggregation by blocking tau-tau interaction, which occurs through the MBR.[101] Methylene blue prevents tau aggregation as confirmed in cells and tau transgenic mice.[102] Methylene blue (Rember or TRx-014) readily crosses the BBB, and its effects in cognitive ability were tested in mild-to-moderate AD subjects in two phase II clinical trials (NCT00515333 and NCT00684944). The drug showed beneficial effects at lower concentrations, but no effects at higher doses because of differences in redox processing and absorption limitations.[103,104] In addition, adverse side effects were reported including dizziness and falls, urinary urgency and painful urination, and diarrhea.[103] This led to the development of second-generation derivatives of methylene blue with a higher stability in reduced conditions, such as LMTX (leucomethylthioninium bis-hydromethanesulfonate, also named TRx-0237). LMTX was confirmed to inhibit aggregation of tau in cells and in mouse tauopathy models, in which it also rescued learning impairment.[105] LMTX has reached phase III clinical trials for the treatment of AD (NCT01689246, NCT01689233, and NCT02245568) (extension study currently ongoing), some of them on subjects under AchE and/or memantine treatments. LMTX showed safety with an increased absorption and improved tolerability as compared with Rember, although it presented gastrointestinal and urinary disorders as common adverse side effects at high doses.[106,107] However, different trials produced mixed results concerning the improvement of cognition. One of these phase III studies did not show significant efficacy in improving cognition in people with mild-to-moderate AD (e.g., NCT01689246)[106]. Yet, post hoc analyses of this trial exhibited some beneficial effects of LMTX in monotherapy, even at low doses, but the use of controversial methodology has been criticized. In contrast, another trial in mild AD patients yielded promising results, providing significant improvement in cognitive and functional outcomes with LMTX monotherapy, even at a control low dose.[107] To further confirm that LMTX is more efficient as a monotherapy than as a multitherapy and to determine safety and efficacy of the drug in early AD, a phase II/III study (LUCIDUTY NCT03446001) is currently recruiting. In this study FDG-PET imaging will be used to monitor LMTX efficacy in delaying AD pathology progression in the brain of patients with early AD, who are not under other treatments.

AZP2006 is another compound that clears tau and β-amyloid aggregates in cells and in tau transgenic mice by activating autophagy.[108] AZP2006 reduced tau phosphorylation and tau pathology progression and reversed already existing neurodegeneration, apparently through its induction of PRGN,[109] a multifunctional growth factor. AZP2006 is thus a drug with both neuroprotective and neurodegenerative potentials.[110] AZP2006 is orally available and has been designated as orphan drug for the treatment of PSP. Phase I clinical trials in healthy subjects indicated that AZP2006 is well absorbed by the intestine and well tolerated. In September 2018, a study was launched on the potential effects of food intake on the pharmacokinetics of AZP2006 administered orally. A phase II clinical trial in PSP patients is planned to start soon after the completion of the former study. If positive results are obtained on PSP patients, the effect of AZP2006 will be evaluated in AD.

Nilotinib (Tasigna) is a c-Abl tyrosine kinase inhibitor used to treat chronic myeloid leukemia.[111] This drug has been shown to degrade phospho-tau and β-amyloid by autophagic clearance, preventing neuronal loss and reversing cognitive decline in AD mouse models.[112] Since preclinical data were strong, Nilotinib was tested in subjects with Parkinson dementia (PD) and Lewi body dementia (LBD), disorders in which CSF total tau and p-tau are increased (NCT02281474). This study showed that Nilotinib significantly reduces the levels of total tau and p-tau in CSF, in correlation with cognitive improvement,[113] paving the way for testing Nilotinib as a potential treatment for AD. Currently, Nilotinib is tested in a phase II clinical trial in subjects with mild-to-moderate AD (NCT02947893). Safety, tolerability, and tau and phospho-tau levels in CSF will be assessed, as well as cognitive and functional abilities.

Immunotherapy

One of the most promising tau-targeting approaches with therapeutic potential to combat AD is immunotherapy, an approach that has been shown to reduce tau pathology and improve cognitive deficits in animal models of AD.[114] Immunotherapy might be passive or active. Passive immunotherapy involves injection of antibodies or antisera against the unwanted element, removing the need of the recipient's body to generate an immune response and create its own antibodies. This has economical drawbacks since the costs of giving every week/month monoclonal antibody infusions for long periods (years to decades) could be substantial. Active immunization, also known as vaccination, is the most common form of immunotherapy and involves antigen administration, usually with coadjuvant, triggering the immune response in the recipient's body. Vaccination is more cost-effective than passive immunotherapy, because it is long-lasting, inducing immunological memory.[115] Furthermore, active vaccines are easy to administer using different methods, and their production is cost-effective. Immunization causes polyclonal response with antibodies that can recognize multiple epitopes on the target protein with different affinities and avidities. However, active vaccine-induced immune response depends on the immune system of the host, leading to an intrinsic variability in the antibody response among patients.

Passive immunotherapy

Therapeutic antibodies have been raised against conformationally altered forms of tau,[116–120] monomeric[116,121,122] or aggregated tau[123] and phosphorylated tau.[114] Diverse features, including epitope selection, isotype, charge, or affinity, influence the action of different types of antibodies.[114] Antitau antibodies also show distinct efficacies depending on the stage of the pathology, which also influences the pools of intra- and extracellular tau.[114] Based on all these considerations, several antibodies have been raised against different tau epitopes and/or conformations and tested as potential immunotherapies for AD in diverse tau transgenic mice. These antibodies have shown a wide range of degrees of efficacy in decreasing tau pathology, with a concomitant improvement in motor and cognitive functions, depending on the location of tau pathology.[114]

Different antibodies recognize different sites of the tau molecule, such as the N-terminal region,[121,122,124–126] proline-rich region (midregion),[120,126] MT-binding region (MBR),[122,127] or C-terminus.[114,117] Antibodies targeting the N-terminus of tau have arisen as promising therapeutic agents by preclinical studies serving as potential inhibitors of tau uptake or seeding, due to the specific characteristics of this protein region.[121,122,124–126] N-terminal fragments of tau have been involved in mitochondrial dysfunction, impairment of synaptic plasticity,[128–131] and increase in β-amyloid production.[132] This is most likely a consequence of early conformational changes that tau N-terminus undergoes during the onset of the pathogenesis of AD that alters the protein function.[133] Antibodies against N-terminal tau have proven to be neuroprotective in diverse preclinical studies performed in transgenic mice (although varied results have been obtained).[121,122,125,126,134] Notably, only N-terminal fragments of tau have been found in CSF of AD patients,[135] and the majority of the toxic tau present in AD brain is truncated, mostly at the N-terminus, including oligomers and fibrils.[136] This implies that a large proportion of pathological tau is not recognized by antibodies against its N-terminus.[136–138] It is worth mentioning that antibodies directed to the N-terminus on tau are not specific to pathological tau, and they most likely decrease as well the level of physiological tau.

A second class of antibodies have been raised against the middomain of tau, since this domain contains some of the residues that become highly phosphorylated during AD (e.g., S202, T205, and T231).[139,140] Although these middomain antibodies (PT51, PT79, PT89, and 6C5) have proven to be effective in the reduction of tau uptake and neutralization of seeding activity to different extents in cell and animal models,[137,138] results obtained in preclinical assays have been inconsistent.

Another group of antitau antibodies are those targeting the MBR,[122,124] a part of the protein involved in MT stabilization and polymerization, as well as in the pathological tau-tau interaction that takes place during aggregation and fibril formation.[141,142] Up to now, only two preclinical studies have been reported using antibodies directed to tau MBR (DC8E8 and HJ93), both providing neuroprotective effects by reducing neurofibrillary pathology in the brain of transgenic tau mice.[122,143]

The C-terminal region of tau contains several residues that undergo hyperphosphorylation in early and late AD pathology,[144] regulates binding to MTs, and influences tau aggregation.[145] As a consequence, antibodies raised against tau C-terminus have been assessed in several preclinical studies in tau transgenic mice.[116,117,146–148] Promising results of these studies include reduction in tau pathology by aggregate clearance, delay of disease progression, or decrease in functional decline.[116,117,146–148]

In addition, antibodies that target conformational changes of tau, as well as tau oligomers, are being used in preclinical studies in mice, since these tau species occur early on during AD pathogenesis.[149–151] An example of these antibodies is MC1 that reacts only against a conformational tau epitope present in pathological tau,[150] decreasing to some extent the levels of insoluble tau and tangles in tau transgenic mouse models.[117,118,123]

Antitau antibodies face the challenge of becoming effective and safe therapeutic drugs for AD. To accomplish this, tau-targeting antibodies should be selective toward pathogenic forms of tau, thus avoiding side effects derived from neutralizing healthy tau. Passive immunotherapy offers several advantages in comparison with active immunotherapy/vaccination. Pharmacologically, antibodies are well-characterized molecules in terms of affinity, specificity, half-life, avidity, single isotype, and concentration. Additionally, passive immunotherapy does not trigger the immune response of the host. Importantly, therapeutic antibodies can target extracellular tau and, in this way, block intercellular tau spreading in a prion-like fashion. Notwithstanding, this type of therapy also bears serious disadvantages, such as the high costs related to the chronic administration and short half-lives of the used therapeutic antibodies. Moreover, administering the antibodies in a chronic way may induce the generation of anti-antibodies that in turn may produce undesired side effects or neutralize the effect of the passive immunotherapy.[152]

Anti-tau antibody therapies being investigated in clinical trials

At present, passive immunotherapy using different types of tau-targeting antibodies is being assessed in clinical trials at different phases as potential therapies for AD. Humanized versions of antibodies against the N-terminal part of tau are mostly tested. One of these antibodies is

C2N-8E12 (a humanized IgG4, also known as ABBV-8E12), which recognizes an aggregated, extracellular form of pathological tau, which makes it potentially capable of blocking tau spreading, since it does not require active uptake into neurons.[153] C2N-8E12 was proved to be safe and well tolerated in clinical trials in PSP subjects (NCT02494024). Positive results prompted the FDA and EMA to grant C2N-8E12 a designation as orphan drug for PSP treatment. The effects of C2N-8E12 are currently being analyzed in a phase II clinical trial (NCT02880956) in subjects with early AD in 6 European countries and 66 sites in North America, plus Australia, and New Zealand. The trial is planned to be completed in 2021. Outcomes include a range of clinical and functional measures.

Another N-terminal antibody being tested as passive immunotherapeutic agent is RO 7105705 (RG 6100). It recognizes the N-termini of all six tau isoforms of human tau, both monomeric and oligomeric tau.[148] As 8E12, RO 7105705 is a IgG4 antibody that targets extracellular, not intracellular, tau, with the potential to stop cell-to-cell spreading of pathological tau in the extracellular space of the brain.[148] Furthermore, it has been shown to exert limited microglial activation.[148] RO 7105705 was initially tested in 2016–17 in a phase I clinical trial in volunteers including both healthy controls and subjects with mild-to-moderate AD, for safety, tolerability, pharmacokinetics, and preliminary activity outcomes (NCT02820896). No severe adverse events were induced by the drug, only minor ones such as pain at the injection site and bruising. After this study, in September 2017, a phase II trial started in subjects with prodromal or probable AD diagnosed on the base of mild symptoms, positive amyloid PET, or CSF Aβ42 peptide (NCT03289143). This study is being conducted at 153 locations in various European countries, North America, and Australia, and it is expected to run until 2020. Outcomes include a number of measures relating to cognitive and functional abilities, as well as serum-drug concentrations and antidrug antibodies. Another phase II clinical trial has just been started in February 2019 in people with probable AD (confirmed by CSF and PET analyses) and moderate dementia. Primary and secondary end points include again a number of measures related to cognitive and functional abilities and serum concentration of the antibody and antidrug antibodies. This study is planned to be completed by 2021 and currently involves seven US locations.

JNJ-63733657 is an antibody that recognizes the mid-region of tau and is effective in the elimination of pathological tau seeds in a cell-based assay and inhibits the propagation of pathology in a mouse model.[7] A phase I clinical trial is currently being conducted to examine safety and tolerability, pharmacokinetics, and pharmacodynamics of JNJ-63733657 in healthy subjects and subjects with AD (NCT03375697). Outcomes include measure of JNJ-63733657 levels in the blood and CSF, presence of antibodies against it, and adverse events. This study was planned to run until February 2019; no results have been reported yet.

Another antibody targeting the mid-region of tau, UCB0107,[114] was also tested in a phase I clinical trial that started in February 2018 (NCT03464227). UCB1017 was reported to bind tau monomers, inhibit tau seeding in a cell-based assay, and further inhibit tau spreading in a tau transgenic mouse model (reported in the congress Advances in Alzheimer's and Parkinson's Therapies Focus Meeting (AAT-AD/PD), held in March 15–18, 2018, in Turin, Italy). Outcomes from this study include adverse events, measure of UCB0107 levels in the blood and CSF, presence of anti-UCB0107 host antibodies, and pharmacokinetic parameters of antibody distribution and clearance. The trial was expected to be completed in November 2018, but no results have been reported yet.

LY3303560 is a conformational antibody,[154] derived from monoclonal MC1 antibody, which recognizes both the MBR and the N-terminus of tau[155] and neutralizes soluble tau aggregates in preclinical studies in mice.[117,118] Effects of LY3303560 are currently tested in phase I clinical assays in healthy volunteers and people with mild-to-moderate AD (NCT02754830 and NCT03019536). Outcomes of both studies include measures of adverse effects, antibody concentration in both serum and CSF, and pharmacokinetic parameters. These outcomes along with others, including several ones related to cognitive and functional abilities, will be measured in an ongoing phase II clinical trial that began in April 2018 (NCT03518073).

To end with, BIIB076 (6C5 huIgG1/l) is another human monoclonal antibody against tau that binds monomeric and fibrillar tau and the human brain tau isolated from both healthy subjects and AD patients.[156] Initial pharmacokinetic and dose-finding data were reported in the AD/PD Meeting 2017 (29th March to 2nd April, Vienna). BIIB076 is currently being tested for safety and tolerability in a phase I clinical trial involving healthy volunteers and subjects with mild-to-moderate AD (NCT03056729). Other outcomes comprise adverse events, neurological exam, pharmacokinetic parameters of antibody exposure, clearance, and immunogenicity.

Active immunotherapy: Vaccination against tau

The development of vaccines against tau has been focused on targeting different tau regions, such as the midregion, the MBR, and the C-terminus (as occurred in passive immunization).[157] The majority of the generated tau vaccines reduced tau pathology[124,134,158–161] and improved cognitive or motor abilities in preclinical studies in mice.[116,124,158,162,163] Antibody responses triggered by these active vaccines have scarcely been characterized in a quantitative way but rather described in a qualitative manner, except for two studies in which the titer of the antibody response was quantified.[124,161] Measurement of antibody responses has to be standardized to allow a direct comparison between antibody responses between different assays and clinical trials. Besides this, the quality of the antibodies generated in response to the vaccines is also decisive for the efficacy of active immunotherapy (e.g., their affinity, target epitope, functional activity, and isotypes).[152]

While a number of antitau antibodies used in passive immunotherapy are under study in clinical trials, only two tau active vaccines are being tested in clinical trials for AD: AADvac1 and ACI-35.

AADvac1 is one of these active tau vaccines under development, which consists of a synthetic peptide derived from amino acids 294 to 305 of the tau sequence (i.e., KDNIKHVPGGGS), coupled to keyhole limpet hemocyanin, a large antigen frequently used as an immune adjuvant.[124] AADvac1 recognizes preferentially pathological truncated and misfolded tau over healthy full-length tau.[164] AADvac1 induces generation of antibodies that target conformational epitopes in tau MBR, preventing tau aggregation and spreading and promoting tau clearance.[164] The potential of the vaccine as a tau-targeted therapeutic agent was assessed in tau transgenic mice and rats. In these experiments, AADvac1 led to the reduction of tau phosphorylation and neurofibrillary tangles and improved its cognitive and behavioral impairment in rodent brain.[124] In humans, AADvac1 proved to be particularly safe and well tolerated, as shown in phase I clinical trials (NCT01850238, followed up in NCT02031198).[124,164,165] Most of the treated patients developed an IgG response to the tau peptide component of AADvac1 and against recombinant pathological tau. AADvac1 led to increasing antibody titers after repeating antibody injections in a majority of participants, stable over 6 months. Delivery of AADvac1 showed to be protective by decreasing tau fibrils and insoluble tau and reducing hippocampal atrophy and cognitive decline in the brains of participants with high titers.[165] At the moment, the therapeutic potential of AADvac1 to treat AD continues to be assessed in a phase II clinical trial in subjects with mild-to-moderate AD (NCT02579252). Outcomes of this assay include safety, cognitive measurements, immunogenicity, and CSF biochemistry constituting exploratory outcomes. This study is being run in different European countries, such as Sweden, Slovakia, Austria, and the Czech Republic. No results have been reported yet.

ACI-35 is a liposome-based vaccine against certain pathological conformers of phosphorylated tau protein.[158] The vaccine encompasses 16 copies of a synthetic fragment of tau that is phosphorylated at the residues S396 and S404, attached into a lipid bilayer, with the MPLA as adjuvant. When injected in tau transgenic mice, ACI-35 elicited rapid and robust immune response with high titers of polyclonal IgG against phosphorylated tau and neurofibrillary tangles and a reduction in tau pathology.[166] ACI-35 was tested in a completed phase I clinical trial for safety and clinical parameters, but results have not been reported (International Clinical Trials Registry Platform, ISRCTN13033912).

Reduction of tau levels

Hyperphosphorylated tau detaches from MTs, leading to an excess of unbound tau that in turn gets mislocalized, undergoes conformational conversion, and forms toxic aggregates and fibrils (especially the most fibrilogenic 4R-tau).[167] Based on this, it has been proposed that decreasing the levels of total tau (or of the extra 4R-tau), by specific antisense oligonucleotides (ASOs) targeting tau, might be a potential therapeutic treatment for AD, as reported in a number of cell and animal models.[168] ASOs are single-stranded short oligos (8–50 nucleotides) designed to downregulate specific genes by specific interaction with complementary messenger RNAs (mRNAs), leading to their degradation or inhibition of their translation. Human therapies involving ASOs are challenging due to questionable uptake, low stability, vulnerability to degradation, and difficulties to cross the BBB.[169] Chemical modifications of the molecules may help to overcome these issues, such as (1) the use of liposomes or cell-penetrating peptides (to enhance ASOs entrance into the cells) or (2) the modification of the DNA or RNA backbones or that of the ribose sugar.[169]

ASOs that reduce tau expression have proven to be neuroprotective as they decrease tau neurofibrillary pathology and reduce neuronal loss in tau transgenic mice.[168,169] Of note, these ASOs have also been tested in cynomolgus monkeys by infusion into the CSF, reducing the levels of tau in CSF and of tau mRNA in different brain regions.[168,169]

BIIB080 (also named IONIS-MAPTRx or ISIS 814907) is the first ASO directed to inhibit tau mRNA translation that have entered clinical trials. BIIB080 effects are being assessed in patients with mild AD (NCT03186989) in a study conducted in five European countries and 10 sites in Canada. This phase I clinical trial is aimed to address safety, pharmacokinetics, and pharmacodynamics of the drug and is planned to be completed by January 2020.

It has to be noted that tau is a multifaceted protein with numerous functions in neurons. Therefore, the reduction of the levels of nonpathological tau may be an adverse side effect and an important drawback of ASOs therapy. In tau knockout mice, there is a developmental compensatory increase in the expression of other MAPs, such as MAP1B, thus overcoming the effect of lacking tau on MT stabilization in neurons.[170] Tau-deficient mice present normal learning, memory, and cognition but display a slight motor phenotype at a later age.[170] In addition, tau knockout mice show brain iron accumulation, brain insulin resistance, and deficits in synaptic plasticity and cognition.[170] Hence, the consequences of tau downregulation in the adult brain have to be analyzed in detail.

Concluding remarks and future directions

For decades, intensive research has been devoted to decrease Aβ burden to combat the progression of AD. A number of compounds and antibodies targeting Aβ have been tested in large clinical trials in patients with early and/or mild-to-moderate AD, but most of them have failed to slow down the progression of AD or improve cognition.[10] On the other hand, recent findings indicate that progression of tau pathology correlates much better than Aβ plaque burden with cognitive decline progression in AD.[14] Therefore, it has been predicted that targeting tau protein may be more effective than targeting Aβ clearance to treat AD.[11,12] As a consequence, tau has lately attracted much attention as a potential therapeutic target to reduce or delay AD progression. This chapter presents an overview of the different approaches currently used to target several aspects of tau biology, including tau phosphorylation, acetylation, *O*-GlcNAcylation, aggregation, and tau-based immunotherapy (as summarized in Fig. 10.1 and Tables 10.1 and 10.2). The objective of all of these approaches is inhibiting or decelerating different aspects of tau pathology, thus potentially delaying the progression of AD. No tau-based drug has been approved yet as a disease-modifying treatment for AD, but a number of novel potential therapeutic agents are currently under study in clinical trials, as discussed in this chapter and summarized in Tables 10.1 and 10.2. The most promising therapeutic approaches range from inhibitors of tau aggregation that have reached phase III in clinical trials, to passive and active immunotherapies currently being tested in a number of ongoing studies, which results are awaited soon. Meanwhile, intensive research for novel approaches that effectively target tau in the human brain, with clear beneficial effects in delaying AD progression and without severe adverse effects, is in progress.

References

1. James BD, Bennett DA. Causes and patterns of dementia: an update in the era of redefining Alzheimer's disease. *Annu Rev Public Health*. 2019;40:65–84.
2. Saxena M, Dubey R. Target enzyme in Alzheimer's disease: acetylcholinesterase inhibitors. *Curr Top Med Chem*. 2019;19(4):264–275.
3. McShane R, Westby MJ, Roberts E, et al. Memantine for dementia. *Cochrane Database Syst Rev*. 2019;3:. CD003154.
4. Wilcock GK, Esiri MM, Bowen DM, Smith CC. Alzheimer's disease. Correlation of cortical choline acetyltransferase activity with the severity of dementia and histological abnormalities. *J Neurol Sci*. 1982;57(2–3):407–417.
5. Whitehouse PJ, Price DL, Clark AW, Coyle JT, DeLong MR. Alzheimer disease: evidence for selective loss of cholinergic neurons in the nucleus basalis. *Ann Neurol*. 1981;10(2):122–126.
6. Yiannopoulou KG, Anastasiou IP, Ganetsos TK, Efthimiopoulos P, Papageorgiou SG. Prevalence of dementia in elderly patients with hip fracture. *Hip Int*. 2012;22(2):209–213.

7. Cummings J, Lee G, Ritter A, Zhong K. Alzheimer's disease drug development pipeline: 2018. *Alzheimers Dement (N Y)*. 2018;4:195–214.
8. Serrano-Pozo A, Frosch MP, Masliah E, Hyman BT. Neuropathological alterations in Alzheimer disease. *Cold Spring Harb Perspect Med*. 2011;1(1):a006189.
9. Hardy JA, Higgins GA. Alzheimer's disease: the amyloid cascade hypothesis. *Science*. 1992;256(5054):184–185.
10. Panza F, Lozupone M, Logroscino G, Imbimbo BP. A critical appraisal of amyloid-beta-targeting therapies for Alzheimer disease. *Nat Rev Neurol*. 2019;15(2):73–88.
11. Holmes BB, Diamond MI. Prion-like properties of Tau protein: the importance of extracellular Tau as a therapeutic target. *J Biol Chem*. 2014;289(29):19855–19861.
12. Stopschinski BE, Diamond MI. The prion model for progression and diversity of neurodegenerative diseases. *Lancet Neurol*. 2017;16(4):323–332.
13. Nelson PT, Alafuzoff I, Bigio EH, et al. Correlation of Alzheimer disease neuropathologic changes with cognitive status: a review of the literature. *J Neuropathol Exp Neurol*. 2012;71(5):362–381.
14. Bejanin A, Schonhaut DR, La Joie R, et al. Tau pathology and neurodegeneration contribute to cognitive impairment in Alzheimer's disease. *Brain*. 2017;140(12): 3286–3300.
15. Novak P, Kontsekova E, Zilka N, Novak M. Ten years of tau-targeted immunotherapy: the path walked and the roads ahead. *Front Neurosci*. 2018;12:798.
16. Morris M, Knudsen GM, Maeda S, et al. Tau post-translational modifications in wild-type and human amyloid precursor protein transgenic mice. *Nat Neurosci*. 2015;18(8):1183–1189.
17. Marcelli S, Corbo M, Iannuzzi F, et al. The involvement of post-translational modifications in Alzheimer's disease. *Curr Alzheimer Res*. 2018;15(4):313–335.
18. Alonso A, Zaidi T, Novak M, Grundke-Iqbal I, Iqbal K. Hyperphosphorylation induces self-assembly of tau into tangles of paired helical filaments/straight filaments. *Proc Natl Acad Sci U S A*. 2001;98(12):6923–6928.
19. Alonso AD, Grundke-Iqbal I, Barra HS, Iqbal K. Abnormal phosphorylation of tau and the mechanism of Alzheimer neurofibrillary degeneration: sequestration of microtubule-associated proteins 1 and 2 and the disassembly of microtubules by the abnormal tau. *Proc Natl Acad Sci U S A*. 1997;94(1):298–303.
20. Alonso AC, Grundke-Iqbal I, Iqbal K. Alzheimer's disease hyperphosphorylated tau sequesters normal tau into tangles of filaments and disassembles microtubules. *Nat Med*. 1996;2(7):783–787.
21. Lee VM, Brunden KR, Hutton M, Trojanowski JQ. Developing therapeutic approaches to tau, selected kinases, and related neuronal protein targets. *Cold Spring Harb Perspect Med*. 2011;1(1):a006437.
22. Takashima A. GSK-3 is essential in the pathogenesis of Alzheimer's disease. *J Alzheimers Dis*. 2006;9(3 Suppl):309–317.
23. Liu SL, Wang C, Jiang T, Tan L, Xing A, Yu JT. The role of Cdk5 in Alzheimer's disease. *Mol Neurobiol*. 2016;53(7):4328–4342.
24. Yang K, Belrose J, Trepanier CH, Lei G, Jackson MF, MacDonald JF. Fyn, a potential target for Alzheimer's disease. *J Alzheimers Dis*. 2011;27(2):243–252.
25. Hanger DP, Noble W. Functional implications of glycogen synthase kinase-3-mediated tau phosphorylation. *Int J Alzheimers Dis*. 2011;2011:352805.
26. Avila J, Hernandez F. GSK-3 inhibitors for Alzheimer's disease. *Expert Rev Neurother*. 2007;7(11):1527–1533.
27. Eldar-Finkelman H, Martinez A. GSK-3 inhibitors: preclinical and clinical focus on CNS. *Front Mol Neurosci*. 2011;4:32.
28. Dessalew N, Patel DS, Bharatam PV. 3D-QSAR and molecular docking studies on pyrazolopyrimidine derivatives as glycogen synthase kinase-3beta inhibitors. *J Mol Graph Model*. 2007;25(6):885–895.

29. Gadakar PK, Phukan S, Dattatreya P, Balaji VN. Pose prediction accuracy in docking studies and enrichment of actives in the active site of GSK-3beta. *J Chem Inf Model.* 2007;47(4):1446–1459.
30. Wiese J, Imhoff JF, Gulder TA, Labes A, Schmaljohann R. Marine fungi as producers of benzocoumarins, a new class of inhibitors of glycogen-synthase-kinase 3beta. *Mar Drugs.* 2016;14(11). E200.
31. Ryves WJ, Harwood AJ. Lithium inhibits glycogen synthase kinase-3 by competition for magnesium. *Biochem Biophys Res Commun.* 2001;280(3):720–725.
32. Chalecka-Franaszek E, Chuang DM. Lithium activates the serine/threonine kinase Akt-1 and suppresses glutamate-induced inhibition of Akt-1 activity in neurons. *Proc Natl Acad Sci U S A.* 1999;96(15):8745–8750.
33. Kirshenboim N, Plotkin B, Shlomo SB, Kaidanovich-Beilin O, Eldar-Finkelman H. Lithium-mediated phosphorylation of glycogen synthase kinase-3beta involves PI3 kinase-dependent activation of protein kinase C-alpha. *J Mol Neurosci.* 2004;24(2):237–245.
34. Hong M, Chen DC, Klein PS, Lee VM. Lithium reduces tau phosphorylation by inhibition of glycogen synthase kinase-3. *J Biol Chem.* 1997;272(40):25326–25332.
35. Caccamo A, Oddo S, Tran LX, LaFerla FM. Lithium reduces tau phosphorylation but not A beta or working memory deficits in a transgenic model with both plaques and tangles. *Am J Pathol.* 2007;170(5):1669–1675.
36. Pasquali L, Busceti CL, Fulceri F, Paparelli A, Fornai F. Intracellular pathways underlying the effects of lithium. *Behav Pharmacol.* 2010;21(5–6):473–492.
37. Machado-Vieira R, Manji HK, Zarate Jr. CA. The role of lithium in the treatment of bipolar disorder: convergent evidence for neurotrophic effects as a unifying hypothesis. *Bipolar Disord.* 2009;11(Suppl 2):92–109.
38. Hampel H, Ewers M, Burger K, et al. Lithium trial in Alzheimer's disease: a randomized, single-blind, placebo-controlled, multicenter 10-week study. *J Clin Psychiatry.* 2009;70(6):922–931.
39. Matsunaga S, Kishi T, Annas P, Basun H, Hampel H, Iwata N. Lithium as a treatment for Alzheimer's disease: a systematic review and meta-analysis. *J Alzheimers Dis.* 2015;48(2):403–410.
40. Forlenza OV, Aprahamian I, de Paula VJ, Hajek T. Lithium, a therapy for AD: current evidence from clinical trials of neurodegenerative disorders. *Curr Alzheimer Res.* 2016;13(8):879–886.
41. Macdonald A, Briggs K, Poppe M, Higgins A, Velayudhan L, Lovestone S. A feasibility and tolerability study of lithium in Alzheimer's disease. *Int J Geriatr Psychiatry.* 2008;23(7):704–711.
42. Bone R, Springer JP, Atack JR. Structure of inositol monophosphatase, the putative target of lithium therapy. *Proc Natl Acad Sci U S A.* 1992;89(21):10031–10035.
43. Arafa RK, Elghazawy NH. Personalized medicine and resurrected hopes for the management of Alzheimer's disease: a modular approach based on GSK-3beta inhibitors. *Adv Exp Med Biol.* 2017;1007:199–224.
44. Dominguez JM, Fuertes A, Orozco L, del Monte-Millan M, Delgado E, Medina M. Evidence for irreversible inhibition of glycogen synthase kinase-3beta by tideglusib. *J Biol Chem.* 2012;287(2):893–904.
45. Sereno L, Coma M, Rodriguez M, et al. A novel GSK-3beta inhibitor reduces Alzheimer's pathology and rescues neuronal loss in vivo. *Neurobiol Dis.* 2009;35(3):359–367.
46. Lovestone S, Boada M, Dubois B, et al. A phase II trial of tideglusib in Alzheimer's disease. *J Alzheimers Dis.* 2015;45(1):75–88.
47. Avrahami L, Licht-Murava A, Eisenstein M, Eldar-Finkelman H. GSK-3 inhibition: achieving moderate efficacy with high selectivity. *Biochim Biophys Acta.* 2013;1834(7):1410–1414.
48. Lee KY, Clark AW, Rosales JL, Chapman K, Fung T, Johnston RN. Elevated neuronal Cdc2-like kinase activity in the Alzheimer disease brain. *Neurosci Res.* 1999;34(1):21–29.

49. Chang KH, Vincent F, Shah K. Deregulated Cdk5 triggers aberrant activation of cell cycle kinases and phosphatases inducing neuronal death. *J Cell Sci*. 2012;125(Pt 21): 5124–5137.
50. Bhounsule AS, Bhatt LK, Prabhavalkar KS, Oza M. Cyclin dependent kinase 5: a novel avenue for Alzheimer's disease. *Brain Res Bull*. 2017;132:28–38.
51. Bain J, McLauchlan H, Elliott M, Cohen P. The specificities of protein kinase inhibitors: an update. *Biochem J*. 2003;371(Pt 1):199–204.
52. Cheung ZH, Ip NY. Cdk5: a multifaceted kinase in neurodegenerative diseases. *Trends Cell Biol*. 2012;22(3):169–175.
53. Mushtaq G, Greig NH, Anwar F, et al. Neuroprotective mechanisms mediated by CDK5 inhibition. *Curr Pharm Des*. 2016;22(5):527–534.
54. Pozo K, Bibb JA. The emerging role of Cdk5 in cancer. *Trends Cancer*. 2016;2(10):606–618.
55. Shupp A, Casimiro MC, Pestell RG. Biological functions of CDK5 and potential CDK5 targeted clinical treatments. *Oncotarget*. 2017;8(10):17373–17382.
56. Corbel C, Zhang B, Le Parc A, et al. Tamoxifen inhibits CDK5 kinase activity by interacting with p35/p25 and modulates the pattern of tau phosphorylation. *Chem Biol*. 2015;22(4):472–482.
57. Sun LM, Chen HJ, Liang JA, Kao CH. Long-term use of tamoxifen reduces the risk of dementia: a nationwide population-based cohort study. *QJM*. 2015;109(2):103–109.
58. Liao KF, Lin CL, Lai SW. Nationwide case-control study examining the association between tamoxifen use and Alzheimer's disease in aged women with breast cancer in Taiwan. *Front Pharmacol*. 2017;8:612.
59. Lee G, Thangavel R, Sharma VM, et al. Phosphorylation of tau by fyn: implications for Alzheimer's disease. *J Neurosci*. 2004;24(9):2304–2312.
60. Li C, Gotz J. Somatodendritic accumulation of Tau in Alzheimer's disease is promoted by Fyn-mediated local protein translation. *EMBO J*. 2017;36(21):3120–3138.
61. Kaufman AC, Salazar SV, Haas LT, et al. Fyn inhibition rescues established memory and synapse loss in Alzheimer mice. *Ann Neurol*. 2015;77(6):953–971.
62. Damiens E, Baratte B, Marie D, Eisenbrand G, Meijer L. Anti-mitotic properties of indirubin-3′-monoxime, a CDK/GSK-3 inhibitor: induction of endoreplication following prophase arrest. *Oncogene*. 2001;20(29):3786–3797.
63. Polychronopoulos P, Magiatis P, Skaltsounis AL, et al. Structural basis for the synthesis of indirubins as potent and selective inhibitors of glycogen synthase kinase-3 and cyclin-dependent kinases. *J Med Chem*. 2004;47(4):935–946.
64. Xie H, Wen H, Zhang D, et al. Designing of dual inhibitors for GSK-3beta and CDK5: virtual screening and in vitro biological activities study. *Oncotarget*. 2017;8(11):18118–18128.
65. Wang Y, Mandelkow E. Tau in physiology and pathology. *Nat Rev Neurosci*. 2016;17(1):5–21.
66. Hwang AW, Trzeciakiewicz H, Friedmann D, et al. Conserved lysine acetylation within the microtubule-binding domain regulates MAP2/tau family members. *PLoS One*. 2016;11(12). e0168913.
67. Min SW, Cho SH, Zhou Y, et al. Acetylation of tau inhibits its degradation and contributes to tauopathy. *Neuron*. 2010;67(6):953–966.
68. Marmorstein R, Roth SY. Histone acetyltransferases: function, structure, and catalysis. *Curr Opin Genet Dev*. 2001;11(2):155–161.
69. Cohen TJ, Friedmann D, Hwang AW, Marmorstein R, Lee VM. The microtubule-associated tau protein has intrinsic acetyltransferase activity. *Nat Struct Mol Biol*. 2013;20(6):756–762.
70. Cook C, Carlomagno Y, Gendron TF, et al. Acetylation of the KXGS motifs in tau is a critical determinant in modulation of tau aggregation and clearance. *Hum Mol Genet*. 2014;23(1):104–116.
71. Ding H, Dolan PJ, Johnson GV. Histone deacetylase 6 interacts with the microtubule-associated protein tau. *J Neurochem*. 2008;106(5):2119–2130.

72. Cohen TJ, Guo JL, Hurtado DE, et al. The acetylation of tau inhibits its function and promotes pathological tau aggregation. *Nat Commun*. 2011;2:252.
73. Tracy TE, Gan L. Acetylated tau in Alzheimer's disease: an instigator of synaptic dysfunction underlying memory loss: increased levels of acetylated tau blocks the postsynaptic signaling required for plasticity and promotes memory deficits associated with tauopathy. *Bioessays*. 2017;39(4). https://doi.org/10.1002/bies.201600224.
74. Lucke-Wold B, Seidel K, Udo R, et al. Role of tau acetylation in Alzheimer's disease and chronic traumatic encephalopathy: the way forward for successful treatment. *J Neurol Neurosurg*. 2017;4(2). 140.
75. Bianchi Porro G, Petrillo M, Ardizzone S. Salsalate in the treatment of rheumatoid arthritis: a double-blind clinical and gastroscopic trial versus piroxicam. II. Endoscopic evaluation. *J Int Med Res*. 1989;17(4):320–323.
76. Goldfine AB, Fonseca V, Jablonski KA, Pyle L, Staten MA, Shoelson SE. The effects of salsalate on glycemic control in patients with type 2 diabetes: a randomized trial. *Ann Intern Med*. 2010;152(6):346–357.
77. Shirakawa K, Wang L, Man N, et al. Salicylate, diflunisal and their metabolites inhibit CBP/p300 and exhibit anticancer activity. *Elife*. 2016;5:. e11156.
78. Gong CX, Liu F, Iqbal K. O-GlcNAcylation: a regulator of tau pathology and neurodegeneration. *Alzheimers Dement*. 2016;12(10):1078–1089.
79. Liu F, Iqbal K, Grundke-Iqbal I, Hart GW, Gong CX. O-GlcNAcylation regulates phosphorylation of tau: a mechanism involved in Alzheimer's disease. *Proc Natl Acad Sci U S A*. 2004;101(29):10804–10809.
80. Liu F, Shi J, Tanimukai H, et al. Reduced O-GlcNAcylation links lower brain glucose metabolism and tau pathology in Alzheimer's disease. *Brain*. 2009;132(Pt 7):1820–1832.
81. Yang X, Qian K. Protein O-GlcNAcylation: emerging mechanisms and functions. *Nat Rev Mol Cell Biol*. 2017;18(7):452–465.
82. Yuzwa SA, Macauley MS, Heinonen JE, et al. A potent mechanism-inspired O-GlcNAcase inhibitor that blocks phosphorylation of tau in vivo. *Nat Chem Biol*. 2008;4(8):483–490.
83. Quattropani A, Beher D, Neny M, et al. Preclinical profiles of drug candidates promoting O-linked glycosilation of tau for the treatment of Alzheimer's disease and tauopathies. *Alzheimers Dement*. 2014;10(4):148–149.
84. Ryan JM, Quattropani A, Abd-Elaziz K, et al. Phase 1 study in healthy volunteers of the O-GlcNase inhibitor ASN120290 as a novel therapy for progressive supranuclear palsy and related tauopathies. *Alzheimers Dement*. 2018;14(7):P251.
85. Cleveland DW, Hwo SY, Kirschner MW. Physical and chemical properties of purified tau factor and the role of tau in microtubule assembly. *J Mol Biol*. 1977;116(2):227–247.
86. Cuchillo-Ibanez I, Seereeram A, Byers HL, et al. Phosphorylation of tau regulates its axonal transport by controlling its binding to kinesin. *FASEB J*. 2008;22(9):3186–3195.
87. Trojanowski JQ, Lee VM. Pathological tau: a loss of normal function or a gain in toxicity? *Nat Neurosci*. 2005;8(9):1136–1137.
88. Ballatore C, Brunden KR, Huryn DM, Trojanowski JQ, Lee VM, Smith 3rd AB. Microtubule stabilizing agents as potential treatment for Alzheimer's disease and related neurodegenerative tauopathies. *J Med Chem*. 2012;55(21):8979–8996.
89. Brunden KR, Ballatore C, Lee VM, Smith 3rd AB, Trojanowski JQ. Brain-penetrant microtubule-stabilizing compounds as potential therapeutic agents for tauopathies. *Biochem Soc Trans*. 2011;40(4):661–666.
90. Brunden KR, Zhang B, Carroll J, et al. Epothilone D improves microtubule density, axonal integrity, and cognition in a transgenic mouse model of tauopathy. *J Neurosci*. 2010;30(41):13861–13866.
91. Zhang B, Carroll J, Trojanowski JQ, et al. The microtubule-stabilizing agent, epothilone D, reduces axonal dysfunction, neurotoxicity, cognitive deficits, and Alzheimer-like pathology in an interventional study with aged tau transgenic mice. *J Neurosci*. 2012;32(11):3601–3611.

92. Barten DM, Fanara P, Andorfer C, et al. Hyperdynamic microtubules, cognitive deficits, and pathology are improved in tau transgenic mice with low doses of the microtubule-stabilizing agent BMS-241027. *J Neurosci*. 2012;32(21):7137–7145.
93. Shi QW, Oritani T, Sugiyama T. A newly rearranged 2(3→20)abeotaxane diterpene from the bark of Chinese yew, Taxus mairei. *Biosci Biotechnol Biochem*. 1998;62(11):2263–2266.
94. Gozes I, Morimoto BH, Tiong J, et al. NAP: research and development of a peptide derived from activity-dependent neuroprotective protein (ADNP). *CNS Drug Rev*. 2005;11(4):353–368.
95. Gozes I. NAP (davunetide) provides functional and structural neuroprotection. *Curr Pharm Des*. 2011;17(10):1040–1044.
96. Oz S, Kapitansky O, Ivashco-Pachima Y, et al. The NAP motif of activity-dependent neuroprotective protein (ADNP) regulates dendritic spines through microtubule end binding proteins. *Mol Psychiatry*. 2014;19(10):1115–1124.
97. Morimoto BH, Schmechel D, Hirman J, Blackwell A, Keith J, Gold M. A double-blind, placebo-controlled, ascending-dose, randomized study to evaluate the safety, tolerability and effects on cognition of AL-108 after 12 weeks of intranasal administration in subjects with mild cognitive impairment. *Dement Geriatr Cogn Disord*. 2013;35(5–6):325–336.
98. Takashima A. Tau aggregation is a therapeutic target for Alzheimer's disease. *Curr Alzheimer Res*. 2010;7(8):665–669.
99. Bulic B, Pickhardt M, Schmidt B, Mandelkow EM, Waldmann H, Mandelkow E. Development of tau aggregation inhibitors for Alzheimer's disease. *Angew Chem Int Ed Engl*. 2009;48(10):1740–1752.
100. Lu G, Nagbanshi M, Goldau N, et al. Efficacy and safety of methylene blue in the treatment of malaria: a systematic review. *BMC Med*. 2018;16(1):59.
101. Wischik CM, Edwards PC, Lai RY, Roth M, Harrington CR. Selective inhibition of Alzheimer disease-like tau aggregation by phenothiazines. *Proc Natl Acad Sci U S A*. 1996;93(20):11213–11218.
102. Hosokawa M, Arai T, Masuda-Suzukake M, et al. Methylene blue reduced abnormal tau accumulation in P301L tau transgenic mice. *PLoS One*. 2012;7(12):e52389.
103. Wischik CM, Staff RT, Wischik DJ, et al. Tau aggregation inhibitor therapy: an exploratory phase 2 study in mild or moderate Alzheimer's disease. *J Alzheimers Dis*. 2015;44(2):705–720.
104. Baddeley TC, McCaffrey J, Storey JM, et al. Complex disposition of methylthioninium redox forms determines efficacy in tau aggregation inhibitor therapy for Alzheimer's disease. *J Pharmacol Exp Ther*. 2015;352(1):110–118.
105. Melis V, Magbagbeolu M, Rickard JE, et al. Effects of oxidized and reduced forms of methylthioninium in two transgenic mouse tauopathy models. *Behav Pharmacol*. 2015;26(4):353–368.
106. Gauthier S, Feldman HH, Schneider LS, et al. Efficacy and safety of tau-aggregation inhibitor therapy in patients with mild or moderate Alzheimer's disease: a randomised, controlled, double-blind, parallel-arm, phase 3 trial. *Lancet*. 2016;388(10062): 2873–2884.
107. Wilcock GK, Gauthier S, Frisoni GB, et al. Potential of low dose leuco-methylthioninium bis(hydromethanesulphonate) (LMTM) monotherapy for treatment of mild Alzheimer's disease: cohort analysis as modified primary outcome in a phase III clinical trial. *J Alzheimers Dis*. 2018;61(1):435–457.
108. Melnyk P, Vingtdeux V, Burlet S, et al. Chloroquine and chloroquinoline derivatives as models for the design of modulators of amyloid Peptide precursor metabolism. *ACS Chem Neurosci*. 2015;6(4):559–569.
109. Medina M. An overview on the clinical development of tau-based therapeutics. *Int J Mol Sci*. 2018;19(4). E1160.
110. Liu CJ, Bosch X. Progranulin: a growth factor, a novel TNFR ligand and a drug target. *Pharmacol Ther*. 2012;133(1):124–132.

111. Saglio G, Kim DW, Issaragrisil S, et al. Nilotinib versus imatinib for newly diagnosed chronic myeloid leukemia. *N Engl J Med*. 2010;362(24):2251–2259.
112. Hebron ML, Lonskaya I, Moussa CE. Nilotinib reverses loss of dopamine neurons and improves motor behavior via autophagic degradation of alpha-synuclein in Parkinson's disease models. *Hum Mol Genet*. 2013;22(16):3315–3328.
113. Pagan F, Hebron M, Valadez EH, et al. Nilotinib effects in Parkinson's disease and dementia with Lewy bodies. *J Parkinsons Dis*. 2016;6(3):503–517.
114. Sigurdsson EM. Tau immunotherapy. *Neurodegener Dis*. 2016;16(1–2):34–38.
115. Wisniewski T, Boutajangout A. Immunotherapeutic approaches for Alzheimer's disease in transgenic mouse models. *Brain Struct Funct*. 2010;214(2–3):201–218.
116. Boutajangout A, Ingadottir J, Davies P, Sigurdsson EM. Passive immunization targeting pathological phospho-tau protein in a mouse model reduces functional decline and clears tau aggregates from the brain. *J Neurochem*. 2011;118(4):658–667.
117. Chai X, Wu S, Murray TK, et al. Passive immunization with anti-Tau antibodies in two transgenic models: reduction of Tau pathology and delay of disease progression. *J Biol Chem*. 2011;286(39):34457–34467.
118. d'Abramo C, Acker CM, Jimenez HT, Davies P. Tau passive immunotherapy in mutant P301L mice: antibody affinity versus specificity. *PLoS One*. 2013;8(4):e62402.
119. Gu J, Congdon EE, Sigurdsson EM. Two novel Tau antibodies targeting the 396/404 region are primarily taken up by neurons and reduce Tau protein pathology. *J Biol Chem*. 2013;288(46):33081–33095.
120. Walls KC, Ager RR, Vasilevko V, Cheng D, Medeiros R, FM LF. p-Tau immunotherapy reduces soluble and insoluble tau in aged 3xTg-AD mice. *Neurosci Lett*. 2014;575:96–100.
121. Yanamandra K, Jiang H, Mahan TE, et al. Anti-tau antibody reduces insoluble tau and decreases brain atrophy. *Ann Clin Transl Neurol*. 2015;2(3):278–288.
122. Yanamandra K, Kfoury N, Jiang H, et al. Anti-tau antibodies that block tau aggregate seeding in vitro markedly decrease pathology and improve cognition in vivo. *Neuron*. 2013;80(2):402–414.
123. Castillo-Carranza DL, Sengupta U, Guerrero-Munoz MJ, et al. Passive immunization with Tau oligomer monoclonal antibody reverses tauopathy phenotypes without affecting hyperphosphorylated neurofibrillary tangles. *J Neurosci*. 2014;34(12):4260–4272.
124. Kontsekova E, Zilka N, Kovacech B, Novak P, Novak M. First-in-man tau vaccine targeting structural determinants essential for pathological tau-tau interaction reduces tau oligomerisation and neurofibrillary degeneration in an Alzheimer's disease model. *Alzheimers Res Ther*. 2014;6(4):44.
125. Agadjanyan MG, Zagorski K, Petrushina I, et al. Humanized monoclonal antibody armanezumab specific to N-terminus of pathological tau: characterization and therapeutic potency. *Mol Neurodegener*. 2017;12(1):33.
126. Dai CL, Chen X, Kazim SF, et al. Passive immunization targeting the N-terminal projection domain of tau decreases tau pathology and improves cognition in a transgenic mouse model of Alzheimer disease and tauopathies. *J Neural Transm (Vienna)*. 2015;122(4):607–617.
127. Croft CL, Moore BD, Ran Y, et al. Novel monoclonal antibodies targeting the microtubule-binding domain of human tau. *PLoS One*. 2018;13(4). e0195211.
128. Borreca A, Latina V, Corsetti V, et al. AD-related N-terminal truncated tau is sufficient to recapitulate in vivo the early perturbations of human neuropathology: implications for immunotherapy. *Mol Neurobiol*. 2018;55(10):8124–8153.
129. Amadoro G, Corsetti V, Florenzano F, et al. AD-linked, toxic NH2 human tau affects the quality control of mitochondria in neurons. *Neurobiol Dis*. 2014;62:489–507.
130. Amadoro G, Corsetti V, Stringaro A, et al. A NH2 tau fragment targets neuronal mitochondria at AD synapses: possible implications for neurodegeneration. *J Alzheimers Dis*. 2010;21(2):445–470.

131. Florenzano F, Veronica C, Ciasca G, et al. Extracellular truncated tau causes early presynaptic dysfunction associated with Alzheimer's disease and other tauopathies. *Oncotarget*. 2017;8(39):64745–64778.
132. Bright J, Hussain S, Dang V, et al. Human secreted tau increases amyloid-beta production. *Neurobiol Aging*. 2015;36(2):693–709.
133. Combs B, Kanaan NM. Exposure of the amino terminus of tau is a pathological event in multiple tauopathies. *Am J Pathol*. 2017;187(6):1222–1229.
134. Asuni AA, Boutajangout A, Quartermain D, Sigurdsson EM. Immunotherapy targeting pathological tau conformers in a tangle mouse model reduces brain pathology with associated functional improvements. *J Neurosci*. 2007;27(34):9115–9129.
135. Johnson GV, Seubert P, Cox TM, Motter R, Brown JP, Galasko D. The tau protein in human cerebrospinal fluid in Alzheimer's disease consists of proteolytically derived fragments. *J Neurochem*. 1997;68(1):430–433.
136. Zhou Y, Shi J, Chu D, et al. Relevance of phosphorylation and truncation of tau to the etiopathogenesis of Alzheimer's disease. *Front Aging Neurosci*. 2018;10:27.
137. Vandermeeren M, Borgers M, Van Kolen K, et al. Anti-tau monoclonal antibodies derived from soluble and filamentous tau show diverse functional properties in vitro and in vivo. *J Alzheimers Dis*. 2018;65(1):265–281.
138. Courade JP, Angers R, Mairet-Coello G, et al. Epitope determines efficacy of therapeutic anti-Tau antibodies in a functional assay with human Alzheimer Tau. *Acta Neuropathol*. 2018;136(5):729–745.
139. Braak H, Braak E. Neuropathological stageing of Alzheimer-related changes. *Acta Neuropathol*. 1991;82(4):239–259.
140. Luna-Munoz J, Chavez-Macias L, Garcia-Sierra F, Mena R. Earliest stages of tau conformational changes are related to the appearance of a sequence of specific phospho-dependent tau epitopes in Alzheimer's disease. *J Alzheimers Dis*. 2007;12(4):365–375.
141. Lee G, Neve RL, Kosik KS. The microtubule binding domain of tau protein. *Neuron*. 1989;2(6):1615–1624.
142. Stohr J, Wu H, Nick M, et al. A 31-residue peptide induces aggregation of tau's microtubule-binding region in cells. *Nat Chem*. 2018;9(9):874–881.
143. Kontsekova E, Zilka N, Kovacech B, Skrabana R, Novak M. Identification of structural determinants on tau protein essential for its pathological function: novel therapeutic target for tau immunotherapy in Alzheimer's disease. *Alzheimers Res Ther*. 2014;6(4):45.
144. Augustinack JC, Schneider A, Mandelkow EM, Hyman BT. Specific tau phosphorylation sites correlate with severity of neuronal cytopathology in Alzheimer's disease. *Acta Neuropathol*. 2002;103(1):26–35.
145. Lathuiliere A, Valdes P, Papin S, et al. Motifs in the tau protein that control binding to microtubules and aggregation determine pathological effects. *Sci Rep*. 2017;7(1):13556.
146. Ittner A, Bertz J, Suh LS, Stevens CH, Gotz J, Ittner LM. Tau-targeting passive immunization modulates aspects of pathology in tau transgenic mice. *J Neurochem*. 2015;132(1):135–145.
147. Umeda T, Eguchi H, Kunori Y, et al. Passive immunotherapy of tauopathy targeting pSer413-tau: a pilot study in mice. *Ann Clin Transl Neurol*. 2015;2(3):241–255.
148. Lee SH, Le Pichon CE, Adolfsson O, et al. Antibody-mediated targeting of tau in vivo does not require effector function and microglial engagement. *Cell Rep*. 2016;16(6):1690–1700.
149. Patterson KR, Remmers C, Fu Y, et al. Characterization of prefibrillar Tau oligomers in vitro and in Alzheimer disease. *J Biol Chem*. 2011;286(26):23063–23076.
150. Weaver CL, Espinoza M, Kress Y, Davies P. Conformational change as one of the earliest alterations of tau in Alzheimer's disease. *Neurobiol Aging*. 2000;21(5):719–727.
151. Braak H, Alafuzoff I, Arzberger T, Kretzschmar H, Del Tredici K. Staging of Alzheimer disease-associated neurofibrillary pathology using paraffin sections and immunocytochemistry. *Acta Neuropathol*. 2006;112(4):389–404.

152. Jadhav S, Avila J, Scholl M, et al. A walk through tau therapeutic strategies. *Acta Neuropathol Commun*. 2019;7(1):22.
153. West T, Hu Y, Verghese PB, et al. Preclinical and clinical development of ABBV-8E12, a humanized anti-tau antibody, for treatment of Alzheimer's disease and other tauopathies. *J Prev Alzheimers Dis*. 2017;4(4):236–241.
154. Alam R, Driver D, Wu S, et al. Preclinical characterization of an antibody [LY3303560] targeting aggregated tau. *Alzheimers Dement*. 2017;13(7):P592–P593.
155. Jicha GA, Bowser R, Kazam IG, Davies P. Alz-50 and MC-1, a new monoclonal antibody raised to paired helical filaments, recognize conformational epitopes on recombinant tau. *J Neurosci Res*. 1997;48(2):128–132.
156. Czerkowicz J, Chen W, Wang Q, et al. Pan-tau antibody BIIB076 exhibits promising safety and biomarker profile in cynomolgus monkey toxicity study. *Alzheimers Dement*. 2017;13(7):P1271.
157. Herline K, Drummond E, Wisniewski T. Recent advancements toward therapeutic vaccines against Alzheimer's disease. *Expert Rev Vaccines*. 2018;17(8):707–721.
158. Theunis C, Crespo-Biel N, Gafner V, et al. Efficacy and safety of a liposome-based vaccine against protein Tau, assessed in tau.P301L mice that model tauopathy. *PLoS One*. 2013;8(8). e72301.
159. Boimel M, Grigoriadis N, Lourbopoulos A, Haber E, Abramsky O, Rosenmann H. Efficacy and safety of immunization with phosphorylated tau against neurofibrillary tangles in mice. *Exp Neurol*. 2010;224(2):472–485.
160. Rajamohamedsait H, Rasool S, Rajamohamedsait W, Lin Y, Sigurdsson EM. Prophylactic active tau immunization leads to sustained reduction in both tau and amyloid-beta pathologies in 3xTg mice. *Sci Rep*. 2017;7(1):17034.
161. Richter M, Mewes A, Fritsch M, Krugel U, Hoffmann R, Singer D. Doubly phosphorylated peptide vaccines to protect transgenic P301S mice against Alzheimer's disease like tau aggregation. *Vaccines (Basel)*. 2014;2(3):601–623.
162. Boutajangout A, Quartermain D, Sigurdsson EM. Immunotherapy targeting pathological tau prevents cognitive decline in a new tangle mouse model. *J Neurosci*. 2010;30(49):16559–16566.
163. Troquier L, Caillierez R, Burnouf S, et al. Targeting phospho-Ser422 by active Tau Immunotherapy in the THYTau22 mouse model: a suitable therapeutic approach. *Curr Alzheimer Res*. 2012;9(4):397–405.
164. Novak P, Schmidt R, Kontsekova E, et al. Safety and immunogenicity of the tau vaccine AADvac1 in patients with Alzheimer's disease: a randomised, double-blind, placebo-controlled, phase 1 trial. *Lancet Neurol*. 2017;16(2):123–134.
165. Novak P, Schmidt R, Kontsekova E, et al. FUNDAMANT: an interventional 72-week phase 1 follow-up study of AADvac1, an active immunotherapy against tau protein pathology in Alzheimer's disease. *Alzheimers Res Ther*. 2018;10(1):108.
166. Pihlgren M, Vukicevic M, Gafner V, et al. Efficacy of ACI-35, a liposomal anti-phospho tau vaccine in two different mouse models of Alzheimer's disease. *Alzheimers Dement*. 2016;12(7):P260–P261.
167. Simic G, Babic Leko M, Wray S, et al. Tau protein hyperphosphorylation and aggregation in Alzheimer's disease and other tauopathies, and possible neuroprotective strategies. *Biomolecules*. 2016;6(1):6.
168. DeVos SL, Miller RL, Schoch KM, et al. Tau reduction prevents neuronal loss and reverses pathological tau deposition and seeding in mice with tauopathy. *Sci Transl Med*. 2017;9(374). eaag0481.
169. DeVos SL, Miller TM. Antisense oligonucleotides: treating neurodegeneration at the level of RNA. *Neurotherapeutics*. 2013;10(3):486–497.
170. Ke YD, Suchowerska AK, van der Hoven J, et al. Lessons from tau-deficient mice. *Int J Alzheimers Dis*. 2012;2012:873270.

CHAPTER

11

Acetylation of tubulin: A feasible protective target from neurodevelopment to neurodegeneration

Cappelletti Graziella[a,b], *Cartelli Daniele*[c]

[a]Department of Biosciences, Università degli Studi di Milano, Milan, Italy, [b]Center of Excellence on Neurodegenerative Diseases, Università degli Studi di Milano, Milan, Italy, [c]Neuroalgology Unit, IRCCS Foundation, "Carlo Besta" Neurological Institute, Milan, Italy

Introduction

Acetylation refers to the transfer of the acetyl group from acetyl-coenzyme A to ε-amino group of lysine residues and is a widespread strategy for controlling proteins in health and disease. The impact on this post-translational modification (PTM) on protein structure and function is considerable and goes far beyond the originally identified control of gene expression due to acetylated histones, even though most acetylated protein appeared for a long time to be nuclear and associated with the regulation of transcription.[1] Up to now, acetylation was more broadly distributed then in the nucleus, and the proteomic survey of protein acetylation opened the door on the cellular acetylome.[2] This includes thousands of nuclear, cytoplasmic, and mitochondrial proteins involved in a range of cellular functions such as chromatin remodeling, cell division, signal transduction, protein folding, and metabolism. In line with the overall role of PTMs, acetylation confers novel properties to the modified proteins including changes in enzymatic activity, protein stability, sub-cellular localization, and the capability to interact with DNA and other proteins.[3]

https://doi.org/10.1016/B978-0-12-814037-6.00009-4

Large macromolecular complexes are preferentially targeted by acetylation, including the cytoskeletal proteins actin[4] and tubulin.[5]

Here, we focus on the building block of microtubules, tubulin, whose extraordinary wide range of properties is strictly dependent on a complex regulation, which includes acetylation. We aim to briefly introduce the so-called tubulin code, which is the strategy for defining the lots of cell-specific properties of tubulin, and to get the gist on tubulin acetylation in physiology. We will then examine the evidence supporting the defective regulation of tubulin acetylation in a spectrum of disorders from neurodevelopmental to neurodegenerative diseases to critically look at the feasibility of targeting this PTM for neuroprotection.

The basis for tubulin code

About 10 years ago the term "tubulin code" was coined to reveal the heterogeneity of PTMs of tubulin that mark and confer structural and functional diversity to subpopulations of microtubules.[6] Interestingly, this was considered to be analogous to the "histone code," which is defined as many PTMs occurring on histone and plays a fundamental regulatory role on gene transcription.[7] The concept of "tubulin code" has evolved over time, and it is now established that it collectively refers to the expression of different tubulin genes and the accumulation of a variety of PTMs on tubulin, which modulate microtubule building and rearrangement giving rise to highly specialized microtubules.[8]

Microtubules are polymers made of α-/β-tubulin heterodimers that associate head to tail and laterally to form hollow tubes. There are eight α-tubulin and nine β-tubulin genes in mammals (https://www.genenames.org/genefamilies/TUB) whose selective restriction or expression in different cell types creates many combinations of α-/β-tubulin heterodimers that profoundly impact the structure and behavior of microtubules and confer peculiar dynamic properties.[9] The second group of crucial actors in generating microtubule diversity includes tubulin PTMs and the modifying enzymes. The large amount of evidence on the wide range of modifications occurring on tubulins has been recently described in several considerable reviews.[10–13] Beyond tubulin acetylation, which we devote the majority of this chapter to, the chemically diverse modifications affecting tubulins include de-tyrosination/tyrosination cycle (the removal and addition of the genetically encoded C-terminal tyrosine on α-tubulin), Δ2- and Δ3-tubulin (the removal of the last two and three amino acids of the α-tubulin C-terminal tail, respectively), phosphorylation, polyglutamylation, poly-amination, (poly)glycylation, and, although less abundant or less common, palmitoylation, ubiquitylation, sumoylation, nitrosylation, nitration, and methylation. All of them can drastically change the biochemical and biophysical properties of microtubules and, most important, have an impact on the complexity of their interplay with organelles in the cytoplasm and microtubule-interacting proteins in the different cell types. A great deal of work has been done to identify the unknown forward and reverse enzymes that "are responsible for writing the tubulin code."[14] For example, after many years of searching, the long-sought existence of the tubulin-tyrosine carboxypeptidases responsible for the reversible de-tyrosination of α-tubulin has been recently proven and its critical role in neuron and brain physiology setup.[15,16]

Focusing on neurons, central in this chapter, microtubules are integral to their function because they are essential for a healthy nervous system from neurodevelopment to neuronal differentiation and for maintenance of the mature neuron's phenotype. This is strictly dependent on the complex and multiple-level regulation due the combinatorial information expressed through tubulin genetic and chemical diversity that generates the neuronal tubulin code and that controls microtubule dynamics, mechanics, and interactions with microtubule effectors.[17]

Acetylation of tubulin

In the 1980s, when acetylation was thought to be a prerogative of nuclear proteins, the discovery of acetylated tubulin in *Chlamydomonas* sp. flagella was completely unexpected.[18] Thanks to the development of a specific monoclonal antibody,[19] the first studies mapped the modified residues on Lysine 40 (K40) in α-tubulin from *Chlamydomonas reinhardtii*[20] to mammals[21] and plants.[22] Pioneering studies reported that microtubule domains that are acetylated turn over slowly[23] and that acetylated α-tubulin marks microtubules associated to particularly stable shafts of both axons and dendrites in cultured neurons[24–26] and are very abundant and widespread in the brain.[27] Paradoxically, although α-tubulin acetylation has been intensively studied and is a highly conserved modification, it remained poorly understood for around three decades, and the impact of this modification on microtubule structure and function is only beginning to emerge[5] (Fig. 11.1). The unusual mapping of this modification on the residue K40 exposed at the luminal microtubule surface raised many questions: first, on the feasibility of enzymatic control on addition and removal of the acetyl moiety and, second, on its biological role in controlling microtubule functions. Many years after its discovery, two de-acetylating enzymes were discovered, the histone de-acetylase 6 (HDAC6)[28] and sirtuin 2 (SIRT2).[29] Later, Mec-17 in *Caenorhabditis elegans*[30] and the mammalian ortholog tubulin acetyl-transferase αTAT1[31] were identified as the enzymes responsible for microtubule acetylation. Explaining how these enzymes work remains an intriguing and partially unsolved question. To date, it's generally accepted that α-tubulin acetylation only takes place on polymerized microtubules and that this requires luminal diffusion

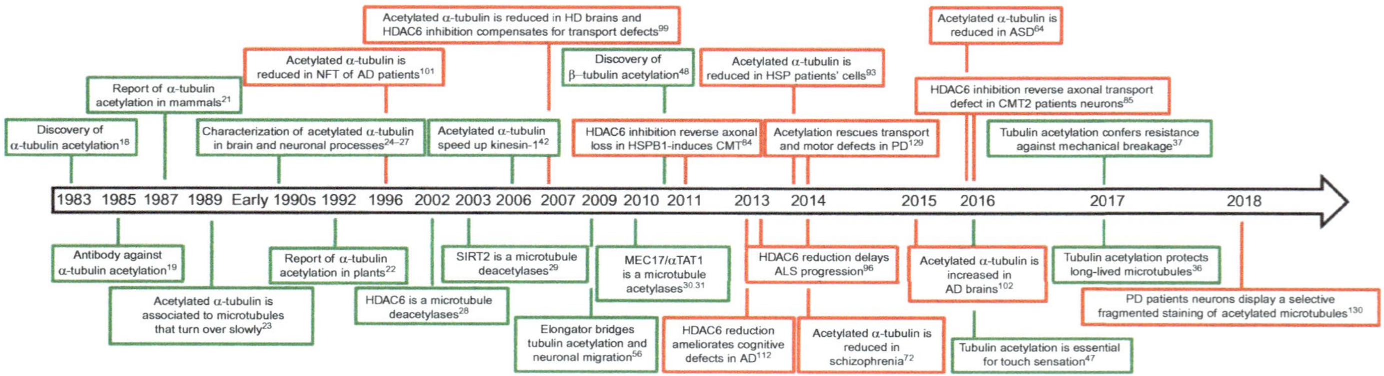

FIG. 11.1 Timeline of tubulin acetylation. Milestones in the history of acetylated tubulin research including findings for its physiological roles *(green boxes)* and studies linking its defects to neurodevelopmental and neurodegenerative diseases *(red boxes)*. *AD*, Alzheimer's disease; *ALS*, amyotrophic lateral sclerosis; *ASD*, autism spectrum disorder; *CMT*, Charcot-Marie-Tooth; *HD*, Huntington's disease; *HSP*, hereditary spastic paraplegia; *PD*, Parkinson's disease.

of αTAT1,[32] which could occur through the opened microtubule ends or through bends and breaks in the lattice.[33]

Acetylated α-tubulin has been found in many cell types, and its localization into "stable" structures as cilia, flagella, and axons was very early described,[20, 34] thus supporting the concept that it is a marker of microtubule stability. However, the question is whether the α-tubulin acetylation induces microtubule stability or whether it occurs exclusively on stable microtubules. For a long time the prevailing model was that α-tubulin acetylation is a consequence of microtubule stabilization on the basis that drugs stabilizing microtubules led to their acetylation whereas artificially increasing tubulin acetylation did not stabilize microtubules.[35] Emerging evidence actually made this model too simplistic. Very recently, it has been demonstrated that α-tubulin acetylation weakens lateral interactions between protofilaments, thus softening and enhancing flexibility of the microtubules and conferring them resilience against repeated mechanical stresses.[36, 37] Surprisingly, this makes acetylated microtubules long lived by making them softer. In addition, acetylated microtubules may be a preferred substrate of the microtubule-severing protein katanin in neurons,[38] thus supporting the view that tubulin acetylation may influence microtubule stability by making them more sensitive to severing, which refers to cutting or breaking of microtubule lattice.[39]

Beyond the earlier described and emerging structural consequences, many intracellular functions depend on or are linked to α-tubulin acetylation. This includes cell migration,[40] autophagy,[41] intracellular trafficking,[42] cell adhesion,[43] and primary cilium disassembly.[44] Tubulin acetylation does affect not only strictly cellular events but also even higher-order processes. In fact *C. elegans* touch receptor neurons contain heavily acetylated 15-protofilament microtubules, and their disruption or mutation of MEC17, the major tubulin acetyltransferase, reduces touch sensation in *C. elegans*.[45,46] More recently, it has been demonstrated that mice lacking αTAT1 in sensory neurons are largely insensitive to mechanical touch and pain possibly due to a reduction of the cellular stiffness; both cellular rigidity and mechanosensitivity can be restored by genetic mimicking of α-tubulin acetylation.[47]

Although α-tubulin acetylation on K40 is the most known acetylated form of tubulin, a whole-proteome mass spectrometry approach has identified a number of potential acetylation sites on α- and β-tubulin suggesting that acetylation events on tubulins are complex.[2] Lastly, acetylation on lysine 252 of β-tubulin was reported to take place on non-polymerized tubulin, in contrast to α-tubulin acetylation on K40; is catalyzed by the San acetyl-transferase; and down-regulates tubulin incorporation into microtubules.[48]

In conclusion, given that acetylation contributes to generating the tubulin code and plays important roles in regulating microtubule architecture, maintaining microtubule integrity, and supporting cell function, any imbalance in the levels of this modified form of tubulin can be detrimental for the cell. An increasing number of human pathologies including neurological disorders, cancer, and heart diseases point to defects of tubulin acetylation and dysfunction of the microtubule cytoskeleton.[12]

Acetylation of tubulin in neurodevelopmental disorders

Neurodevelopmental disorders are a heterogeneous group of diseases characterized by defects in the development of the central nervous system. They can occur as neuropsychiatric problems, impaired motor function, learning disabilities, language development, or non-verbal communication. The fine-tuning of microtubules plays a fundamental role in the physiological development of the nervous system from proliferation and neurogenesis to neuronal migration, differentiation, and synapse formation[49, 50] and, consequently, in neurodevelopmental diseases.[51]

Specific stages of brain development require the diverse functional repertoire of the microtubule cytoskeleton, which is strictly dependent on the tubulin code. First the developmental profile and the subcellular organization of the neuronal tubulin isotypes actually change and dictate the microtubule behavior throughout the different cellular processes,[52] which come in succession and build the nervous system. For example, *Tuba1* is expressed almost exclusively in post-mitotic neurons and controls neurogenesis and neuronal migration, as revealed by the defects in hippocampal development and lamination observed in knock-out mice[53] and by the discovery that mutations on its human homolog are linked to lissencephaly.[54] Besides tubulin genetics, PTMs of tubulin are importantly involved in neurodevelopment as well. Neuronal microtubules are functionalized with abundant PTMs that are asymmetrically distributed in the neuron: microtubules in dendrites are enriched in tyrosination, acetylation, and short-chain glutamylation, whereas axonal microtubules are enriched in long-chain glutamylation, acetylation, poly-amination, de-tyrosination, and Δ2-tubulin.[10] Insights into the contribution of individual tubulin PTM to the development and differentiation of the nervous system come from mouse models. For instance, the importance of the cycle tyrosination/detyrosination for in vivo neuronal differentiation was disclosed by studies on (i) tubulin-tyrosine ligase knockout mice in which the levels of de-tyrosinated and Δ2-tubulin in the brain were up-regulated showing dramatic defects in brain development due to improper neurite extention[55] and (ii) the recently discovered tubulin-tyrosine carboxy-peptidases vasohibins, whose deficiency impairs neuronal migration in mouse neo-cortex.[15]

Focusing on acetylation of α-tubulin, a significant amount of data supports its crucial role in neuronal development. The migration and differentiation of cortical neurons strictly depend on tight regulation of tubulin acetylation being the processes impaired when the levels of acetylated α-tubulin decrease.[56, 57] Noticeably the imbalance of acetylated α-tubulin in mouse lacking αTAT1 has an impact on hippocampus development leading to dentate gyrus distortion.[58] Next, we focus on the emerging data linking changes in this tubulin PTM to autism spectrum disorders and schizophrenia.

Autism spectrum disorders

Neurodevelopmental disorders characterized by defective social interaction and communication, repetitive behaviors, and restrictive interests are collectively called autism spectrum disorders. Although their pathogenesis is not yet fully understood, increasing data link the mechanisms that underline this disease to microtubule dysfunction.[51] Genomic studies have revealed rare copy number variation linked to risk for autism spectrum disorder and converging on networks related to neuronal signaling and development[59]; notably, genetic mutations identified in autism spectrum disorders include mutations in microtubule-associated genes. One of these genes is activity-dependent neuroprotective protein (ADNP),[60] which is vital for brain development. Interestingly, ADNP directly interacts with microtubule end-binding proteins (EB1 and EB3), thus promoting axonogenesis and dendritic spine formation.[61] In addition, rare deletions of Janus kinase and microtubules interacting protein 1 (JAKMIP1), an RNA-binding protein able to regulate microtubule dynamics,[62] have been identified in patients affected by autism spectrum disorders.[63] Although the potential and direct involvement of microtubules in autism spectrum disorder pathogenesis is emerging, there are only few works that address this issue. Intriguing data come from a study in a mouse model that exhibits delayed neuronal migration and defects in the formation of neural circuits in the cortex, circuits responsible for social behavior typically observed in autism spectrum disorders.[64] In this model the increased activity of the tubulin deacetylase HDAC6 leads to reduced levels of acetylated α-tubulin. The treatment with the HDAC6 inhibitor tubacin A rescues both defects in neuronal migration and social behavior suggesting that this could be a novel approach to treating psychiatric diseases. In addition, a very recent paper shows that the autism-related protein Ube3A/E6AP induces defects in dendritic arborization and synapse formation via caspase-dependent microtubule destabilization and that the expression of E6AP along with α-tubulin K40A, which is a non-acetylatable mutant tubulin variant, deeply interferes with this process blocking the E6AP-induced reduction in dendritic arborization.[65] Interestingly a previous study investigated post-mortem brain samples from patients by histone acetylome-wide association study (HAWAS) and uncovered a distinct histone acetylome pattern and epigenetic signature,[66] thus opening the way to future molecular analyses on cellular acetylome in autism spectrum disorders and in other diseases. A similar approach could have the potentiality of getting novel insights into the investigation of non-histone proteins, including cytoskeletal proteins and tubulin, in multiple pathologies.

Schizophrenia

Schizophrenia is a psychiatric disorder that manifests itself in the adulthood. It is well accepted that it has a strong neurodevelopmental component.[67] Experimental evidence from mouse models and human studies (more than 200 papers listed in Pubmed at February 2019) has linked this common and debilitating mental illness to defects in the microtubule functions. After it was discovered that altered expression and distribution of the microtubule-associated proteins MAP2 and MAP5 in post-mortem brain of patients affected by schizophrenia[68] and the evidence that mutations in DISC1, encoded by *Disrupted-In-Schizophrenia 1* gene, cause susceptibility to schizophrenia, affect a variety of microtubule-based functions,[69] and perturb the development of cerebral cortex,[70] many laboratories in the world addressed this issue. The beneficial effect of pharmacological microtubule stabilization on synaptic function and behavior in a mouse model of schizophrenia,[71] generated by the deletion of microtubule-associated protein 6, forced the interest into the deep understanding of molecular mechanisms regulating microtubule dysfunction in this disorder. More recent data show that deletions in unc-51-like kinase (*ULK4*) gene, which are present in schizophrenia patients, compromise neuritogenesis and cell motility in cultured cells, modulate microtubule integrity, and reduce the levels of acetylated α-tubulin, thus suggesting a causative link between microtubule alteration and defective neurite elaboration.[72] Since *ULK4* null mice showed very severe phenotype, the same authors moved to in utero gene transfer for silencing Ulk4 in subsets of neuronal progenitors in the developing mouse cortex and validated the effects of Ulk4 on corticogenesis and on the acetylation of α-tubulin.[73] Alterations of α-tubulin acetylation were also found in olfactory neurosphere-derived cells isolated from schizophrenia patients.[74] In detail, this study revealed that patient-derived olfactory cells, which are considered an excellent disease model as they encompass the polygenic nature of schizophrenia, express less filamentous actin and less acetylated α-tubulin than cells derived from healthy subjects. These defects could be fundamental in regulating cell migration and leading to altered brain development observed in schizophrenia patients. A further interesting study on olfactory neuronal precursors derived from schizophrenia patients showed that the formation of primary cilia was dramatically reduced, which suggests that the molecular pathways involved in their formation are defective in this disease.[75] Primary cilia are non-motile organelles whose disruption in cortical and hippocampal neurons impairs dendritic arborization[76] and whose role is fundamental in brain development.[77] Intriguingly the core of primary cilium structure is a radial array of nine pairs of peripheral microtubules, which are enriched in tubulin PTMs. Acetylation of α-tubulin is the best studied PTM on ciliary axoneme,[78] and the de-acetylase activity of HDAC6 is necessary for ciliary microtubule disassembly.[79] To conclude the emerging data showing defects in the acetylation of tubulin in both schizophrenia mouse models and patient cells suggest that the unbalancing in this tubulin modification could contribute to the disease pathogenesis.

Acetylation of tubulin in neurodegenerative diseases

The large amount of data coming from in vitro and in vivo studies disclosed the involvement of defects in microtubule structure and dynamics[80] and in microtubule-dependent functions, including axonal transport,[81, 82] in a spectrum of disorders ranging from peripheral to central nervous system.[39, 83] Little is known about the role of tubulin PTMs, but because of the well-studied and easily available antibody, acetylation of α-tubulin and its fine regulation are the most intensively investigated PTM and notably suggested to be relevant in the pathogenic pathways (Fig. 11.1).

Starting from the peripheral nervous system, studies on the most common hereditary neuropathy Charcot-Marie-Tooth disease, linked to mutations in more than 80 different genes, are compelling. First, in mouse models, the expression of the mutated gene encoding heat shock protein β1 (*HSPB1*) linked to Charcot-Marie-Tooth disease leads to a severe reduction of tubulin acetylation in peripheral nerves that provokes axonal transport defects, which are successfully rescued by both trichostatin A and the specific HDAC6 inhibitor tubastatin A together with the complete reversal of the defective motor performance.[84] This was the first of a series of evidence showing that defects in axonal transport and reduced α-tubulin acetylation, both of which are restored by HDCA6 inhibitors, occur in experimental models of Charcot-Marie-Tooth disease including (i) human motor neurons derived from induced pluripotent stem cells of patients carrying mutations in *HSBP1*[85]; (ii) in vitro motor neurons isolated from mouse lacking *GDAP1* gene, whose loss of function recessive mutations cause different forms of the disease[86]; and (iii) sciatic nerves and dorsal root ganglia from mouse expressing dominant mutations in glycyl-tRNA synthetase that cause a subtype of the disease.[87, 88] Collectively, these studies demonstrate that α-tubulin deacetylation and disrupted axonal transport may represent a common pathogenic mechanism underlying Charcot-Marie-Tooth disease, despite the genetic heterogeneity of its etiology.

Moving to central nervous system, additional evidence comes from murine to human cellular models. The most common form of hereditary spastic paraplegia is due to mutations in the *SPAST* gene, which encode for spastin, a well-characterized microtubule-severing protein,[89, 90] and exhibits the degeneration of corticospinal tracts. The levels of acetylated α-tubulin are significantly modulated by the expression of spastin in murine cultured neurons,[91] decreased in neural progenitor cells derived from biopsies of the olfactory mucosa from patients with *SPAST* mutations.[92] This decrease is related to peroxisome trafficking deficits in olfactory neurosphere-derived cells from patients and is rescued by microtubule-targeting drugs.[93] Furthermore, increasing evidence concentrates on the idea that HDAC6 impairment and, consequently, acetylated α-tubulin unbalance might be altered in various subtypes of amyotrophic lateral sclerosis.[94, 95] In a mouse model of this disease, the deletion of HDAC6 delays disease progression, and this protective effect is associated with increased α-tubulin acetylation.[96] Notably, human motor neurons derived from induced pluripotent stem cells of patients affected by amyotrophic lateral sclerosis and carrying mutations in *FUS* gene show defects in the axonal transport that are restored by increasing the levels of acetylated α-tubulin through pharmacological inhibition and genetic silencing of HDAC6.[97]

Finally, before moving to the most diffuse neurodegenerative disorders, Alzheimer's and Parkinson's disease, we would like to cite interesting but almost completely neglected data supporting the role of acetylated α-tubulin in Huntington's disease. Beyond the initial demonstration that the protein huntingtin, whose abnormal polyglutamine (polyQ) expansion results in Huntington's disease, binds to microtubules,[98] some evidence revealed that changes on microtubule system could play a role in the disease. Noteworthy, tubulin acetylation is reduced in post-mortem brains from patients with Huntington's disease, and the specific inhibition of HDAC6 increases the acetylation of tubulin, the recruitment of molecular motors to microtubules, and intraneuronal transport of vesicles in cultured neurons obtained from knock-in mouse over-expressing polyQ huntingtin.[99] On the contrary, HDAC6 deletion failed to correct the transport impairment in a different mouse model of the disease.[100] In previous years, few papers have been published about the contribution of tubulin acetylation to Huntington's disease, thus leaving the debated question unsolved.

Alzheimer's disease

More than 20 years ago a basic but informative approach revealed that in the post-mortem brain of Alzheimer's disease patients, the immunoreactivity of acetylated α-tubulin was reduced in the neurofibrillary tangle-bearing neurons.[101] More recently the analysis of tubulin PTMs by western blotting has led to conflicting results as both the increase[102] and the decrease of acetylated α-tubulin[103] were found in patients' brain. In addition, the discoveries that tau is acetylated both in vitro and in vivo,[104] that this modified form promotes tau aggregation,[105] and that the aggregates accumulate in post-mortem brain as a "pathological signature in Alzheimer's disease"[106] make the picture even more complex. Intriguingly, acetylated α-tubulin is the first identified and the major substrate for HDAC6,[28] and tau interacts with, impacts on, and regulates the activity of HDAC6,[107] suggesting that acetylation of α-tubulin and tau could collaborate in maintaining the equilibrium between neuronal health and disease. Thus, in the attempt to clear up whether and how acetylated tubulin plays a role in Alzheimer's disease, we will focus on the large amount of evidence coming from experimental models. Cultured hippocampal neurons exposed to extra-cellular beta-amyloid peptide (Aβ), one of the major components of the neurofibrillary tangles in Alzheimer's disease, offer controversial results as both increased levels of acetylated α-tubulin and disrupted microtubule dynamics via inhibition of HDAC6[108] and impaired mitochondrial transport through HDAC6 activation[109] have been reported.

The suggested Alzheimer's disease animal model, aged tau knock-out mice, shows the significant reduction in acetylated α-tubulin that is linked to deficits in spatial memory and loss of hippocampal synapse.[110] Similar results have been obtained in mouse model of amyloid precursor protein/presenilin 1 (APP/PS1), which exhibits severe amyloid pathology and a lack of associative and spatial memory formation other than unbalanced acetylated α-tubulin in the hippocampus and cortex.[111] Interesting data also come from HDAC6 knockout mice that are viable and show the loss of de-acetylase function resulting in the increases of α-tubulin acetylation levels in the hippocampus with no changes in histone acetylation.[112] Crossing HDAC6 knock-out mice with APP/PS1-21 mice rescues all the defects observed in the amyloid pathology model and suggests that the reduction of endogenous HDAC6 levels could be a reasonable strategy to counteract cognitive decline in Alzheimer's disease.[112] This concept has been supported by further evidences including the recovery of short-term learning and memory deficits, hyper-activity, and modifications of tau and tubulin in APP/PS1 mouse models following the treatment with potent HDAC6 inhibitors ACY-738[113] and tubastatin A/ACY-1215.[114] Although there are many mouse models employed for studying molecular mechanisms in Alzheimer's disease, emerging non-mammals models have also been developed. Interestingly, acetylated α-tubulin has been investigated in a *Drosophila melanogaster* model expressing human tau[115] showing that pharmacological and genetic control of α-tubulin acetylated rescues human Tau-induced defects in microtubules and neuromuscular junction abnormalities.[116]

Parkinson's disease

For a long time the multiple hit hypotheses to explain the molecular pathogenesis of Parkinson's disease ignored the potential role of microtubule dysfunction. The picture has changed, and the overall contribution of microtubule defects to the neurodegenerative processes that lead to Parkinson's disease is emerging in terms of microtubule stability and microtubule-dependent intracellular trafficking.[117–120] Acetylation of tubulin has been linked to both these aspects of microtubule biology, as reported earlier (point 1.2), and, in the last 10 years, it has been investigated in many cellular and animal models of Parkinson's disease up to human cells obtained from patients.

Starting from studies with toxins inducing experimental parkinsonism, the unbalance of tubulin PTMs, including acetylated α-tubulin, triggers cell death in cultured neuronal cells exposed to the neurotoxin *N*-methyl-1-4-phenylpyridium,[121, 122] in neuroblastoma cells exposed to 6-hydroxydopamine,[123] and in the brain of mice injected with *N*-methy-4-phenyl-1,2,3,6-tetrahydropyridine.[124] Lately, three lines of evidences link acetylation of α-tubulin to molecular mechanisms underlying physiological and pathologic roles of the proteins encoded by genes linked to familial and non-familial forms of Parkinson's disease. First, relevant data come from studies on Leucine-rich repeats kinase 2 (LRRK2) whose mutations cause the most common familial forms of the disease.[125] LRRK2 directly interacts with three isoforms of β-tubulin, and the binding site is mapped to the proximity of acetylatable K40 residues of α-tubulin, thus contributing to maintain the low levels of acetylated α-tubulin in cell.[126] Interestingly, embryonic fibroblasts derived from *LRRK2* knock-out mice[126] and skin fibroblasts isolated from patients harboring *LRRK2* mutation[127] display increased tubulin acetylation and, together with the increased levels of acetylated tubulin reported in peripheral blood mononuclear cells from both patients affected by sporadic Parkinson's disease and patients bearing *LRRK2* mutation,[128] suggest that the control of this tubulin PTM can be relevant for the pathology. However, the question is in what way does the unbalance of tubulin acetylation impair neuronal functions. Godena and colleagues[129] showed that the expression of pathogenic forms of LRRK2 disrupts axonal transport in rat cortical neuron cultures, causes locomotor deficits in *D. melanogaster* model, and genetic and pharmacological strategies for increasing tubulin acetylation restore axonal transport and motor behavior. In line with the results from *LRRK2* models, we found that acetylation of tubulin changes in *PARK2* knock-out mice and in human neurons from patients harboring *PARK2* mutations.[130] Loss of function caused by mutations in *PARK2* gene is linked to early-onset familial Parkinson's disease and impacts on multiple intracellular pathways including microtubule system.[131, 132] We found that parkin absence induces (over)acetylation of α-tubulin in mouse brain, where it precedes the alteration of mitochondria transport, and we observed a significant change in acetylated microtubules in midbrain neurons derived from induced pluripotent stem cells: axonal microtubules are homogenously acetylated in controls, whereas a reduced and discontinuous staining for acetylated α-tubulin distinguishes axonal microtubules of patients neurons.[130] Intriguingly, these changes in acetylation status of microtubules could play a relevant pathologic role in causing the loss of mechanical resistance and/or the impairment of axonal transport. Further evidence comes from studies on α-synuclein, which is the major component of Lewy bodies,[133] the cellular marker of the pathology, and whose mutations and allele multiplication are linked to familial forms of Parkinson's disease.[134] Increasing evidence suggests that α-synuclein interacts with and modulates microtubule system and that this could be important for its biology.[135] Notably, pharmacological inhibition and modulation of the deacetylase SIRT2 by gene silencing increase tubulin acetylation, reduce the formation of α-synuclein inclusions, and alleviate neurotoxicity in cell and in *D. melanogaster* models,[136] supporting the idea that this could be a promising strategy for blocking α-synuclein aggregation, one of the putative triggering molecular events underlying Parkinson's disease. When using as an ex-vivo model cybrid cells, which are created through the fusion of teratocarcinoma (NT2) cells depleted from their own mitochondrial DNA, with mitochondria isolated from platelets of patients affected by idiopathic Parkinson's disease, the loss of microtubule mass and the concomitant α-synuclein aggregation are reported,[137] and this is related to the decrease of acetylated α-tubulin and the impairment of mitochondria trafficking.[138] In conclusion a large amount of studies on cellular and animal models of Parkinson's disease converges on looking at the unbalance in tubulin acetylation as a pathogenic mechanism of the disease.

Open questions and future perspective

The concept that acetylation of α-tubulin is merely a passive marker of microtubule stability has been definitively overcome by recent seminal papers that demonstrated its direct and broad role as a regulator of microtubule properties and functions in many cell types including neurons.[36, 37] Given this statement, any change in the levels of acetylated α-tubulin can likely be detrimental for cells and, consequently, contribute to the pathogenesis of many disorders. Undeniably the increasing amount of data from basic science summarized in this chapter (Fig. 11.1) suggests that neurodevelopmental and neurodegenerative diseases are the perfect examples for confirming the fundamental pathogenic role of the unbalance in acetylated α-tubulin and for looking at this as a potential target for therapy. Dissecting the experimental data that support this hypothesis, however, we must admit, leaves many open questions unanswered.

The first debatable point is the specificity of the enzyme involved in regulating tubulin acetylation and, in turn, the feasibility of designing efficient drugs. The outcome from the majority of the studies reported earlier converges on the idea that increasing the levels of acetylated α-tubulin throughout the inhibition of the de-acetylating enzymes could be a promising strategy for the rescue of many neuronal defects. The most investigated enzyme is HDAC6, but, actually, it is known that this enzyme catalyzes the removal of acetyl groups from the lysine residues of a series of non-histone proteins besides α-tubulin, thus affecting multiple cellular processes.[139] The crucial point then is to get increasingly more molecular details on the mechanisms at the basis of its catalysis and inhibition. Interestingly a recent paper aimed at supplying structural insights into HDAC6 tubulin de-acetylation shows that only one of the two catalytic domains of this enzyme de-acetylates K40 in α-tubulin[140]; this makes the study of selective inhibitors extremely important and challenging to validate HDAC6 as a drug target. To date, starting from the initial discovery of HDAC6 selectivity of tubastatin A, revealed by its efficacy in elevating the levels of acetylated α-tubulin in primary cortical neuron cultures,[141] new selective and activity-specific HDAC6 inhibitors are developing,[142] and interesting data are emerging. Among the others a bicyclic-capped tubastatin A has been described, and its activity validated in a cell model of Charcot-Marie-Tooth disease.[143] To make the picture even more complex, HDAC6 could simultaneously regulate more than one protein playing a pathogenic role. Indeed a very recent paper reports the development of a series of novel selective inhibitors of HDAC6, which induce changes in the acetylation levels of both α-tubulin and tau and that ameliorate Alzheimer's disease phenotype in a mouse model of the disease.[144] At last, we cannot forget that emerging data suggest strategies to modulate tubulin acetylation that could be alternative to HDAC6 inhibition. Those include the discovery of a potent inhibitor of the de-acetylating enzyme SIRT1/2,[145] identified in coffee with moderate consumption revealing several positive effects on human health, and the evidence that lithium chloride regulates both primary cilia length and acetylation of α-tubulin via the activation of αTAT1.[146] Although the regulation of αTAT1 is much less clear than its role in acetylation of α-tubulin, very recent data get novel insights into cell signaling mechanisms that control this enzyme, thus opening the way to a more efficient modulation of its activity with the potential of controlling microtubule acetylation in health and disease.[147]

The second, not negligible question arises from the evidence that, beyond the well-characterized K40, many residues in α-tubulin have been reported to be potentially acetylated.[2] To date the extensive study of a K40 acetylated α-tubulin is undoubtedly due to the excellent antibody discovered about 30 years ago,[19] and, thanks to this tool, a large and even more accurate amount of data on its functional role in regulating tubulin biology has been collected. Now, we move to the actual questions of whether the acetylation of additional residues on α-tubulin potentiates, hinders, or induces alternative effects if compared with acetylated K40 and whether HDAC6 is in charge for de-acetylating all these other residues. To our knowledge, no functional data have been reported on the role of the additional acetylation residues on α-tubulin. Otherwise the proteomic analysis performed on tubulin purified from wild-type and HDAC6 knock-out mice mapped 12 acetylation sites of α- and β-tubulin and revealed six lysine residues to be putative for de-acetylation mediated by HDAC6.[79] The design of a pharmacological strategy based on HDAC6 inhibition has to take into account the need for specifically targeting tubulin and the existence of the multi-faceted interplay of the enzyme with tubulin.

Finally, although most of the basic and pre-clinical studies in a wide range of neurodevelopmental and neurodegenerative disorders that focused on the inhibition of HDAC6 provided positive clues and suggest the feasibility of this strategy, few steps have been taken in the direction of clinical trials. Currently almost all of trials with inhibitors of HDAC6 are recruiting patients affected by different tumors. The exception is one Phase I trial started in May 2018 recruiting patients affected by Huntington's disease whose estimated study completion date is in February 2020 (https:// ClinicalTrials.gov identifier NCT03713892). The goal of this study is to assess the safety, tolerability, pharmacokinetics, and pharmacodynamics of the specific HDAC6 inhibitor CKD-504 compared with placebo in healthy Korean and Caucasian adult subjects.

To conclude, in the future, a more comprehensive analysis of molecular details at the base of tubulin acetylation and, most important, the consequent determination of novel compounds targeting this process are the "conditio sine qua non" for moving toward therapeutic interventions. Obviously the idea that this approach could be working in so many different diseases from neurodevelopment to neurodegeneration might weaken its potential. We are firmly convinced of the opposite and indeed believe that this could be an advantage as the step forward in a specific pathology could lead to new discoveries in different diseases, thus providing a solid starting point and challenge for the future research.

Acknowledgments

The authors are grateful to Fondazione Grigioni per il Morbo di Parkinson, Milano, Italy, for the long-lasting support to G.C. and to Susan Morrison for reading and editing the manuscript. We thank all the members of our teams that have exceptionally contributed to the role of microtubules and tubulin acetylation in disease models. We apologize to all colleagues whose work we were unable to include due to space limitations.

References

1. Verdin E, Ott M. 50 years of protein acetylation: from gene regulation to epigenetics, metabolism and beyond. *Nat Rev Mol Cell Biol*. 2015;16(4):258–264.
2. Choudhary C, Kumar C, Gnad F, et al. Lysine acetylation targets protein complexes and co-regulates major cellular functions. *Science*. 2009;325:834–841.
3. Narita T, Weinert BT, Choudhary C. Functions and mechanisms of non-histone protein acetylation. *Nat Rev Mol Cell Biol*. 2019;20:156–174.
4. Varland S, Vandekerckhove J, Drazic A. Actin post-translational modifications: the Cinderella of cytoskeletal control. *Trends Biochem Sci*. 2019;44(6):502–516.
5. Janke C, Montagnac G. Causes and consequences of microtubule acetylation. *Curr Biol*. 2017;27(23):R1287–R1292.
6. Verhey KJ, Gaertig J. The tubulin code. *Cell Cycle*. 2007;6(17):2152–2160.
7. Strahl BD, Allis D. The language of covalent histone modications. *Nature*. 2000;403:41–45.
8. Gadadhar S, Bodakuntla S, Natarajan K, Janke C. The tubulin code at a glance. *J Cell Sci*. 2017;130:1347–1353.
9. Vemu A, Atherton J, Spector JO, Moores CA, Blanchoin L. Tubulin isoform composition tunes microtubule dynamics. *Mol Biol Cell*. 2017;28:3564–3572.
10. Song Y, Brady ST. Post-translational modifications of tubulin: pathways to functional diversity of microtubules. *Trends Cell Biol*. 2015;25(3):125–136.
11. Wloga D, Joachimiak E, Fabczak H. Tubulin post-translational modifications and microtubule dynamics. *Int J Mol Sci*. 2017;18:2207.
12. Magiera MM, Singh P, Gadadhar S, Janke C. Tubulin posttranslational modifications and emerging links to human disease. *Cell*. 2018;173(6):1323–1327.
13. Roll-Mecak A. How cells exploit tubulin diversity to build functional cellular microtubule mosaics. *Curr Opin Cell Biol*. 2019;56:102–108.
14. Yu I, Garnham CP, Roll-Mecak A. Writing and reading the tubulin code. *J Biol Chem*. 2015;290(28):17163–17172.
15. Aillaud C, Bosc C, Peris L, et al. Vasohinibins/SVBP are tubulin carboxypeptidases (TCPs) that regulate neuron differentiation. *Science*. 2017;358:1448–1453.
16. Nieuwenhuis J, Adamopoulos A, Bleijerveld OB, et al. Vasohibins encode tubulin detyrosinating activity. *Science*. 2017;358(6369):1453–1456.
17. Park JH, Roll-Mecak A. The tubulin code in neuronal polarity. *Curr Opin Neurobiol*. 2018;51:95–102.
18. Hernault SWL, Rosenbaum JL. Chlamydomonas α-tubulin is posttranslationally modified in the flagella during flagellar assembly. *J Cell Biol*. 1983;97:258–263.
19. Piperno G, Fuller MT. Monoclonal antibodies specific for an acetylated form of α-tubulin recognize the antigen in cilia and flagella from a variety of organisms. *J Cell Biol*. 1985;101(37):2085–2094.
20. LeDizet M, Piperno G. Identification of an acetylation site of Chlamydomonas α-tubulin. *Proc Natl Acad Sci*. 1987;84:5720–5724.
21. Piperno G, LeDizet M, Chang X. Microtubules containing acetylated α-tubulin in mammalian cells in culture. *J Cell Biol*. 1987;104:289–302.
22. Astrom H. Acetylated alpha-tubulin in the pollen tube. *Cell Biol Int Rep*. 1992;16(9):871–881.
23. Webster DR, Borisy GG. Microtubules are acetylated in domains that turn over slowly. *J Cell Sci*. 1989;92:57–66.
24. Falconer M, Vielkind U, Brown D. Establishment of a stable, acetylated microtubule bundle during neuronal commitment. *Cell Motil Cytoskeleton*. 1989;12(3):169–180.
25. Baas PW, Black MM. Individual microtubules in the axon consist of domains that differ in both composition and stability. *J Cell Biol*. 1990;111:495–509.

26. Brown A, Li Y, Slaughter T, Black MM. Composite microtubules of the axon: quantitative analysis of tyrosinated and acetylated tubulin along individual axonal microtubules. *J Cell Sci*. 1993;104:339–352.
27. Morales M, Fifkovà E. Distribution of acetylated α-tubulin in brain. *Cell Tissue Res*. 1991;265:415–423.
28. Hubbert C, Guardiola A, Shao R, et al. HDAC6 is a microtubule-associated deacetylase. *Nature*. 2002;417(6887):455–458.
29. North BJ, Marshall BL, Borra MT, Denu JM, Verdin E. The human Sir2 ortholog, SIRT2, is an NAD^+-dependent tubulin deacetylase. *Mol Cell*. 2003;11(2):437–444.
30. Akella JS, Wloga D, Kim J, et al. MEC-17 is an α-tubulin acetyltransferase. *Nature*. 2010;467(7312):218–222.
31. Shida T, Cueva JG, Xu Z, Goodman MB, Nachury MV. The major α-tubulin K40 acetyltransferase αTAT1 promotes rapid ciliogenesis and efficient mechanosensation. *Proc Natl Acad Sci*. 2010;107(50):21517–21522.
32. Szyk A, Deaconescu AM, Spector J, et al. Molecular basis for age-dependent microtubule acetylation by tubulin acetyltransferase. *Cell*. 2014;157(6):1405–1415.
33. Coombes C, Yamamoto A, McClellan M, et al. Mechanism of microtubule lumen entry for the α-tubulin acetyltransferase enzyme αTAT1. *Proc Natl Acad Sci*. 2016;113(46): E7176–E7184.
34. Chambray-Deakin M, Burgoyne RD. Posttranslational modifications of α-tubulin: acetylated and detyrosinated forms in axons of rat cerebellum. *J Cell Biol*. 1987;104: 1569–1574.
35. Palazzo A, Ackerman B, Gundersen GG. Tubulin acetylation and cell motility. *Nature*. 2003;421:230.
36. Portran D, Schaedel L, Xu Z, Thèry M, Nachury MV. Tubulin acetylation protects long-lived microtubules against mechanical aging. *Nat Cell Biol*. 2017;19(4):391–398.
37. Xu Z, Schaedel L, Portran D, et al. Microtubules acquire resistance from mechanical breakage through intralumenal acetylation. *Science*. 2017;356(6335):328–332.
38. Sudo H, Baas PW. Acetylation of microtubules influences their sensitivity to severing by Katanin in neurons and fibroblasts. *J Neurosci*. 2010;30(21):7215–7226.
39. Baas PW, Rao AN, Matamoros AJ, Leo L. Stability properties of neuronal microtubules. *Cytoskeleton (Hoboken)*. 2016;73(9):442–460.
40. Montagnac G, Meas-Yedid V, Irondelle M, et al. αTAT1 catalyses microtubule acetylation of clathrin-coated pits. *Nature*. 2013;502(7472):567–570.
41. Geeraert C, Ratier A, Pfisterer SG, et al. Starvation-induced hyperacetylation of tubulin is required for the stimulation of autophagy by nutrient deprivation. *J Biol Chem*. 2010;285(31):24184–24194.
42. Reed NA, Cai D, Blasius TL, et al. Microtubule acetylation promotes kinesin-1 binding and transport. *Curr Biol*. 2006;16(21):2166–2172.
43. Aguilar A, Becker L, Tedeschi T, Heller S, Iomini C, Nachury MV. α-Tubulin K40 acetylation is required for contact inhibition of proliferation and cell–substrate adhesion. *Mol Biol Cell*. 2014;25:1854–1866.
44. Pugacheva EN, Jablonski SA, Hartman TR, Henske EP, Golemis EA. HEF1-dependent Aurora A activation induced disassembly of the primary cilium. *Cell*. 2007;129(7):1351–1363.
45. Bounoutas A, O'Hagan R, Chalfie M. The multipurpose 15-protofilament microtubules in *C. elegans* have specific roles in mechanosensation. *Curr Biol*. 2009;19(16):1362–1367.
46. Topalidou I, Keller C, Kalebic N, et al. Enzymatic and non-enzymatic activities of the tubulin acetyltransferase MEC-17 are required for microtubule organization and mechanosensation in *C. elegans*. *Curr Biol*. 2012;22(12):1057–1065.
47. Morley SJ, Qi Y, Iovino L, et al. Acetylated tubulin is essential for touch sensation in mice. *elife*. 2016;5:e20813.

48. Chu C, Hou F, Zhang J, et al. A novel acetylation of β-tubulin by San modulates microtubule polymerization via down-regulating tubulin incorporation. *Mol Biol Cell.* 2011;22(4):448–456.
49. Sakakibara A, Ando R, Sapir T, Tanaka T. Microtubule dynamics in neuronal morphogenesis. *Open Biol.* 2013;3(7):130061.
50. Kapitein LC, Hoogenraad CC. Building the neuronal microtubule cytoskeleton. *Neuron.* 2015;87(3):492–506.
51. Lasser M, Tiber J, Lowery LA. The role of the microtubule cytoskeleton in neurodevelopmental disorders. *Front Cell Neurosci.* 2018;12:165.
52. Breuss MW, Leca I, Gstrein T, Hansen AH, Keays DA. Tubulins and brain development—the origins of functional specifi cation. *Mol Cell Neurosci.* 2017;84:58–67.
53. Keays DA, Cleak J, Huang G-J, et al. The role of *Tuba1a* in adult hippocampal neurogenesis and the formation of the dentate gyrus. *Dev Neurosci.* 2010;32:268–277.
54. Keays DA, Tian G, Poirier K, et al. Mutations in α-tubulin cause abnormal neuronal migration in mice and lissencephaly in humans. *Cell.* 2007;128:45–57.
55. Erck C, Peris L, Andrieux A, et al. A vital role of tubulin-tyrosine-ligase for neuronal organization. *Proc Natl Acad Sci.* 2005;102(22):7853–9858.
56. Creppe C, Malinouskaya L, Volvert ML, et al. Elongator controls the migration and differentiation of cortical neurons through acetylation of α-tubulin. *Cell.* 2009;136(3):551–564.
57. Li L, Wei D, Wang Q, et al. MEC-17 deficiency leads to reduced α-tubulin acetylation and impaired migration of cortical neurons. *J Neurosci.* 2012;32(37):12673–12683.
58. Kim G, Li L, Gorbani M, You L, Yang X. Mice lacking α-tubulin acetyltransferase 1 are viable but display α-tubulin acetylation deficiency and dentate gyrus distortion. *J Biol Chem.* 2013;288(28):20334–20350.
59. Pinto D, Delaby E, Merico D, et al. Convergence of genes and cellular pathways dysregulated in autism spectrum disorders. *Am J Hum Genet.* 2014;94:677–694.
60. Helsmoortel C, Vulto-van Silfhout AT, Coe BP, et al. A SWI/SNF-related autism syndrome caused by de novo mutations in ADNP. *Nat Genet.* 2014;46(4):380–384.
61. Oz S, Kapitansky O, Ivashco-Pachima Y, et al. The NAP motif of activity-dependent neuroprotective protein (ADNP) regulates dendritic spines through microtubule end binding proteins. *Mol Psychiatry.* 2014;19(10):1115–1124.
62. Steindler C, Li Z, Algarte M, et al. Jamip1 (marlin-1) defines a family of proteins interacting with janus kinases and microtubules. *J Biol Chem.* 2004;279(41):43168–43177.
63. Kaminsky EB, Kaul V, Paschall J, et al. An evidence-based approach to establish the functional and clinical significance of CNVs in intellectual and developmental disabilities. *Genet Med.* 2011;13(9):777–784.
64. Fukuda T, Nagashima S, Abe T, Kiyonari H, Inatome R, Yanagi S. Rescue of CAMDI deletion-induced delayed radial migration and psychiatric behaviors by HDAC6 inhibitor. *EMBO Rep.* 2016;17(12):1785–1798.
65. Khatri N, Gilbert JP, Huo Y, et al. The autism protein Ube3A/E6AP remodels neuronal dendritic arborization via caspase-dependent microtubule destabilization. *J Neurosci.* 2018;38(2):363–378.
66. Sun W, Poschmann J, Cruz-Herrera del Rosario R, et al. Histone Acetylome-wide Association Study of Autism Spectrum Disorder. *Cell.* 2016;167(5):1385–1397. e11.
67. Owen MJ, O'Donovan MC, Thapar A, Craddock N. Neurodevelopmental hypothesis of schizophrenia. *Br J Psychiatry.* 2011;198:173–175.
68. Arnold SE, Lee VM, Gur RE, Trojanowski JQ. Abnormal expression of two microtubule-associated proteins (MAP2 and MAP5) in specific subfields of the hippocampal formation in schizophrenia. *Proc Natl Acad Sci.* 1991;88:10850–10854.
69. Morris JA, Kandpal G, Ma L, Austin CP. DISC1 (disrupted-in-schizophrenia 1) is a centrosome-associated protein that interacts with MAP1A, MIPT3, ATF4/5 and NUDEL: regulation and loss of interaction with mutation. *Hum Mol Genet.* 2003;12(13):1591–1608.

70. Kamiya A, Kubo K, Tomoda T, et al. A schizophrenia-associated mutation of DISC1 perturbs cerebral cortex development. *Nat Cell Biol.* 2005;7(12):1167–1178.
71. Andrieux A, Salin P, Schweitzer A, et al. Microtubule stabilizer ameliorates synaptic function and behavior in a mouse model for schizophrenia. *Biol Psychiatry.* 2006;60(11):1224–1230.
72. Lang B, Pu J, Hunter I, et al. Recurrent deletions of *ULK4* in schizophrenia: a gene crucial for neuritogenesis and neuronal motility. *J Cell Sci.* 2014;127:630–640.
73. Lang B, Zhang L, Jiang G, et al. Control of cortex development by *ULK4*, a rare risk gene for mental disorders including schizophrenia. *Sci Rep.* 2016;6:31126.
74. Tee JY, Sutharsan R, Fan Y, Mackay-Sim A. Schizophrenia patient-derived olfactory neurosphere-derived cells do not respond to extracellular reelin. *NPJ Schizophr.* 2016;2:16027.
75. Muñoz-Estrada J, Lora-Castellanos A, Meza I, Elizalde SA, Benítez-King G. Primary cilia formation is diminished in schizophrenia and bipolar disorder: a possible marker for these psychiatric diseases. *Schizophr Res.* 2018;195:412–420.
76. Guadiana SM, Semple-Rowland S, Daroszewski D, et al. Arborization of dendrites by developing neocortical neurons is dependent on primary cilia and type 3 adenylyl cyclase. *J Neurosci.* 2013;33(6):2626–2638.
77. Youn YH, Han Y-G. Primary cilia in brain development and diseases. *Am J Pathol.* 2018;188(1):11–22.
78. Janke C, Chloë Bulinski J. Post-translational regulation of the microtubule cytoskeleton: mechanisms and functions. *Nat Rev Mol Cell Biol.* 2011;12(12):773–786.
79. Ran J, Yang Y, Li D, Liu M, Zhou J. Deacetylation of α-tubulin and cortactin is required for HDAC6 to trigger ciliary disassembly. *Sci Rep.* 2015;5:12917.
80. Penazzi L, Bakota L, Brandt R. Microtubule dynamics in neuronal development, plasticity, and neurodegeneration. *Int Rev Cell Mol Biol.* 2016;321:89–169.
81. Millecamps S, Julien J-P. Axonal transport deficits and neurodegenerative diseases. *Nat Rev Neurosci.* 2013;14(3):161–176.
82. Brady ST, Morfini GA. Regulation of motor proteins, axonal transport deficits and adult-onset neruodegenerative diseases. *Neurobiol Dis.* 2017;105:273–282.
83. Cappelletti G, Cartelli D, Christodoulou MS, Passarella D. Microtubule-directed therapeutic strategy for neurodegenerative disorders: starting from the basis and looking on the emergences. *Curr Pharm Des.* 2017;23:784–808.
84. d'Ydewalle C, Krishnan J, Chiheb DM, et al. HDAC6 inhibitors reverse axonal loss in a mouse model of mutant HSPB1-induced Charcot-Marie-Tooth disease. *Nat Med.* 2011;17(8):968–974.
85. Kim J-Y, Woo S-Y, Bin HY, et al. HDAC6 inhibitors rescued the defective axonal mitochondrial movement in motor neurons derived from the induced pluripotent stem cells of peripheral neuropathy patients with *HSPB1* mutation. *Stem Cells Int.* 2016;9475981.
86. Barneo-Munoz M, Juárez P, Civera-Tregón A, et al. Lack of GDAP1 induces neuronal calcium and mitochondrial defects in a knockout mouse model of Charcot-Marie-Tooth neuropathy. *PLoS Genet.* 2015;1–27.
87. Mo Z, Zhao X, Liu H, et al. Aberrant GlyRS-HDAC6 interaction linked to axonal transport deficits in Charcot-Marie-Tooth neuropathy. *Nat Commun.* 2018;9:1007.
88. Benoy V, Van Helleputte L, Prior R, et al. HDAC6 is a therapeutic target in mutant GARS-induced Charcot-Marie-Tooth disease. *Brain.* 2018;141:673–687.
89. Evans KJ, Gomes ER, Reisenweber SM, Gundersen GG, Lauring BP. Linking axonal degeneration to microtubule remodeling by Spastin-mediated microtubule severing. *J Cell Biol.* 2005;168(4):599–606.
90. Roll-Mecak A, Vale RD. Structural basis of microtubule severing by hereditary spastic paraplegia protein spastin. *Nature.* 2008;451(7176):363–367.
91. Riano E, Martignoni M, Mancuso G, et al. Pleiotropic effects of spastin on neurite growth depending on expression levels. *J Neurochem.* 2009;108:1277–1288.

92. Abrahamsen G, Fan Y, Matigian N, et al. A patient-derived stem cell model of hereditary spastic paraplegia with SPAST mutations. *Dis Model Mech*. 2013;6:489–502.
93. Fan Y, Wali G, Sutharsan R, et al. Low dose tubulin-binding drugs rescue peroxisome trafficking deficit in patient-derived stem cells in hereditary spastic paraplegia. *Biol Open*. 2014;3:494–502.
94. Gal J, Chen J, Barnett KR, Yang L, Brumley E, Zhu H. HDAC6 regulates mutant SOD1 aggregation through two SMIR motifs and tubulin acetylation. *J Biol Chem*. 2013;288(21):15035–15045.
95. Lee J-Y, Kawaguchi Y, Li M, et al. Uncoupling of protein aggregation and neurodegeneration in a muse ALS model. *Neurodegener Dis*. 2015;15(6):339–349.
96. Taes I, Timmers M, Hersmus N, et al. *Hdac6* deletion delays disease progression in the sod1^{g93a} mouse model of ALS. *Hum Mol Genet*. 2013;22(9):1783–1790.
97. Guo W, Naujock M, Fumagalli L, et al. HDAC6 inhibition reverses axonal transport defects in motor neurons derived from FUS-ALS patients. *Nat Commun*. 2017;8:861.
98. Hoffner G, Kahlem P, Djian P. Perinuclear localization of huntingtin as a consequence of its binding to microtubules through an interaction with β-tubulin relevance to Huntington's disease. *J Cell Sci*. 2002;115:941–948.
99. Dompierre JP, Godin JD, Charrin BC, et al. Histone deacetylase 6 inhibition compensates for the transport deficit in Huntington's disease by increasing tubulin acetylation. *J Neurosci*. 2007;27(13):3571–3583.
100. Bobrowska A, Paganetti P, Matthias P, Bates GP. *Hdac6* knock-out increases tubulin acetylation but does not modify disease progression in the R6/2 mouse model of Huntington's disease. *PLoS ONE*. 2011;6:e20696.
101. Hempen BBS, Brion J-P. Reduction of acetylated α-tubulin immunoreactivity in neurofibrillary tangle-bearing neurons in Alzheimer's disease. *J Neuropathol Exp Neurol*. 1996;55(9):964–972.
102. Zhang F, Su B, Wang C, et al. Posttranslational modifications of α-tubulin in Alzheimer disease. *Transl Neurodegener*. 2015;4:9.
103. Ferrero H, Larrayoz IM, Martisova E, et al. Increased levels of brain adrenomedullin in the neuropathology of Alzheimer's disease. *Mol Neurobiol*. 2018;55:5177–5183.
104. Min S-W, Cho S-H, Zhou Y, et al. Acetylation of tau inhibits its degradation and contributes to tauopathy. *Neuron*. 2010;67(6):953–966.
105. Cohen TJ, Guo JL, Hurtado DE, et al. The acetylation of tau inhibits its function and promotes pathological tau aggregation. *Nat Commun*. 2011;2:252.
106. Irwin DJ, Cohen TJ, Grossman M, et al. Acetylated tau neuropathology in sporadic and hereditary tauopathies. *Am J Pathol*. 2013;183(2):344–351.
107. Simões-Pires C, Zwick V, Nurisso A, Schenker E, Carrupt P-A, Cuendet M. HDAC6 as a target for neurodegenerative diseases: what makes it different from the other HDACs? *Mol Neurodegener*. 2013;8:7. https://doi.org/10.1186/1750-1326-8-7.
108. Tsushima H, Emanuele M, Polenghi A, et al. HDAC6 and RhoA are novel players in Abeta-driven disruption of neuronal polarity. *Nat Commun*. 2015;6:7781.
109. Rui Y, Zheng JQ. Amyloid β oligomers elicit mitochondrial transport defects and fragmentation in a manner. *Mol Brain*. 2016;9(1):79.
110. Ma Q, Zuo X, Yang F, et al. Loss of MAP function leads to hippocampal synapse loss and deficits in the Morris Water Maze with aging. *J Neurosci*. 2014;34(21):7124–7136.
111. Radde R, Bolmont T, Kaeser SA, et al. Aβ42-driven cerebral amyloidosis in transgenic mice reveals early and robust pathology. *EMBO Rep*. 2006;7(9):940–946.
112. Govindarajan N, Rao P, Burkhardt S, et al. Reducing HDAC6 ameliorates cognitive deficits in a mouse model for Alzheimer' s disease. *EMBO Mol Med*. 2013;5(1):52–63.
113. Majid T, Griffin D, Criss Z, Jarpe M, Pautler RG. Pharmocologic treatment with histone deacetylase 6 inhibitor (ACY-738) recovers Alzheimer's disease phenotype in amyloid precursor protein/presenilin 1 (APP/PS1) mice. *Alzheimers Dement (N Y)*. 2015;1(3):170–181.

114. Zhang L, Liu C, Wu J, et al. Tubastatin A / ACY-1215 improves cognition in Alzheimer's disease transgenic mice. *J Alzheimers Dis*. 2014;41(4):1193–1205.
115. Xiong Y, Zhao K, Wu J, Xu Z, Jin S, Zhang YQ. *HDAC6* mutations rescue human tau-induced microtubule defects in *Drosophila*. *Proc Natl Acad Sci*. 2013;110(12):4604–4609.
116. Mao C-X, Wen X, Jin S, Zhang YQ. Increased acetylation of microtubules rescues human tau-induced microtubule defects and neuromuscular junction abnormalities in *Drosophila*. *J Cell Sci*. 2017;10:1245–1252.
117. Hunn BHM, Cragg SJ, Bolam JP, Spillantini M-G, Wade-Martins R. Impaired intracellular trafficking defines early Parkinson's disease. *Trends Neurosci*. 2015;38(3):178–188.
118. Abeliovich A, Gitler AD. Defects in trafficking bridge Parkinson's disease pathology and genetics. *Nature*. 2016;539(7628):207–216.
119. Pellegrini L, Wetzel A, Granno S, Heaton G, Harvey K. Back to the tubule: microtubule dynamics in Parkinson's disease. *Cell Mol Life Sci*. 2017;74:409–434.
120. Cartelli D, Cappelletti G. Microtubule destabilization paves the way to Parkinson's disease. *Mol Neurobiol*. 2017;54:6762–6774.
121. Cartelli D, Ronchi C, Maggioni MG, Rodighiero S, Giavini E, Cappelletti G. Microtubule dysfunction precedes transport impairment and mitochondria damage in MPP^+-induced neurodegeneration. *J Neurochem*. 2010;115(1):247–258.
122. Kim-Han JS, Antenor-Dorsey JA, O'Malley KL. The parkinsonian mimetic, MPP^+, specifically impairs mitochondrial transport in dopamine axons. *J Neurosci*. 2011;31(19):7212–7221.
123. Patel VP, Chu CT. Decreased SIRT2 activity leads to altered microtubule dynamics in oxidative-stressed neuronal cells: implications for Parkinson's disease. *Exp Neurol*. 2014;257:170–181.
124. Cartelli D, Casagrande F, Busceti CL, et al. Microtubule alterations occur early in experimental parkinsonism and the microtubule stabilizer epothilone D is neuroprotective. *Sci Rep*. 2013;3:1837.
125. Zimprich A. Genetics of Parkinson's disease and essential tremor. *Curr Opin Neurol*. 2011;24:318–323.
126. Law BMH, Spain VA, Leinster VHL, et al. A direct interaction between leucine-rich repeat kinase 2 and specific β-tubulin isoforms regulates tubulin acetylation. *J Biol Chem*. 2014;289(2):895–908.
127. Cartelli D, Goldwurm S, Casagrande F, Pezzoli G, Cappelletti G. Microtubule destabilization is shared by genetic and idiopathic Parkinson's disease patient fibroblasts. *PLoS ONE*. 2012;7(5):e37467.
128. Esteves AR, Fernandes MG, Santos D, Januario C, Cardoso SM. The upshot of LRRK2 inhibition to Parkinson's disease paradigm. *Mol Neurobiol*. 2015;52(3):1804–1820.
129. Godena VK, Brookes-Hocking N, Moller A, et al. Increasing microtubule acetylation rescues axonal transport and locomotor deficits caused by LRRK2 Roc-COR domain mutations. *Nat Commun*. 2014;5:5245.
130. Cartelli D, Amadeo A, Calogero AM, et al. Parkin absence accelerates microtubule aging in dopaminergic neurons. *Neurobiol Aging*. 2018;61:66–74.
131. Corti O, Lesage S, Brice A. What genetics tells us about the causes and mechanisms of Parkinson's disease. *Physiol Rev*. 2011;91(4):1161–1218.
132. Cappelletti G, Casagrande F, Calogero A, De Gregorio C, Pezzoli G, Cartelli D. Linking microtubules to Parkinson's disease: the case of parkin. *Biochem Soc Trans*. 2015;43:292–296.
133. Spillantini MG, Schmidt ML, Lee VM-Y, Trojanowski JQ, Goedert M. α-Synuclein in Lewy bodies. *Nature*. 1997;388:839–840.
134. Singleton AB, Farrer M, Johnson J, et al. α-Synuclein locus triplication causes Parkinson's disease. *Science*. 2003;302:841.
135. Cartelli D, Aliverti A, Barbiroli A, et al. α-Synuclein is a novel microtubule dynamase. *Sci Rep*. 2016;6:33289.

136. Outeiro TF, Kontopoulos E, Altmann SM, et al. Sirtuin 2 inhibitors rescue α-synuclein-mediated toxicity in models of Parkinson's disease. *Science*. 2007;317(5837):516–519.
137. Esteves AR, Arduíno DM, Swerdlow RH, Oliveira CR, Cardoso SM. Microtubule depolymerization potentiates alpha-synuclein oligomerization. *Front Aging Neurosci*. 2010;1:5.
138. Esteves AR, Gozes I, Cardoso SM. The rescue of microtubule-dependent traffic recovers mitochondrial function in Parkinson' s disease. *Biochim Biophys Acta Mol Basis Dis*. 2014;1842(1):7–21.
139. Li L, Yang X-J. Tubulin acetylation: responsible enzymes, biological functions and human diseases. *Cell Mol Life Sci*. 2015;72(22):4237–4255.
140. Miyake Y, Keusch JJ, Wang L, et al. Structural insights into HDAC6 tubulin deacetylation and its selective inhibition. *Nat Chem Biol*. 2016;12(9):748–754.
141. Butler KV, Kalin J, Brochier C, Vistoli G, Langley B, Kozikowski AP. Rational design and simple chemistry yield a superior, neuroprotective HDAC6 inhibitor, tubastatin A. *J Am Chem Soc*. 2010;132(31):10842–10846.
142. Wang X-X, Wan R-Z, Liu Z-P. Recent advances in the discovery of potent and selective HDAC6 inhibitors. *Eur J Med Chem*. 2018;143:1406–1418.
143. Shen S, Benoy V, Bergman JA, et al. Bicyclic-capped histone deacetylase 6 inhibitors with improved activity in a model of axonal Charcot-Marie-Tooth disease. *ACS Chem Neurosci*. 2016;7(2):240–258.
144. Lee H, Fan S, Huang F, et al. 5-Aroylindoles act as selective histone deacetylase 6 inhibitors ameliorating Alzheimer's disease phenotypes. *J Med Chem*. 2018;61:7087–7102.
145. Park JB. Finding potent Sirt inhibitor in coffee: isolation, confirmation and synthesis of javamide-II (N-Caffeoyltryptophan) as Sirt1/2 inhibitor. *PLoS ONE*. 2016;11(3):e0150392.
146. Nakakura T, Asano-Hoshino A, Suzuki T, et al. The elongation of primary cilia via the acetylation of α-tubulin by the treatment with lithium chloride in human fibroblast KD cells. *Med Mol Morphol*. 2015;48(1):44–53.
147. Shah N, Kumar S, Zaman N, et al. TAK1 activation of alpha-TAT1 and microtubule hyperacetylation control AKT signaling and cell growth. *Nat Commun*. 2018;9:1696.

Index

Note: Page numbers followed by *f* indicate figures and *t* indicate tables.

A
AADvac1, 262
ABA. *See* Applied behavioral analysis (ABA)
Abeotaxane, 255
Abnormal sleep cycle, 85
Acetylation, definition, 273–274
Acetylation of tubulin, 275–278
  neurodegenerative diseases, 281–285
  neurodevelopmental disorders, 278–281
ACI-35, 262
Activity-dependent neuroprotective protein (ADNP), 279–280
  Alzheimer's disease (AD) connection, 13
  epilogue, 13–14
  macromolecular interactions, 4, 5–6*f*
  schizophrenia connection, 11–13
  syndrome, in patients and models, 7–11, 7*t*
  vasoactive intestinal peptide (VIP), 3–4
Adeno-associated virus (AAV), 154
Adnp-deficient mouse, 4
AL-108, 255
Alzheimer's disease (AD), 173–174, 218–221, 282–283
  AchE inhibitors, 245–246
  amyloid plaques, 246
  clinical characteristics and symptomatology, 21–23
  clinical convergence, 23
  complex I and IV, 183–184
  diagnosis, 245–246
  heme metabolism, 184–185
  human clinical studies, 219–221
  mitochondria, 182–184
  neurofibrillary tangles (NFTs), 246
  risk, 219
  social cognition, 35–37
  tau pathology, 246
  tissue and animal studies, 218–219
American Academy of Pediatrics, 80
AMP-dependent tau protein kinase (AMPK), 183–184
Amyloid protein precursor (APP), 182–183
Amyotrophic lateral sclerosis (ALS), 224–226
  human studies, 225–226
  nonhuman studies, 225
Aneuploidies, 62–63
Angelman's syndrome, 85
Animal models, in autism spectrum disorder (ASD), 111, 113–114
Antiinflammatory cytokine, 112
Antipsychotic drugs, in schizophrenia, 109
Antisense oligonucleotides (ASOs), 263
Anti-tau antibody therapies, 259–261
Apoptosis, 215
Applied behavioral analysis (ABA), 91
Arc mRNA expression, 159
ATP-competitive GSK-3β inhibitors, 250–251
Attention-deficit hyperactivity disorder (ADHD), 28, 59
α-Tubulin acetylation, 277
Autism spectrum disorder (ASD), 59, 279–280
  brain-derived neurotropic factor (BDNF) dysregulation, 116–117
  clinical characteristics and symptomatology, 21–23
  clinical convergence, 23
  comprehensive evaluation, 89–93
  cytokine alterations, 112–114
  diagnosis, 79, 86–88
    assessment tools, 89–90
    methods, 86–88
  early diagnosis, 80–85
    age of diagnosis, 83–84
    atypical behaviors, 83
    developmental evaluation during first 6 months, 80–81, 82*t*
    developmental evaluation up to first year and transition to second year, 81–83, 82*t*
    early regulatory difficulties, 84–85
    empathy, 85
    general movements, 85
    parental concern, 83–84
    significance, 93
    sleep disturbances, 85
    social behaviors, 85

Autism spectrum disorder (ASD) *(Continued)*
etiology, 86
evaluation, 79
features, 101
immune-modulating drugs, 114–115
inflammatory hypothesis, 101
metaanalyses, 26–27
with attention-deficit/hyperactivity disorder (ADHD), 28
genetic diagnosis, 27
with schizophrenia, 27–28
social cognition, 28–30
microglial dysfunction, 111–112
oxidative stress, 117–118
oxytocin treatment, 38–40
red flags, 85–86
social cognition, 26–30
treatment
and intervention, 90–93
pharmacological, 92–93
planning, 89–93, 89*f*
Autoimmune lymphoproliferative syndrome (ALPS), 114
Autophagy, 210–211
Axonal microtubules, 284–285
AZP2006, 256–257

B
BDNF. *See* Brain-derived neurotrophic factor (BDNF)
BIIB076, 261
BIIB080, 263
Bipolar disorders (BD), 60–61, 209
BMS-241027. *See* Epothilone D (Epo D)
Brain-derived neurotrophic factor (BDNF), 131–133
autism spectrum disorder (ASD), 116–117
depression, 135–136
lithium, 211–212
major depressive disorder (MDD), 147–148
antidepressant efficacy regulation, 153
depression-like behavior, 152–153
genetics, 148–149
genotype, 149–150
mouse models, 151–152
peripheral level, 149–150
proBDNF/p75NTR signaling, 152–153
Val66Met polymorphism, 150–151
mood disorders
activity-dependent secretion, 157–159, 158*f*
excitatory-inhibitory balance regulation, 156–157
neuronal plasticity regulation, 155–156
neuroprotective effects, 154–155
rapid-acting antidepressants, 157–159
signaling, 159–160
schizophrenia, 107–109, 108*f*
Brain imaging, 87
Brainstem auditory evoked response assay (BAERA), 87–88

C
cAMP-response element-binding protein (CREB), 135
CDK5, 251
Celecoxib, 115
Charcot-Marie-Tooth disease, 281
Chemokines, 129
Chromosome 9 open reading frame 72 (*C9orf72*), 225
Chronic stress, 126, 132–133
C2N-8E12, 259–260
Communication disorders, 59–60
Coproporphyrinogen III dehydrogenase. *See* Coproporphyrinogen III (CPFG) oxidase
Coproporphyrinogen III (CPFG) oxidase, 175–178
Copy number variation (CNV), 63–64
complex structural variation, 66
molecular mechanisms, 64
NHAR-induced CNVs, 64–65
nonhomology-driven mechanisms, 65–66
triplications, 66
Corticosteroids, 114
CP201 (NAP), 7–11, 255
Cyclic AMP (cAMP)-response element-binding protein (CREB), 153, 179–180
Cyclin-dependent kinase 5 (CDK5), 250
Cyclooxygenase-2 inhibitor, 115
Cytokines
autism spectrum disorder (ASD), 112–114
schizophrenia, 103–104, 105*f*

D
Davunetide, 255
Depression
altered brain volume, 127–128
antidepressant treatments, 134–135
brain-derived neurotrophic factor (BDNF), 135–136

cellular atrophy, 127–128
connectivity, 127–128
current successful therapeutic strategies, 134–136
etiology, 126
impaired neurogenesis, 128–129
neuroinflammation and neurotoxicity, 129–130
neurotrophic factor signaling, 127–134
neurotrophic factor signaling alteration, 130–134
genetic studies, 131–132
in mouse and rat models, 132–134
reduced levels, 131
stress and stress-induced depression, 132–134
symptoms, 126
treatment strategies, 126
Developmental, Individual Difference, Relationship-based model (DIR), 91–92
Diacylglycerol (DAG), 130–131
DIR. *See* Developmental, Individual Difference, Relationship-based model (DIR)
Drosophila FXS model, 223
Dynamin-1-like protein (Drp1), 182–183

E
Early Start Denver Model (ESDM), 91
Electroconvulsive shock therapy, 134–135
Electroconvulsive therapy (ECT), 149–150, 155
Empathy, 85
Epilepsy, 61
Epothilone D (Epo D), 254
ESDM. *See* Early Start Denver Model (ESDM)
Eukaryotic translation initiation factor 2-alpha (eIF2α), 174–175
Extensive neurodevelopmental workup, 87–88

F
Fetal alcohol syndrome, 10–11
Fission 1 protein (FIS1), 182–183
Flavin adenine dinucleotide (FAD), 175–178
Floortime model, 91–92
Forced swim test (FST), 159–160
Forkhead-BOX P2 (Foxp2), 12
Fork stalling and template switching (FoSTeS) mechanism, 64
Fragile X mental retardation 1 gene *(FMR1)*, 63
Fragile X syndrome (FXS), 63, 85, 223–224
animal studies, 223–224
human studies, 224
Fyn phosphorylates tau, 252

G
Gamma-amino butyric acid (GABA)ergic neurotransmission, 156
Gene networks
from disease to networks, 71–72
gene-regulatory networks, 70–71
metabolic networks, 70–71
from networks to disease, 72–73
protein-protein interaction networks, 70–71
Gene-regulatory networks, 70–71
Glial cell line-derived neurotrophic factor (GDNF), 131
Glial fibrillary acidic protein/heme oxygenase 1 GFAP.HMOX), 186–187
Glutathione (GSH), 106–107
Glycogen synthase kinase-3 (GSK-3), 250
signaling pathway, 215
Glycogen synthase kinase-3 β (GSK-3β), 250

H
Heat shock protein β1 *(HSPB1)*, 281
Heme
Alzheimer's' disease, 184–185
biosynthesis pathway, 175–178, 177*f*
catabolic pathway, 178
in neurons, 179–180
in neuropsychiatric disorders, 180–182
Parkinson's disease, 186–187
pyrrole ring, 181–182
structure, 175*f*
Heme precursor aminolevulinic acid (Δ-ALA), 175–178
Heme-regulated inhibitor (HRI), 174–175, 176*f*
Hippocampal BDNF, 145–147, 146*f*
Histone acetylome-wide association study (HAWAS), 279–280
Histone deacetylase 6 (HDAC6), 275–277, 282–283
Human Phenotype Ontology (HPO), 71–72
Huntington's disease (HD), 217–218
human studies, 218
nonhuman studies, 217–218
Hyperphosphorylated tau, 250

I

ID. *See* Intellectual disability (ID)
Idiopathic autism, 87
Immune-modulating drugs, 114–115
Immunotherapy, 257–262
  active immunotherapy, 261–262
  passive immunotherapy, 258–261
Inflammation, 212–213
Inflammatory processes, 129
Inositol depletion, 211
Inositol monophosphatase (IMPase), 210–211
Intellectual development disorder (IDD), 58
Intellectual disability (ID), 8, 58
Interleukin-6, 104
Interleukin-17a, 113–114
Intravenous immunoglobulin (IVIG), 114–115
Isolated active normal mitochondria (IAN-MIT), 187–188

J

Janus kinase and microtubules interacting protein 1 (JAKMIP1), 279–280
JNJ-63733657, 260

K

KANLS1, 65–66
The Keshet model, 92
Koolen-de Vries syndrome, 65–66

L

Learning disorder, 59–60
Leucine-rich repeats kinase 2 (LRRK2), 284–285
Leuco-methylthioninium bis-hydromethanesulfonate (LMTX), 256
Light chain 3 (LC3), 12
Lithium, 250
  Alzheimer's disease (AD), 218–221
    human clinical studies, 219–221
    risk, 219
    tissue and animal studies, 218–219
  amyotrophic lateral sclerosis (ALS), 224–226
    human studies, 225–226
    nonhuman studies, 225
  apoptosis, 215
  autophagy, 210–211
  for bipolar disorders, 209
  brain-derived neurotrophic factor (BDNF), 211–212
  cell signaling-related enzymes, 210
  fragile X syndrome (FXS), 223–224
    animal studies, 223–224
    human studies, 224
  Huntington's disease (HD), 217–218
    human studies, 218
    nonhuman studies, 217–218
  inflammation, 212–213
  mitochondria, 213–215
  multiple sclerosis (MS), 226–227
    animal studies, 227
    human studies, 227
  neuropsychiatric disorders, 215–227
  Parkinson's disease (PD), 221–223
    animal studies, 222
    human studies, 222–223
    neuronal cells in culture studies, 221
  stroke, 216–217
    animal studies, 216
    human studies, 216–217
Lithium-induced neuroprotection, 210
L1-mediated retrotransposition, 64
Long-term depression (LTD), 107–108
Long-term potentiation (LTP), 107–108
LY3303560, 261
Lysine 40 (K40), 275–277

M

Macroautophagy. *See* Autophagy
Magnetic resonance imaging (MRI), 127
Major depressive disorder (MDD), 126, 128–129, 131
  brain-derived neurotrophic factor (BDNF), 147–148
    antidepressant efficacy regulation, 153
    depression-like behavior, 152–153
    genetics, 148–149
    genotype, 149–150
    mouse models, 151–152
    peripheral level, 149–150
    proBDNF/p75NTR signaling, 152–153
    Val66Met polymorphism, 150–151
  neurotrophin levels, 147–148
Mammalian target of rapamycin (mTOR), 155–157, 210
Maternal immune activation (MIA), 112
Mauve Factor, 181–182
Memantine, 245–246
Metabolic networks, 70–71
Methyl-CpG-binding protein 2 (Mecp2), 111–112
Methylene blue, 256
1-Methyl-4-phenylpyridinium (MPP), 221

Microarrays, 61–62
Microglia
  autism spectrum disorder (ASD), 111–112
  schizophrenia, 102–103, 102*f*
Microhomology-mediated break-induced replication (MMBIR) mechanism, 64
Microtubule (MT), 274–275
  health, 13–14
  stabilizers, 254–255
Microtubule associated protein 6 (Map6), 11
Mild cognitive impairment (MCI), 35–36
Minor inconsistent asymmetry, 80–81
Missense variant, 66–67
Mitochondria
  Alzheimer's' disease, 182–184
  lithium, 213–215
  Parkinson's disease, 185
Mitochondrial dysfunction, 173–174
Mood disorders, 125–126
  brain-derived neurotrophic factor (BDNF)
    activity-dependent secretion, 157–159, 158*f*
    excitatory-inhibitory balance regulation, 156–157
    neuronal plasticity regulation, 155–156
    neuroprotective effects, 154–155
    rapid-acting antidepressants, 157–159
    signaling, 159–160
  disease-modifying therapies for, 136–137
  pathophysiology, 147–153
Motor disorders, 59–60
Multiple sclerosis (MS), 226–227
  animal studies, 227
  human studies, 227

N

N-Acetylcysteine, 107
NAHR, 64
NDDs. *See* Neurodevelopmental disorders (NDDs)
Nerve growth factor (NGF), 145–147
Neural stem cells (NSCs), 154
Neurodegenerative diseases, 125–126
  acetylation of tubulin, 281–285
Neurodegenerative disorders, 137
Neurodevelopmental disorders (NDDs), 57
  acetylation of tubulin, 278–281
  aneuploidies, 62–63
  attention-deficit hyperactivity disorder (ADHD), 59
  autism spectrum disorders (ASD), 59
  bipolar disorders (BD), 60–61
  communication disorders, 59–60
  comorbidity rates, 57–58
  copy number variation (CNV), 63–64
    complex structural variation, 66
    molecular mechanisms, 64
    NHAR-induced CNVs, 64–65
    nonhomology-driven mechanisms, 65–66
    triplications, 66
  definition, 57–58
  epilepsy, 61
  gene networks
    from disease to networks, 71–72
    gene-regulatory networks, 70–71
    metabolic networks, 70–71
    from networks to disease, 72–73
    protein-protein interaction networks, 70–71
  genetic causes, 61–70
  intellectual disability (ID), 58
  learning disorder, 59–60
  motor disorders, 59–60
  repeat expansions, 63
  schizophrenia spectrum, 60
  single-gene defects, 66–70
  single-nucleotide variants (SNVs), 66–67
    convergence to overlapping networks, 70
    gene identification strategies, 69–70
    mutational effects, 67
    variants from mutations, 68–69
  structural variation (SV), 63
  symptoms, 57–58
Neurofibrillary tangles (NFTs), 246
Neurogenesis, 128–129
Neurological abnormalities, 81
Neuronal microtubules, 278
Neuronal pruning defect, 103
Neuropeptide Y (NPY), 156
Neuroporphyria, 180–181
Neuropsychiatric disorders, 215–227. *See also Specific diseases*
Neurotrophin 3 (NT-3), 145–147
Neurotrophin 4/5 (NT-4/5), 145–147
Neurotrophins, 130–131
Next-generation sequencing (NGS) techniques, 61–62
Nilotinib (Tasigna), 257
N-methyl-D-aspartate (NMDA) receptor, 107–108, 157
Nonallelic homologous recombination (NAHR), 64

Non-ATP competitive GSK-3 inhibitors, 250–251
Nonhomologous end joining (NHEJ), 64
Noninvasive brain stimulation techniques, 42–43
Nonsense-mediated decay (NMD) process, 67
Nonsteroidal antiinflammatory drug (NSAID), 253
Nonsynonymous, 66–67

O

O-GlcNAcylation, 253–254
Oxidative stress
  autism spectrum disorder (ASD), 117–118
  schizophrenia, 104–107, 106*f*
Oxygen metabolites, 104
Oxytocin, 38
  autism spectrum disorder (ASD), 38–40
  schizophrenia, 40–41
Oxytocin receptor (OXTR), 39

P

Parkinson's disease (PD), 173–174, 221–223, 284–285
  animal studies, 222
  complex I, 185–186
  heme metabolism, 186–187
  human studies, 222–223
  mitochondria, 185
  neuronal cells in culture studies, 221
  physiopathology, 185
Passive immunotherapy, 258–261
Pharmacotherapy, 114
Phospholipase C-γ (PLC-γ), 130–131
Phosphorylated eIF2α (peIFf2α), 174–175
Pioglitazone, 115
Pituitary adenylate cyclase-activating polypeptide (PACAP), 3–4
Plaques, 246
Porphobilinogen (PBG), 175–178
Porphyria, 180–181
Positron emission tomography (PET), 156, 182–183
Posttranslational modification (PTM), 273–274, 278
Primitive reflexes, 81
Proinflammatory cytokines, 112
Protein-protein interaction (PPI) networks, 70–72
Psychotic disorders, 60
PTEN-induced kinase 1 (PINK1), 185

R

Reactive oxygen species (ROS), 182–183
Rett syndrome, 111–112
Risperidone, 12
Rodents, 223–224

S

Salsalate, 253
Schizophrenia (SZ), 60, 173–174, 280–281
  brain-derived neurotrophic factor (BDNF), 107–109, 108*f*
  clinical characteristics and symptomatology, 21–23
  clinical convergence, 23
  complex I, 188
  cytokines, 103–104, 105*f*
  features, 101
  heme metabolism, 188–189
  inflammatory hypothesis, 101
  metaanalysis, 32
    bipolar disorder, 34
    neurocognition and social cognition, 33–34
    of social cognition domains, 32–33
  microglia, 102–103, 102*f*
  mitochondria, 187–188
  oxidative stress, 104–107, 106*f*
  oxytocin treatment, 40–41
  social cognition, 30–35
Schizotypal (personality) disorder, 60
Seizures, 61
Single nucleotide polymorphism (SNP), 147–149
Single-nucleotide variants (SNVs), 61–62, 66–67
  convergence to overlapping networks, 70
  gene identification strategies, 69–70
  mutational effects, 67
  variants from mutations, 68–69
Sleep disturbances, 85
Social cognition
  Alzheimer's disease (AD), 35–37
  attribution style, 24
  dysfunctional domains, 43
  emotional processing, 24
  functional magnetic resonance imaging (fMRI), 25
  NMDA glutamate receptor, 41–42
  oxytocin, 38
  pharmacological treatment, 37–43
  psychostimulant, 42
  schizophrenia, 30–35

social perception, 24
theory of mind (ToM), 24
Somatostatin (SST), 156–157
Somerville-Van der Aa syndrome, 64–65
SPAST, 282
Stable tubule-only polypeptide (STOP) model, 11
Stroke, 216–217
animal studies, 216
human studies, 216–217
Structural variation (SV), 63
Subgranular zone (SGZ), 154
Subventricular zone (SVZ), 154
Synaptogenesis, 155–156
Synonymous, 66–67

T

Tau acetylation inhibitors, 252–253
Tau aggregation inhibitors (TAIs), 255–257
Tau-kinase inhibitors, 252
Tau-targeting therapies
approaches, 247*f*
clinical trials, 248–249*t*
kinases inhibitors, 250–252
MT stabilizers, 254–255
posttranslational modifications, 247–254
reduction, 263
tau aggregation inhibitors (TAIs), 255–257
T-helper 17 (T_H17), 113–114
Threonine 485 (Thr485), 174–175
Tocilizumab, 104
TP1287. *See* Abeotaxane
TrkB signaling pathways, 145–147, 146*f*
Tropomyosin receptor kinase (Trk), 145–147
Tubulin code, 274–275

U

UCB1017, 261
Unc-51-like kinase *(ULK4)* gene, 280–281
Uroporphyrinogen III (UPG), 175–178
US Food and Drug Administration (FDA), 115

V

Val66Met BDNF polymorphism, 148–151
Valproic acid (VPA), 113–114
Vasoactive intestinal peptide (VIP), 3–4
Ventral tegmental area (VTA), 135–136

W

Whole exome sequencing (WES), 69–70
Williams-Beuren syndrome, 64–65

Z

Zinc transporter 3 (ZnT3), 116

FSC
www.fsc.org
MIX
Papier | Fördert
gute Waldnutzung
FSC® C083411

Kolibri 360 GmbH
Ferdinand-Jühlke-Straße 7
99095 Erfurt, Deutschland
produktsicherheit@kolibri360.de